Functional Surfaces in Biology

Functional Surfaces in Biology

Adhesion Related Phenomena

Volume 2

edited by

Stanislav N. Gorb
Functional Morphology and Biomechanics,
University of Kiel, Germany

Springer

Editor
Prof. Stanislav N. Gorb
Department of Functional Morphology and Biomechanics
Zoological Institute
University of Kiel
Am Botanischen Garten 1–9
D-24098 Kiel
Germany
sgorb@zoologie.uni-kiel.de

ISBN 978-1-4020-6694-8 e-ISBN 978-1-4020-6695-5
DOI 10.1007/978-1-4020-6695-5
Springer Dordrecht Heidelberg London New York

Library of Congress Control Number: 2009927034

Cover picture: Japanese gecko (*Gekko japonicus*). Background: scanning electron micrograph of the terminal contact elements of the gecko toe. See chapter M. Johnson, A. Russell, and S. Delannoy on surface characteristics of locomotor substrata and their relationship to gekkonid adhesion.
Background picture: Stanislav N. Gorb (University of Kiel, Germany); Front picture: Naoe Hosoda (National Institute of Materials Science, Tsukuba, Japan)

Printed on acid-free paper

Springer is part of Springer Science+Business Media (www.springer.com)

Abstract

Biological surfaces represent the interface between living organisms and the environment and serve many different functions: (1) They delimit the organism, give the shape to the organism, and provide mechanical stability of the body. (2) They are barriers against dry, wet, cold or hot environments. (3) They take part in respiration and in the transport of diverse secretions, and serve as a chemical reservoir for the storage of metabolic waste products. (4) A variety of specialised surface structures are parts of mechano- and chemoreceptors. (5) The coloration and chemical components of surfaces are important components for thermoregulation, and are often involved in diverse communication systems. (6) A number of specialised surface structures may serve a variety of other functions, such as air retention, food grinding, body cleaning, etc. In spite of a huge number of publications, describing biological surfaces by the use of light and electron microscopy, exact working mechanisms have been clarified only for a few systems, because of the structural and chemical complexity of biological surfaces. However, biological surfaces hide a virtually endless potential for technological ideas for the development of new materials and systems. Because of the broad diversity of functions of biological surfaces, inspirations from biology may be interesting for a broad range of topics in engineering sciences: adhesion, friction, wear, lubrication, filtering, sensorics, wetting phenomena, self-cleaning, anti-fouling, thermoregulation, optics, etc. Since the majority of biological surfaces are multifunctional, it makes them even more interesting from the point of view of biomimetics. In the present book, some structural aspects of biological surfaces in relation to their function are reported. The editor and contributors believe that such a functional approach to biological surfaces will make this book interesting not only for biologists, but also to physicists, engineers and materials scientists.

Editor

Stanislav Gorb is Professor and Director at the Zoological Institute of the University of Kiel, Germany. He received his PhD degree in zoology and entomology at the Schmalhausen Institute of Zoology of the National Academy of Sciences in Kiev, Ukraine. Gorb was a postdoctoral researcher at the University of Vienna, Austria; a research assistant at University of Jena, Germany; a group leader at the Max Planck Institute for Developmental Biology in Tübingen, Germany and at the Max Planck Institute for Metals Research in Stuttgart, Germany. He was a visiting professor at the University of Washington, USA and University of Astronautics and Aeronautics, Nanjing, China. Gorb's research focuses on morphology, structure, biomechanics, physiology, and evolution of surface-related functional systems in animals and plants, as well as the development of biologically inspired technological surfaces and systems. He received the Schlossmann Award in Biology and Materials Science in 1995 and was the 1998 BioFuture Competition winner for his works on biological attachment devices as possible sources for biomimetics. Since 2007, he is the member of the board of the German biomimetics competence network BioKon. Gorb has authored three books, including Attachment Devices of Insect Cuticle and Biological Micro- and Nanotribology, more than 150 papers in peer-reviewed journals, and four patents. He is editor and member of editorial boards of several entomological, zoological and biomimetical journals.

Acknowledgments

The editor would like to thank all contributors of both volumes for excellent collaboration and patience. Zuzana Bernhart's (Springer, Holland) belief in this topic and her personal help in management are greatly acknowledged. Victoria Kastner from Max Planck Institute for Developmental Biology, Tübingen, Germany has helped in polishing English of contributors from 14 countries. Further, the editor would like to express his appreciation to the following colleagues, who took the time to review different book chapters. Their contribution has significantly improved the quality of the book. The individuals who served as reviewers of chapters in this two volume set are:

Zsolt Bálint	Department of Zoology, Hungarian Natural History Museum, Budapest, Hungary
Alison Barker	CABI-Bioscience Centre Switzerland, Delémont, Switzerland
Aaron Bauer	Biology Department, Villanova University, Pennsylvania, PA, USA
Serge Berthier	Institut des Nanosciences de Paris, Campus Boucicaut, Paris, France
Oliver Betz	Department of Evolutionary Biology of Invertebrates, University of Tübingen, Tübingen, Germany
László Péter Biró	Nanotechnology Department, Research Institute for Technical Physics and Materials Science of the Hungarian Academy of Sciences, Budapest, Hungary
Jerome Casas	Institut de Recherche sur la Biologie de l'Insecte (IRBI)/CNRS, University of Tours, Tours, France
Walter Federle	Department of Zoology, University of Cambridge, Cambridge, UK
Laurence Gaume	UMR CNRS 5120 AMAP, Botanique et Bioinformatique de l'architecture des plantes, Montpellier, France
Helen Ghiradella	Department of Biological Sciences, State University of New York at Albany, Albany NY, USA
Elena V. Gorb	Evolutionary Biomaterials Group, Max Planck Institute, Stuttgart, Germany

Hannelore Hoch	Museum für Naturkunde, Humboldt-Universität zu Berlin, Berlin, Germany
Michael Land	Department of Biology and Environmental Science, School of Life Sciences, University of Sussex, Falmer, Brighton, UK
Chris Lawrence	Smart Materials Group (FST), QinetiQ, Cody Technology Park, Farnborough, Hants, UK
Jay McPherson	Department of Zoology, Southern Illinois University, Carbondale, IL, USA
Paul Maderson	Department of Biology, Brooklyn College, New York, NY, USA
John P. Miller	Graduate Group in Neurobiology and Department of Molecular and Cell Biology, University of California, Berkeley, CA, USA
Tommy Nyman	Department of Biology, University of Oulu, Oulu, Finland
Hanno Richter	University of Natural Resources and Applied Life Sciences, Vienna, Austria
George Rogers	School of Molecular and Biomedical Science, University of Adelaide, Adelaide, Australia
Jerome S. Rovner	Department of Zoology, Ohio University, Athens, OH, USA
Andrew M. Smith	Department of Biology, School of Humanities and Sciences, Ithaca College, Ithaca, NY, USA
Annemarie Surlykke	Institute of Biology, University of Southern Denmark, Odense, Danmark
Cameron Tropea	Biotechnik-Zentrum Technical University of Darmstadt, Darmstadt, Germany
Michael Varenberg	Faculty of Mechanical Engineering, Technion – Israel Institute of Technology, Haifa, Israel
Jean-Pol Vigneron	Laboratoire de Physique du Solide, Facultés Universitaires Notre-Dame de la Paix, Namur, Belgium
Dagmar Voigt	Evolutionary Biomaterials Group, Max Planck Institute, Stuttgart, Germany
Tom Weir	Tom Weir, ANIC, CSIRO Entomology, Canberra, Australia
Yehuda Werner	Department of Evolution, Systematics and Ecology, The Hebrew University of Jerusalem, Jerusalem, Israel
Iain Wilkie	School of Life Sciences, Glasgow Caledonian University, Glasgow, Scotland, UK
Giacomo Zaccone	Department of Animal Biology and Marine Ecology, Section of Cell and Evolutionary Biology, Messina University, Messina, Italy

Contents

Volume 2

Part II Adhesion Reduction

Contents

Volume 1

Contributors

Nicholas Aldred University of Newcastle upon Tyne, School of Marine Science and Technology, Newcastle, UK, nicholas.aldred@ncl.ac.uk

Tony Clare University of Newcastle upon Tyne, School of Marine Science and Technology, Newcastle, UK, a.s.clare@ncl.ac.uk

Garrett Clevenger Department of Entomology, Washington State University, Pullman WA, USA

Ana Varela Coelho Instituto de Tecnologia Química e Biológica, Universidade Nova de Lisboa, Mass Spectrometry Laboratory, Oeiras, Portugal

Debasish Das Department of Zoology, Presidency College, Kolkata, India

Sonia Delannoy Department of Biological Sciences, University of Calgary, Calgary, Canada

Bruno Di Giusto UMR CNRS 5120 AMAP, Botanique et Bioinformatique de l'architecture des plantes, Montpellier, France

Hongjian Ding Department of Plant, Soil and Entomological Sciences, University of Idaho, Moscow ID, USA

Sanford D. Eigenbrode Department of Plant, Soil and Entomological Sciences, University of Idaho, Moscow ID, USA, sanforde@uidaho.edu

Patrick Flammang Laboratoire de Biologie Marine, Université de Mons-Hainaut, Mons, Belgium, Patrick.Flammang@umh.ac.be

Leonid I. Frantsevich Department of Ethology and Social Biology of Insects, Schmalhausen Institute of Zoology, Kiev, Ukraine, leopup@izan.kiev.ua

Laurence Gaume UMR CNRS 5120 AMAP, Botanique et Bioinformatique de l'architecture des plantes, Montpellier, France, lgaume@cirad.fr

Dmytro Gladun Department of Ethology and Social Biology of Insects, Schmalhausen Institute of Zoology, Kiev, Ukraine, gladun.dmytro@gmail.com

Elena V. Gorb Evolutionary Biomaterials Group, Max Planck Institute, Stuttgart, Germany, o.gorb@mf.mpg.de

Stanislav N. Gorb Department of Functional Morphology and Biomechanics, Zoological Institute of the University of Kiel, Kiel, Germany, sgorb@zoologie.uni-kiel.de

Michaël Guéroult UMR CNRS 5120 AMAP, Botanique et Bioinformatique de l'architecture des plantes, Montpellier, France

Elise Hennebert Laboratoire de Biologie Marine, Université de Mons-Hainaut, Mons, Belgium, e.hennebert@lasolue.be

Megan Johnson Department of Biological Sciences, University of Calgary, Calgary, Canada

Tapas C. Nag Department of Anatomy, All India Institute of Medical Sciences, New Delhi, India, tapas_nag@hotmail.com

Nick Rowe UMR CNRS 5120 AMAP, Botanique et Bioinformatique de l'architecture des plantes, Montpellier, France, nrowe@cirad.fr

Anthony Russell Department of Biological Sciences, University of Calgary, Calgary, Canada, arussell@ucalgary.ca

Romana Santos Instituto de Tecnologia Química e Biológica, Universidade Nova de Lisboa, Mass Spectrometry Laboratory, Oeiras, Portugal, rsantos@itqb.unl.pt

William E. Snyder Department of Entomology, Washington State University, Pullman WA, USA, wesnyder@wsu.edu

Introduction

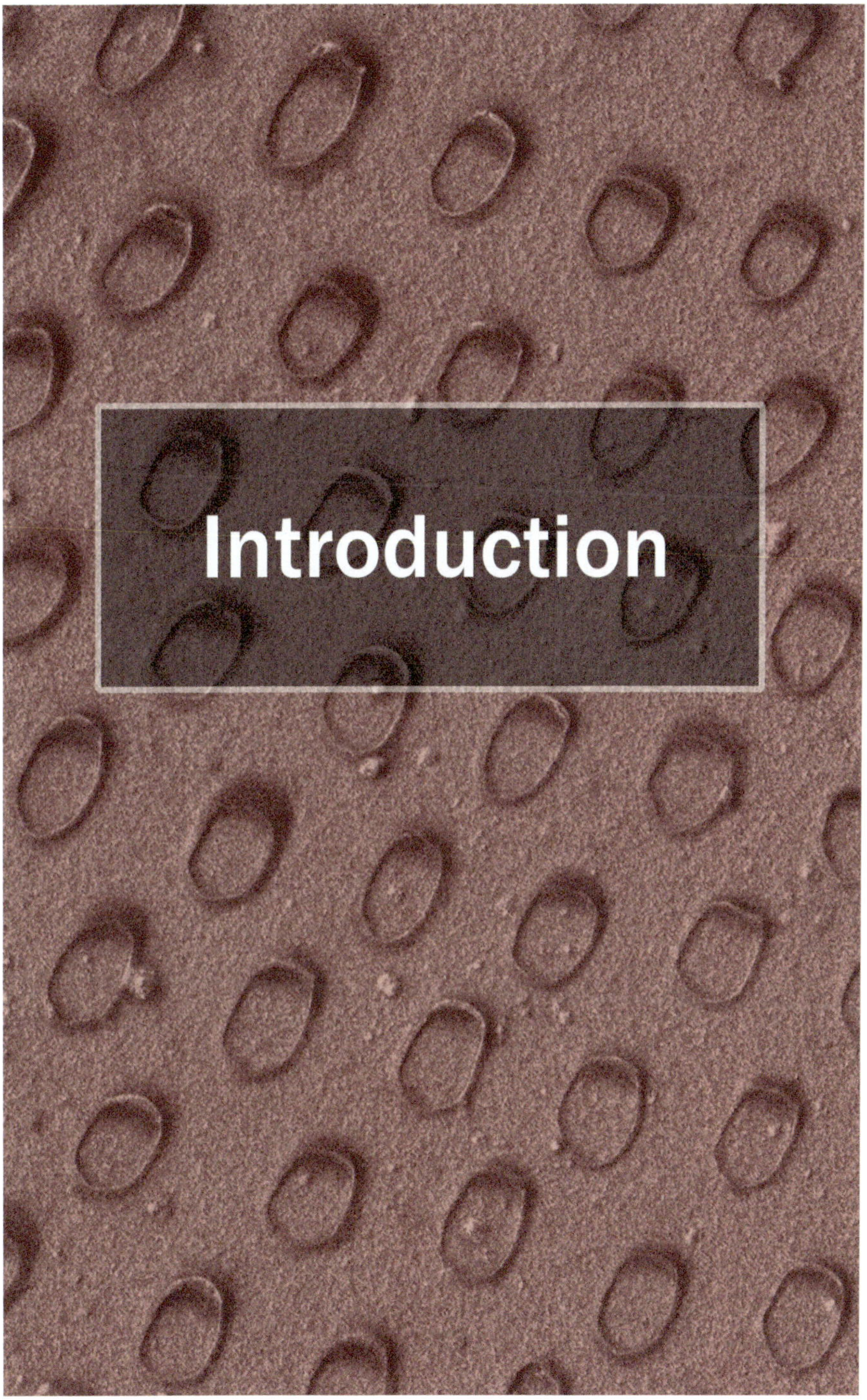

Introduction Adhesion Enhancement and Reduction in Biological Surfaces

Stanislav Gorb

There are numerous publications on cell adhesion 5phenomena, but only few references devoted to the non-specific adhesion of living organisms (Nachtigall, 1974; Gorb, 2001; Scherge and Gorb, 2001; Smith and Callow, 2006). Even less information is published on anti-adhesive surfaces in biology. Because of the structural and chemical complexity of biological surfaces related to adhesion, exact working mechanisms have been clarified only for some systems. The present volume is a continuation of Volume 1: *Functional Surfaces in Biology: Small Structures with Big Effects*. In the present volume, we collected two sets of papers showing biological surfaces and systems specialised for adhesion enhancement (Chapters 16, 17, 18, 19 and 20) and adhesion reduction (Chapters 21, 22, and 23). These contributions discuss adhesive and non-adhesive functions of biological surfaces and their relationship with the structure.

Materials and systems preventing the separation of two surfaces may be defined as adhesives. There are a variety of natural adhesive devices based on entirely mechanical principles, while others additionally rely on the chemistry of polymers and colloids. Adhesive organs are functional systems, the purpose of which is either temporary or permanent attachment of an organism to the substrate surface, to another organism, or temporary interconnection of body parts within an organism. Their design varies enormously and is subject to different functional loads. There is no doubt that many functional solutions have evolved independently in different lineages. Many species of animals and plants are supplied with diverse adhesive surfaces, the morphology of which depends on the species' biology and the particular function, in which the adhesive device is involved. There are eight fundamental classes of attachment principles: (1) hooks, (2) lock or snap, (3) clamp, (4) spacer, (5) suction, (6) expansion anchor, (7) adhesive secretions (glue), and (8) friction. However, different combinations of these principles also occur in existing biological structures. Three types of adhesion at the organism level are known: (1) temporary adhesion allowing an organism to attach strongly to the substrate and detach quickly when necessary; (2) transitory adhesion permitting simultaneous attachment and

S. Gorb (✉)
Zoological Institute of the University of Kiel, Germany
e-mail: sgorb@zoologie.uni-kiel.de

S.N. Gorb (ed.), *Functional Surfaces in Biology*, Vol. 2,
DOI 10.1007/978-1-4020-6695-5_1, © Springer Science+Business Media B.V. 2009

movement along the substrate; (3) permanent adhesion involving the secretion of cement (Flammang, 1996). These three types of adhesion do not have the same purpose and are often based on different physical principles.

This volume begins with Chapter 16 by R. Santos, E. Hennebert, A.V. Coelho and P. Flammang on the echinoderm tube feet adhesion. This adhesion allows these animals to interact with their environment by manipulating items for burrowing or feeding, or to cope structurally with their environment by attaching strongly to the substratum, thus withstanding the action of waves. The tube foot disc often has an enlarged flattened surface, ideal for enhancing contact with the substratum, the ability to deform in order to replicate the surface profile, and possesses internal support structures to bear the tensions associated with adhesion. Moreover, the disc epidermis possesses a duo-glandular adhesive system that produces adhesive secretions for fastening the tube foot to the substratum as well as de-bonding secretions allowing easy detachment.

Chapter 17 by N. Aldred and T. Clare reviews the adhesion mechanism of barnacle cyprids. This contribution discusses the importance of adhesion in barnacle biology and highlights new perspectives for further research on cyprid adhesive mechanism. This chapter brings novel arguments based on analogous systems. It seems that cyprid adhesion relies neither entirely on a viscous mechanism, facilitated by the antennular secretion, or on van der Waals interactions, but presumably requires a combination of the two.

D. Gladun, S.N. Gorb and L.I. Frantsevich in Chapter 18 have reviewed structure and adhesive function of insect arolia, which belong to the smooth type of insect adhesive devices. There are two main types of arolia: those with constant shape and those able to fold and spread. Thysanoptera, Diptera-Tipulomorpha, Mecoptera, Hymenoptera, and Trichoptera possess the second, more complex type of arolia. Contact material of the arolium represents the nanocomposite of elastic cuticular dendrites and a viscous fluid accommodating the dendrites. Tight contact of the arolium bottom with the 3D profile of the ground is ensured by a fluid of rather complex and, probably, even biphasic (emulsion) composition. Complex movements of arolium structures during unfolding, spreading and folding are provided by a multi-sclerite mechanism, in which simple neural control is compensated for by a complex structure and material properties.

Adhesive organs, by which the teleost fish species adhere to submerged rocks and stones of streams, are investigated by D. Das and T.C. Nag in Chapter 19. Adhesive organs are located ventrally at the thoracic region behind the opercular openings or encircle the mouth opening. In mountain-stream catfish (Sisoridae), additional adhesive devices are present on the ventral surface of the paired pectoral and pelvic fins. In these fish, the skin of the outer rays is transformed into a series of alternate ridges and grooves. In their contribution, the authors have discussed material composition of these structures and their functional mechanism.

Gecko adhesion mechanism has been recently stressed in numerous publications (Autumn, 2006). However, many questions about ecological significance of the gecko toe micro- and nanostructures and their operation under natural conditions still remain unresolved. Chapter 20 written by M. Johnson, A. Russell and

S. Delannoy discuss questions of (1) what precise locomotory advantage do setae provide that cannot be adequately met by claws or digital flexure?, and (2) is there any evidence that some or all seta-bearing taxa live on surfaces that have certain physical characteristics that make setae particularly advantageous? The authors have investigated animals and their preferred locomotor substratum and concluded that maximization of contact and adhesive force generation is of major importance, and extensive contact splitting will assist this. Exploitation of smooth surfaces, using such an elaborate surface, is likely a secondary eventuality and not a driving force in the evolution of the configuration and dimensions of the gecko adhesion system.

Some biological systems have developed surfaces covered with micro- and nanostructures having anti-adhesive property. It is known that wax crystalloids on the flowering shoots of plants are adaptations to prevent crawling insects from robbing nectar and other resources (Eigenbrode and Kabalo, 1999). The wax blooms of ant-plants from the genus *Macaranga* seem to be an ecological isolation mechanism for the symbiotic ants. This mechanism is based exclusively on the influence of the ant attachment abilities, but not on the repellent effects of the wax (Federle et al., 1997). To explain the anti-adhesive properties of plant surfaces covered with waxes, several hypotheses have been previously proposed: the roughness-hypothesis, the contamination-hypothesis, the wax-dissolving-hypothesis, and the fluid-absorption-hypothesis (Gorb and Gorb, 2002). The second part of the volume is devoted to anti-adhesive plant surfaces having different insect-related biological functions.

Chapter 21 by S.D. Eigenbrode, W.E. Snyder, G. Clevenger, H. Ding and S.N. Gorb provides specific evidence for the plant surface impact on the foraging of insect predators and parasitoids. This contribution focuses on the role of crystalline waxes on plant surfaces in mediating these types of interactions through their effects on insect attachment. Finally it illustrates the implications of the variability of plant surface waxes and insect responses to surface waxes through a case study examining the attachment and performance of five species of predatory beetle on plants differing in surface wax.

The last two chapters are devoted to the anti-adhesive *Nepenthes* pitcher plant surfaces involved in insect trapping and retention. Chapter 22 by B. Di Giusto, M. Guéroult, N. Rowe and L. Gaume aimed at characterizing the polymorphism of the waxy layer in *N. rafflesiana*, testing whether this layer plays a preponderant role in insect trapping in the species. The authors ask whether the non-adhesive waxy layer of the pitcher varies with plant ontogeny and whether the waxy layer provides a substantial benefit to the plant. Waxy and non-waxy phenotypes were compared for their retention effect on ants and flies and for prey abundance. Based on the literature data and their own results, the authors discuss how and why some *Nepenthes* taxa could have lost their waxy layer during evolution.

The aim of Chapter 23 by E.V. Gorb and S.N. Gorb was to re-examine the structure and microtopography of epidermal surfaces in functional zones of *N. alata* pitchers, using a new cryo-SEM method allowing high-resolution imaging of frozen and fractured samples. The existing electron microscopy studies of *Nepenthes* pitchers have used the conventional technique of sample preparation, including treatment in strong solvents. Since surfaces are covered with layers of waxes or fluids, it was previously impossible to combine high-resolution imaging with keeping samples in a native condition. The new technique allowed obtaining a fresh insight into the ultrastructure of superficial layers in different pitcher zones under native conditions at high resolution. Based on new results and literature data, the role of different functional pitcher surfaces in prey capturing and retention is discussed.

The two volumes on *Functional Surfaces in Biology,* taken together, present an overview of the field. They provide a reference for a novice in the field. The chapters generally have an overview along with new research data. The volumes are also intended for use by researchers who are active, or intend to become active, in the field. The appeal of this topic is expected to be broad, ranging from classical biology, biomechanics and physics to surface engineering.

References

Autumn, K. (2006) How gecko toes stick. Am. Sci. 94: 124–132.

Eigenbrode, S.D., and Kabalo, N.N. (1999) Effects of *Brassica oleracea* waxblooms on predation and attachment by *Hippodomia convergens*. Ent. Exp. Appl. 91: 125–130.

Federle, W., Maschwitz, U., Fiala, B., Riederer, M., and Hölldobler, B. (1997) Slippery ant-plants and skilful climbers: selection and protection of specific ant partners by epicuticular wax blooms in *Macaranga* (Euphorbiaceae). Oecologia 112: 217–224.

Flammang, P. (1996) Adhesion in echinoderms. Echinoderm Stud. 5: 1–60.

Gorb, S.N. (2001) *Attachment Devices of Insect Cuticle*. Dordrecht, Boston, London: Kluwer Academic Publishers.

Gorb, S.N. (2006) Functional surfaces in biology: mechanisms and applications. In: *Biomimetics: Biologically Inspired Technologies*, edited by Y. Bar-Cohen, Boca Raton: CRC Press, pp. 381–397.

Gorb, E.V., and Gorb, S.N. (2002) Attachment ability of the beetle *Chrysolina fastuosa* on various plant surfaces. Ent. Exp. Appl. 105: 13–28.

Nachtigall, W. (1974) *Biological Mechanisms of Attachment*. Berlin, Heidelberg, New York: Springer-Verlag.

Scherge, M., and Gorb, S.N. (2001) *Biological Micro- and Nanotribology*. Berlin: Springer-Verlag.

Smith, A.M., and Callow, J.A. (2006) *Biological Adhesives*. Berlin, Heidelberg, New York: Springer-Verlag.

Part I

Adhesion Enhancement

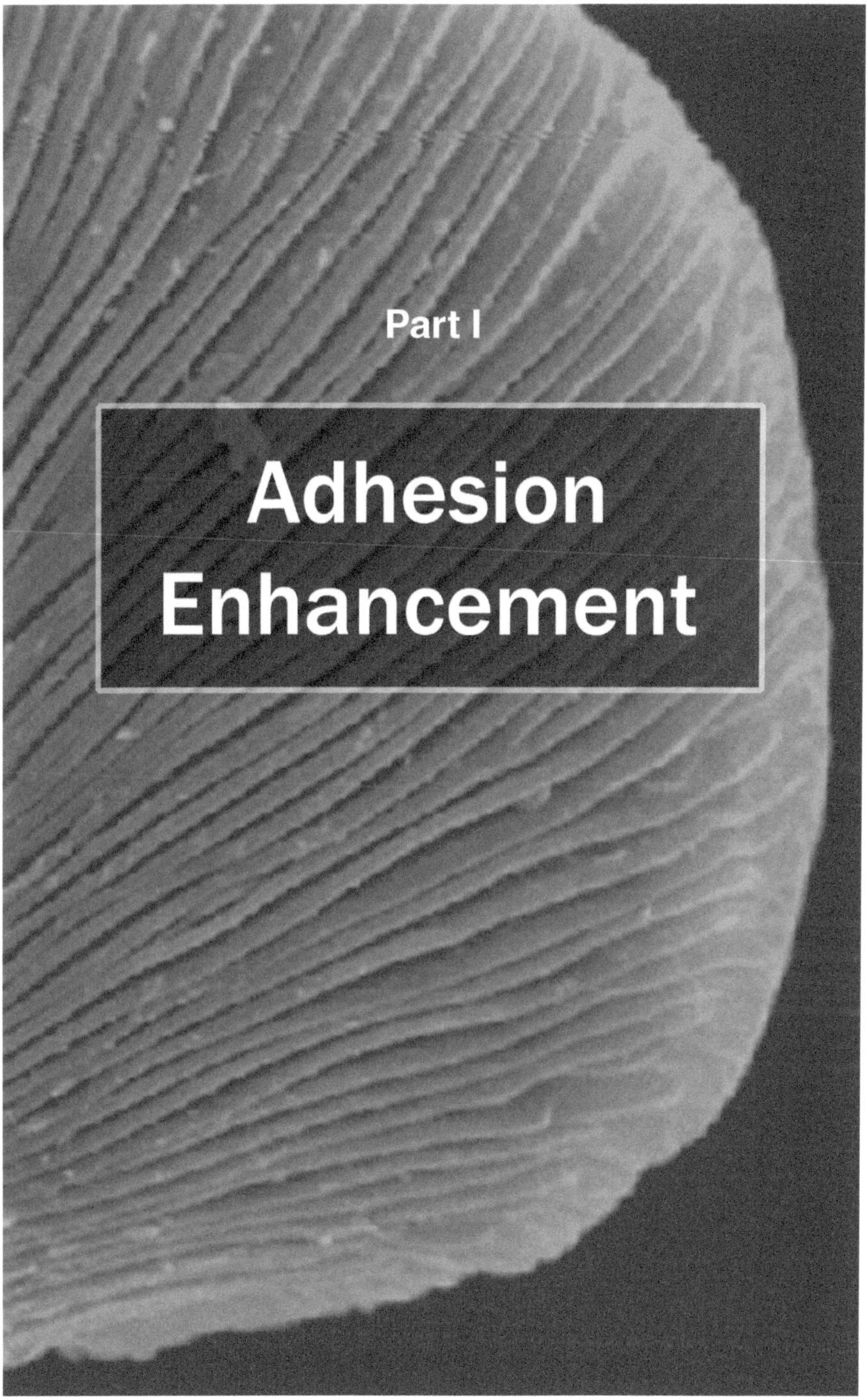

Chapter 1
The Echinoderm Tube Foot and its Role in Temporary Underwater Adhesion

Romana Santos, Elise Hennebert, Ana Varela Coelho and Patrick Flammang

1.1 Introduction

1.1.1 Adhesion in the Sea

Adhesion (attachment with adhesive secretions) is a way of life in the sea (Waite, 1983). Indeed, representatives of bacteria, protoctists (including macroalgae), and all animal phyla, living in the sea attach to natural or artificial surfaces. Adhesion ability is particularly developed and diversified in invertebrates, which adhere during their larval and adult life (Walker, 1987; Smith and Callow, 2006). It is involved in various functions such as the handling of food, the building of tubes or burrowing and, especially, the attachment to the substratum (Walker, 1987; Tyler, 1988; Whittington and Cribb, 2001; Flammang et al., 2005). Indeed, seawater, being a dense medium, denies gravity to hold organisms to the bottom. Thus, to withstand the hydrodynamic forces, marine organisms rely on specialised adhesive mechanisms. Adhesion to the substratum may be permanent, transitory or temporary (Tyler, 1988; Whittington and Cribb, 2001; Flammang et al., 2005). Permanent adhesion involves the secretion of a cement and is characteristic of sessile organisms staying at the same place throughout their adult life (such organisms have representatives among sponges, hydrozoan cnidarians, cirripede crustaceans, bivalve molluscs, tubicolous polychaetes, bryozoans or tunicates) (Walker, 1987). Transitory adhesion allows simultaneous adhesion and locomotion: the animals attach by a viscous film laid down between their body and the substratum, and creep on the film which is left behind as animals move. This type of adhesion is characteristic of invertebrates, mostly small, soft-bodied ones, such as turbellarians, nemertines, gastrotrichs, or polychaetes, moving along the substratum by ciliary

R. Santos (✉)
Instituto de Tecnologia Química e Biológica, Universidade Nova de Lisboa, Mass Spectrometry Laboratory, Oeiras, Portugal
e-mail: romana_santos@yahoo.com

P. Flammang (✉)
Université de Mons, Laboratoire de Biologie Marine, Mons, Belgium
e-mail: Patrick.Flammang@umons.ac.be

S.N. Gorb (ed.), *Functional Surfaces in Biology*, Vol. 2,
DOI 10.1007/978-1-4020-6695-5_2, © Springer Science+Business Media B.V. 2009

gliding (Tyler, 1988; Whittington and Cribb, 2001). Larger animals, such as sea anemones and gastropod molluscs, also use transitory adhesion, but they move by means of waves of muscular contractions running along their foot (Walker, 1987). Temporary adhesion allows organisms to attach firmly but momentarily to a substratum. This type of adhesion is common in small invertebrates inhabiting the interstitial environment, e.g. in turbellarians, gastrotrichs, nematodes, and polychaetes (Tyler, 1988). A few macro-invertebrates, such as some cnidarians and most echinoderms, can also attach and detach repeatedly (Flammang, 1996).

Chemical adhesion is the joining together of two dissimilar materials, the adherends, using a sticky material, the adhesive. The surface properties of the adherends and the properties of the adhesive determine the strength of adhesion (Waite, 1983). For marine invertebrates, the adherends are the animal's integument on one side and an exogenous substratum on the other. The latter may be abiotic (e.g., rock) or biotic (e.g., algal blades or the integument of another animal); it may also be an artificial substratum, such as a ship hull or a pier pillar. In addition, these surfaces may be covered by a biofilm consisting of adsorbed macromolecules and bacteria, whose composition may be variable spatially as well as temporally (Characklis, 1981; Fletcher, 1994). It is clear therefore that, for their attachment, marine invertebrates must be able to cope with substrata varying greatly in their chemical and physical characteristics. For sessile invertebrates, using permanent adhesion, the problem of the adaptation to the substratum occurs only once, at the time of fixation. In these organisms, a copious amount of the adhesive is secreted as a fluid, filling the gap between the animal base and the substratum and then gradually solidifying to form a cement with high adhesive and cohesive strengths (Walker, 1987; Kamino, 2006). Once attached, these organisms are not able to move anymore or self-detach, even though the holdfast may break, be abandoned, or deteriorate in time. On the other hand, motile invertebrates using non-permanent adhesion (either transitory or temporary) can move around and attach themselves strongly but momentarily to the substratum (Walker, 1987; Tyler, 1988; Flammang et al., 2005). They may, therefore, encounter various substrata during their displacements, and their adhesive organs must be able to sense quality of substrata, adapt to the selected substratum to spare adhesive material, provide enough tensile strength, and finally possess means of detachment to allow movement. Thus, adhesive organs, involved in non-permanent adhesion form a multifunctional interface for animal-substratum interactions.

1.1.2 Echinoderm Tube Feet

Being exclusively benthic animals, echinoderms have activities and adaptations that are correlated with their relationship to the sea bottom. Most of these activities, such as attachment to the substratum, locomotion, handling of food and burrow-building, rely on adhesive secretions allowing the animal to stick to or to manipulate a substratum. In post-metamorphic echinoderms, these adhesive secretions are always produced by specialized organs, the podia or tube feet. These are the external appendages of the water-vascular system and are also probably the most advanced hydraulic organs in the animal kingdom (Nichols, 1966). Tube foot attachment is typically temporary adhesion. Indeed, although tube feet can adhere very strongly to the substratum, they are also able to detach easily and voluntarily from the substratum before reinitiating another attachment-detachment cycle (Thomas and Hermans, 1985; Flammang, 1996). Attachment takes place at the level of adhesive areas, which are specialized sensory-secretory areas of the tube foot epidermis.

From their presumed origin as simple respiratory evaginations of the ambulacral system (Nichols, 1966), tube feet have diversified into the wide range of specialized structures found in extant echinoderms. This morphological diversity reflects the variety of functions of tube feet (Lawrence, 1987). Indeed, they take part in locomotion, burrowing, feeding, sensory reception and respiration. In some groups, a single type of tube foot fulfils different functions; in others, different types of tube feet are specialised, each in one particular function. Based on their external morphology only, tube feet can be classified into six broad types: disc-ending (Fig. 1.1A), penicillate, knob-ending (Fig. 1.1B), lamellate, digitate (Fig. 1.1C), and ramified (Fig. 1.1D) (Flammang, 1996). Adhesive areas are organized differently according to the morphotype, and this organization represents the first stage of specialization of the tube feet. Tube feet that capture or manipulate small particles present an adhesive area fragmented into small, discrete zones (Flammang, 1996). This is the case, for example, in the adhesive papillae scattered on the tube feet of filter-feeding ophiuroids (Fig. 1.1C). The fragmented adhesive area reaches its highest development in the dendritic buccal tentacles occurring in some holothuroids in which the adhesive zones are located at the tip of each ramification (Fig. 1.1D and E). Discrete adhesive zones are presumably more efficient in the handling of small particles; conversely, a large adhesive area provides a strong attachment site for tube feet involved in locomotion or in maintaining position (Flammang, 1996). Such large adhesive areas occur, therefore, on the distal surface of the disc in disc-ending tube feet (Fig. 1.1A), or on the surface of the knob in knob-ending tube feet (Fig. 1.1B). Thus, the shape of the adhesive area strongly correlates with the tube foot function: attachment to solid substrata and digging in soft substrata, respectively.

For practical reasons (relatively large size and high adhesion force of the tube feet), only disc-ending tube feet of both asteroids and regular echinoids have been studied in detail in terms of their adhesive strength and mechanical properties. Data on echinoderm attachment strengths are scarce in the literature, and experimental results are usually reported as attachment forces (Table 1.1). There are more data on sea urchins than on sea stars, presumably because it is easier to grab and pull a spherical and rigid echinoid than a multi-armed, soft-bodied sea star. Although sea urchins appear to adhere generally more strongly than sea stars, both can attach with forces up to about 100 N (Märkel and Titschack, 1965; Siddon and Witman, 2003; Santos and Flammang, 2007). Attachment force may even reach 250 N in the echinoid *Colobocentrotus atratus* (Gallien, 1986; Santos and Flammang, 2008). Some measurements have been performed in the field (Märkel and Titschack, 1965; Gallien, 1986; Siddon and Witman, 2003; Santos and Flammang, 2007), other data come from aquarium measurements (Sharp and Gray, 1962; Yamasaki et al., 1993; Berger and Naumov, 1996; Guidetti and Mori, 2005; Santos and Flammang, 2007).

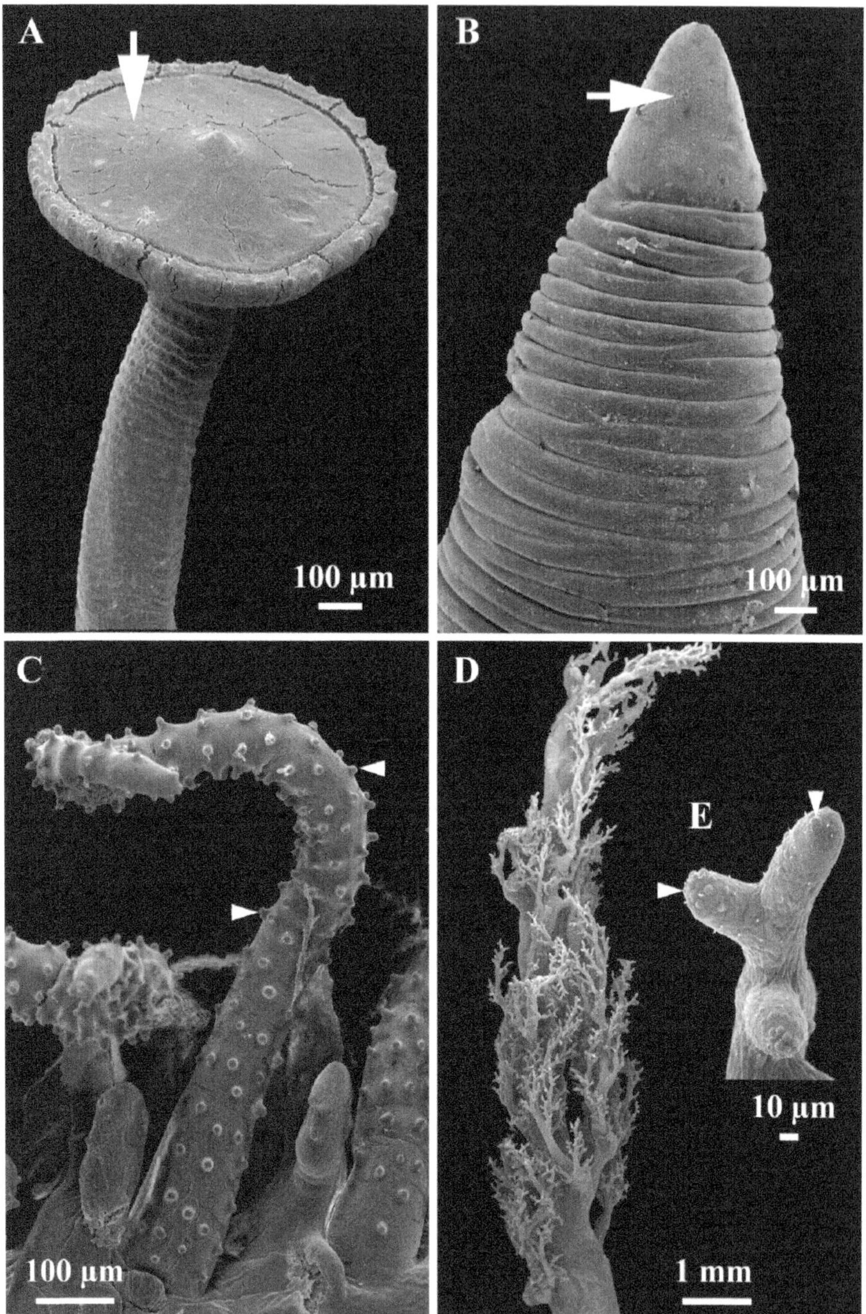

Fig. 1.1

Table 1.1 Critical detachment forces measured for sea stars and sea urchins attached by their disc-ending tube feet

Species	Experimental conditions	Substratum	Force (N)	Refs
Asteroids				
Asterias rubens	Laboratory	Glass	4.8 ± 0.3 (M ± 95%CI)	1
	Laboratory	Rock	5.7 ± 0.5 (M ± 95%CI)	1
Asterias forbesi	Field	Rock	30.8 ± 1.4 (M ± SE)	2
Echinoids				
Arbacia lixula	Field	Rock	Up to 45	3
	Laboratory	Rock	2.8 – 33	4
	Laboratory	Glass	3.5 ± 2.1 (M ± SD)	5
Colobocentrotus atratus	Field	Rock	Up to 250	6
Paracentrotus lividus	Field	Rock	Up to 82	3
	Laboratory	Rock	1.1 – 14.5	4
	Field	Rock	9 – 87	3
	Laboratory	Glass	13.2 ± 9.1 (M ± SD)	5
Sphaerechinus granularis	Laboratory	Glass	33.8 ± 17.9 (M ± SD)	5
Strongylocentrotus droebachiensis	Field	Rock	42.4 ± 1.8 (M ± SE)	2
Strongylocentrotus intermedius	Laboratory	Glass	3 – 28	7
Strongylocentrotus nudus	Laboratory	Glass	2 – 24	7

1, Berger and Naumov, 1996; 2, Siddon and Witman, 2003; 3, Märkel and Titschack, 1965; 4, Guidetti and Mori, 2005; 5, Santos and Flammang, 2007; 6, Gallien, 1986; 7, Yamasaki et al., 1993. M, mean; SD, standard deviation; SE, standard error; CI, confidence interval.

For both sea stars and sea urchins, measurements taken in the field are usually higher than those taken in the laboratory (Table 1.1). Berger and Naumov (1996) also reported that, in the sea star *Asterias rubens*, ability to attach strongly to the substratum quickly drops when animals are held in aquaria.

Attachment strength may also be evaluated by measuring the tenacity, which is the adhesive force per unit area and is expressed in Pascals (Pa). Tenacity of single tube feet has been quantified in several species of asteroids and echinoids under different conditions. Mean tenacity ranges from 0.17 to 0.21 MPa in asteroids (Paine, 1926; Flammang and Walker, 1997; Santos et al., 2005a), and from 0.09 to 0.47 MPa in echinoids (Santos et al., 2005a; Santos and Flammang, 2006). Tenacity was shown to be dependent on the chemical and physical characteristics of the surface to which the tube foot adheres (see Section 2.3. below). All these values are in the same range as those observed in other marine invertebrates using

Fig. 1.1 Morphological and functional diversity in echinoderm tube feet (original SEM pictures; for comparison, all tube feet have been oriented distal end up). (**A**) Disc-ending tube foot of the echinoid *Heterocentrotus trigonarius*. (**B**) Knob-ending tube foot of the asteroid *Astropecten irregularis*. (**C**) Digitate tube foot of the ophiuroid *Ophiothrix fragilis*. (**D**) Ramified tube foot (buccal tentacle) of the holothuroid *Phyllophorus spiculata*, with a close-up on the tip of the smallest branches (**E**). Arrows indicate large adhesive areas and arrow heads small adhesive zones (see text for details) non-permanent adhesion (0.1 to 0.5 MPa) and approach the adhesive strength of permanent adhesives, which is typically 0.5–1 MPa (Smith, 2006). As a comparison, current technological requirements for underwater adhesives are in the range of 0.2–0.7 MPa (see Waite, 2002). Tube foot temporary adhesion compares well therefore to both natural and synthetic adhesives.

1.2 Functional Organisation of Disc-Ending Tube Feet

1.2.1 Overview

Disc-ending tube feet consist of a basal hollow cylinder, the stem, and an enlarged and flattened apical extremity, the disc (Fig. 1.2). The different constituents making up these two parts act cooperatively to make tube feet an efficient holdfast, allowing sea stars and sea urchins to resist hydrodynamically generated forces, but also to perform rather elaborate tasks such as climbing, righting, covering or shell opening (Lawrence, 1987). A generalized model of component functions of a typical disc-ending tube foot is illustrated in Fig. 1.3. The disc makes contact with the substratum (Figs. 1.2 and 1.3). It adapts to the surface profile, produces the adhesive secretion that fastens the tube foot to the substratum, and encloses support structures that bear the tensions associated with adhesion. It also produces the de-adhesive secretion that allows detachment of the tube foot. The stem, on the other hand, acts as a tough tether connecting the disc to the animal's body. It is also mobile and flexible, and thus gives the tube foot the capacity to perform various movements.

1.2.2 The Stem

The stem wall of echinoderm tube feet consists of four tissue layers: (1) an outer epidermis, (2) a basiepidermal nerve plexus, (3) a connective tissue layer, and (4) an inner mesothelium that surrounds the water-vascular lumen (Fig. 1.4A)

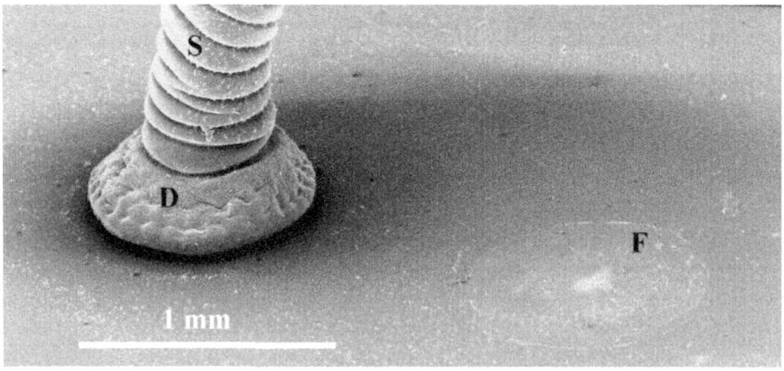

Fig. 1.2 SEM photograph of a disc-ending tube foot of the echinoid *Paracentrotus lividus* attached to a smooth glass substratum (original). The picture also shows an adhesive footprint (F) left by another tube foot after detachment. D, disc; S, stem

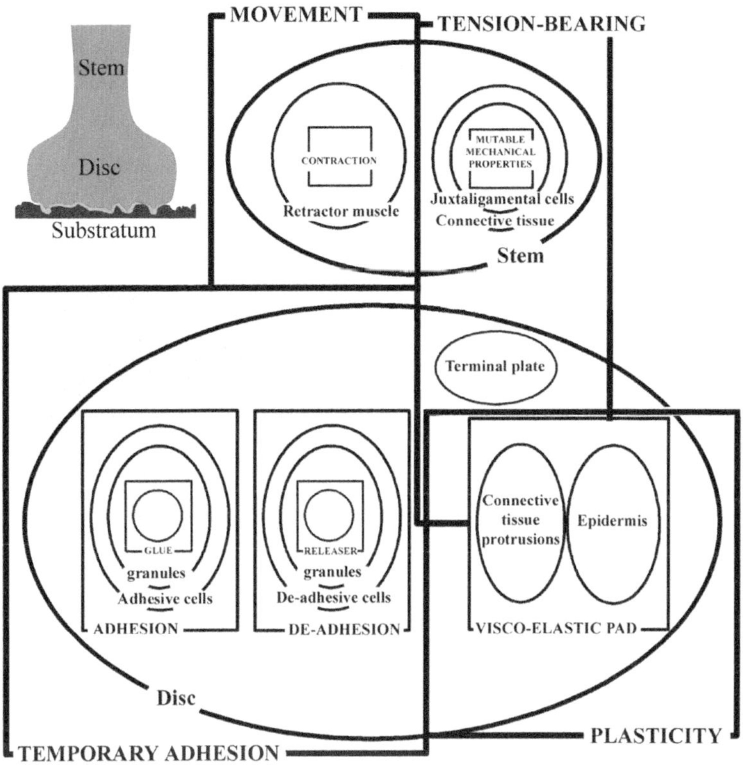

Fig. 1.3 Schematic representation of component functions in a typical disc-ending tube foot adhering to a rough substratum. Functions marked by *boxes*; structures performing them as *ovals* (according to the model of Tyler, 1988)

(Flammang, 1996). The epidermis is a monostratified epithelium which covers the tube foot externally (Flammang, 1996). It encloses support cells, sensory cells and various types of secretory cells (Holland, 1984). The epidermis is coated by a well-developed, multi-layered glycocalyx, the so-called cuticle (Ameye et al., 2000). The nerve plexus is a cylindrical sheath of ectoneural nervous tissue (Cobb, 1987). It is thickened on one side of the tube foot to form the longitudinal nerve, and also at the proximal and distal extremities to form two nerve rings (Nichols, 1966). Nerve cell bodies occur in the longitudinal podial nerve and in the nerve rings; the rest of the plexus consists of a criss-cross of nerve processes (Flammang, 1996). The connective tissue layer is made up of an amorphous ground substance that encloses bundles of collagen fibrils (i.e. fibres), elongated cells with an electron-dense cytoplasm which may be fibrocytes and various types of mesenchymal cells (macrophages, spherulocytes, etc.) (Flammang, 1996; Fig. 1.4A, B). The connective tissue layer is organized into an outer and an inner sheath. Collagen fibres are oriented helicoidally in the inner sheath while they are oriented longitudinally in the outer sheath. The outer sheath also contains juxtaligamental cells filled with electron-dense granules

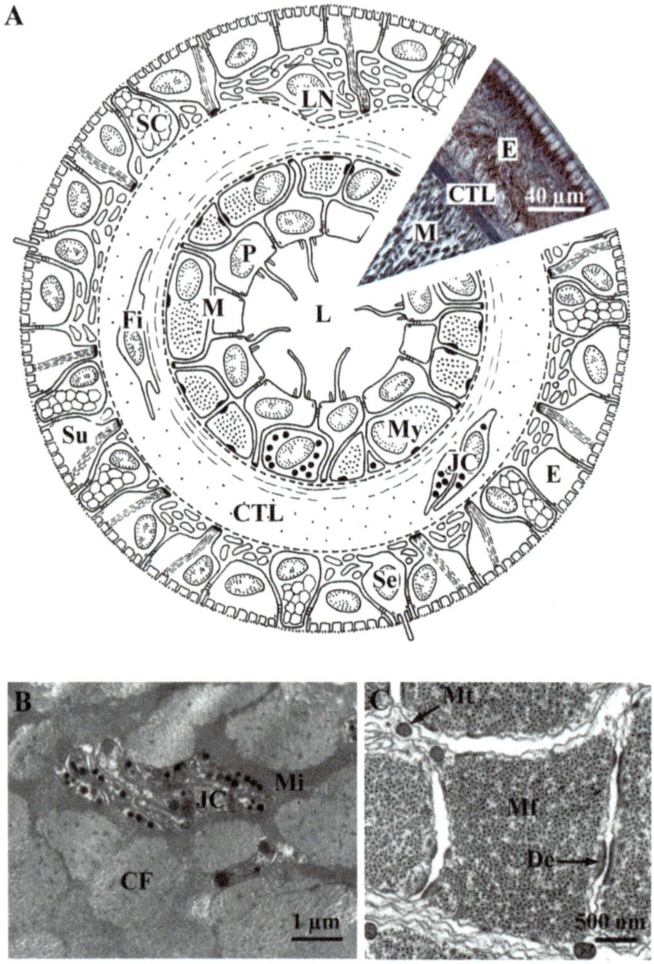

Fig. 1.4 Morphology and ultrastructure of the echinoderm tube foot stem. (**A**) Reconstruction of a transverse section through a typical stem of asteroid or echinoid disc-ending tube foot (modified from Flammang, 1996). The inset is a light micrograph of a histological section through the stem wall in the sea star *Asterias rubens* (original). (**B**) Transverse section through the outer sheath of the connective tissue layer (original TEM picture from the asteroid *Marthasterias glacialis*). (**C**) Transverse section through a myocyte of the mesothelium (original TEM picture from the holothuroid *Holothuria forskali*). Abbreviations: CF, collagen fiber; CTL, connective tissue layer; De, desmosome; E, epidermis; Fi, fibrocyte; JC, juxtaligamental cell; L, ambulacral lumen; LN, longitudinal nerve; M, mesothelium; Mf, myofibril; Mi, microfibrillar network; Mt, mitochondria; My, myocyte; P, peritoneocyte; SC, secretory cell; Se, sensory cell; Su, support cell

and, in sea stars, a well developed microfibrillar network surrounding collagen fibres (Fig. 1.4B) (Santos and Flammang, 2005, Santos et al., 2005b, Hennebert and Flammang, unpubl. obs.). The mesothelium is a myoepithelium comprising two main cell types: peritoneocytes and myocytes (Fig. 1.4A). The myocytes contain a bundle of myofilaments associated with numerous mitochondria, and are connected one to another by spot desmosomes (Fig. 1.4C) (Wood and Cavey, 1981; Flammang, 1996). The myofibrils are always oriented longitudinally and together they form an extensive longitudinal muscle layer (viz. the retractor muscle of the tube foot; Flammang, 1996).

One important function of tube foot stems in both sea stars and sea urchins is to bear tensions applied to the animal by external forces. This load-bearing function may be critical for survival. Indeed, when asteroids and echinoids are subjected to a constant pull, a considerable proportion of the tube feet rupture before the disc is detached from the substratum (Märkel and Titschack, 1965; Smith, 1978; Yamasaki et al., 1993; Berger and Naumov, 1996; Flammang and Walker, 1997; Santos and Flammang, 2007). Among the stem tissues, only the connective tissue and the retractor muscle contain fibrillar elements (i.e., collagen fibres, microfibrils, myofilaments) oriented in parallel to the tube foot axis in the direction of the tensile stress. The comprehension of their respective mechanical properties is therefore important to understand the functioning of disc-ending tube feet.

1.2.2.1 Connective Tissue

Stem mechanical properties of disc-ending tube feet have been investigated in a few species of asteroids and echinoids, always by tensile testing (Leddy and Johnson, 2000; Santos and Flammang, 2005; Santos et al., 2005b; Hennebert and Flammang, unpubl. obs.). Measurements were done on tube feet either in the so-called standard state (i.e. in artificial sea water, ASW; Table 1.2) or in the relaxed state (i.e., in a $MgCl_2$ solution). This last solution prevents muscle contraction and was therefore used to measure the passive material properties of the stem (Leddy and Johnson, 2000). However, it was later demonstrated that the $MgCl_2$ solution also influences the mechanical state of the connective tissue layer (Santos and

Table 1.2 Tensile mechanical properties of asteroid and echinoid tube foot stems measured in sea water. Values of extensibility and strength are expressed as true strain [$\varepsilon=\ln(L/L_0)$] and true stress [$\sigma=(F/S)(L/L_0)$], respectively, because of the high extensions observed for echinoderm tube feet (Shadwick, 1992). Stiffness was calculated on the true stress-true strain curve; toughness was estimated from the force-extension curve

Species	Extensibility	Strength (MPa)	Stiffness (MPa)	Toughness (MJ m^{-3})	Refs
Asteroids					
Asterias rubens	1.6	21	122	3.5	1
Marthasterias glacialis	1.8	23	170	3.7	1
	1.3	13	68	2.6	3
Echinoids					
Arbacia lixula	0.83	23	152	2.9	2
Paracentrotus lividus	0.87	29	328	2.5	2
	1.1	35	273	4.7	3
Sphaerechinus granularis	0.93	24	183	2.9	2

1, Hennebert and Flammang, unpubl. obs.; 2, Santos and Flammang, 2005; 3, Santos et al., 2005b.

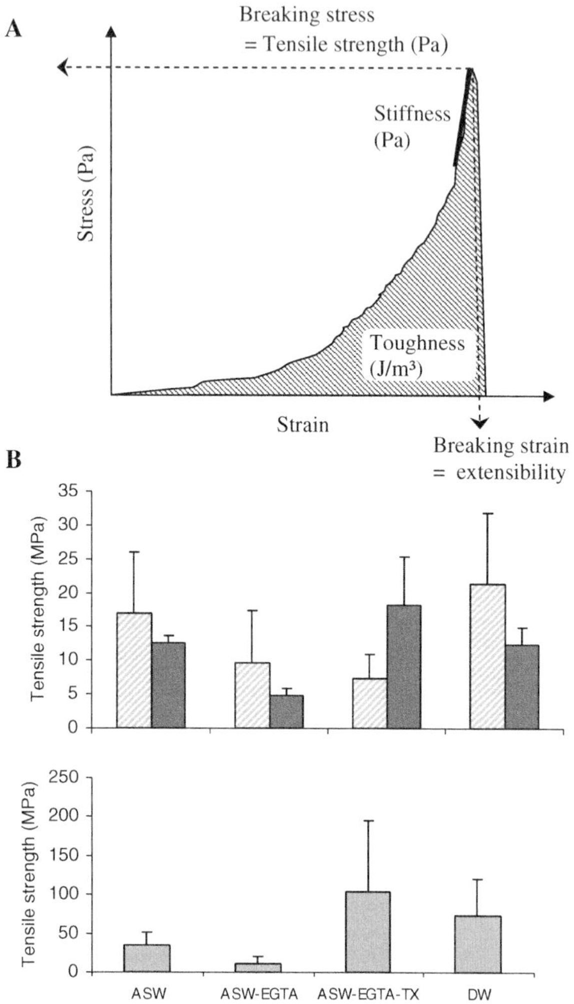

Fig. 1.5 Tensile mechanical properties of the echinoderm tube foot stem. (**A**) Typical J-shaped stress-strain curve for the stem of asteroid and echinoid tube feet, showing the different material properties measured. (**B**) Variation of the tensile strength of the tube foot stem (mean ± SD, n≥3) of the asteroids *Asterias rubens* and *Marthasterias glacialis* (light and dark *grey* bars in top graph, respectively) and of the echinoid *Paracentrotus lividus* (bottom graph) in different bathing solutions. See text for abbreviations

Flammang, 2005; Santos et al., 2005b; see below). Whatever the solution in which they are pulled, tube feet always present a similar complex stress-strain curve in which stress increases with strain, first slowly, and then more rapidly until the stem ruptures (Fig. 1.5A). This results in a typical J-shaped stress-strain curve, which is characteristic of shock-absorbing materials, such as mammalian skin and artery (Vincent, 1990; Vogel, 2003). Four material properties can be calculated from this curve (Fig. 1.5A): (1) extensibility, which is the value of strain when the stem fails (breaking strain); (2) strength, which is the maximal value of stress (breaking stress); (3) stiffness, which is calculated as the slope of the last portion of the stress-strain curve; and (4) toughness, which is a measure of the energy required to extend and break the stem (it corresponds to the area under the stress-strain curve) (see e.g. Santos et al., 2005b). The values of these parameters in ASW are presented in

Table 1.2. Sea star tube foot stems are more extensible (270–500%) than those of sea urchins (130–200%), but they are less stiff than those in echinoids. This makes the tube foot stems in both groups roughly equivalent in terms of toughness (Table 1.2). Tube feet bathed in a $MgCl_2$ solution are slightly more extensible but much less strong, stiff and tough than tube feet left in ASW (Santos and Flammang, 2005). In the $MgCl_2$solution, the material properties measured for the tube feet of the echinoid *Strongylocentrotus droebachiensis* are very close to those of *Arbacia lixula* (Leddy and Johnson, 2000; Santos and Flammang, 2005). These authors have also reported that the stem material properties are strain rate dependent (extensibility, strength, stiffness, and toughness increase as strain rate increases). In the natural environment, this means that tube foot stems possess a higher resistance to rapid loads (such as waves) than to slower, self-imposed loads (such as natural extension) (Leddy and Johnson, 2000; Santos and Flammang, 2005).

Recently, it has been demonstrated that the connective tissue of the stem wall of the tube feet from the echinoid *Paracentrotus lividus* and the asteroids *Asterias rubens* and *Marthasterias glacialis* is a "mutable collagenous tissue" (MCT) (Santos et al., 2005b; Hennebert and Flammang, unpubl. obs.). MCTs which are characteristic of echinoderms, can undergo rapid changes in their passive mechanical properties under nervous control via a specialized cell type, the juxtaligamental cells (Trotter, et al., 2000; Wilkie, 1996, 2005). These cells are neurosecretory cells which are believed to control the mechanical properties of the connective tissue by the secretion of compounds modulating interactions between collagen fibrils (Wilkie, 2005). The mutable character of the connective tissue is usually demonstrated by mechanical tests in solutions depleted of calcium (e.g., ASW in which $CaCl_2$ is replaced by EGTA, a chelator of calcium [ASW-EGTA]), or in solutions causing destabilisation of cell membranes by the action of surfactants (e.g., ASW-EGTA containing Triton-X100 [ASW-EGTA-TX]) or by osmotic shock (e.g., deionised water [DW]). These treatments are known to influence the physiological state of MCTs, the former mimicking its soft state and the latter inducing its stiff state (Trotter and Koob, 1995; Szulgit and Shadwick, 2000). The outer connective tissue sheath of the stem wall in asteroid and echinoid tube feet contain numerous juxtaligamental cell processes with electron-dense granules, as in other known echinoderm MCTs (Fig. 1.4B; Santos et al., 2005b; Hennebert and Flammang, unpubl. obs.). Moreover, tube feet from sea urchins and sea stars show a decrease in tensile strength when placed in a calcium-free solution (Fig. 1.5B) (Santos et al., 2005b; Hennebert and Flammang, unpubl. obs.). In sea urchins, all the values of measured material properties (except extensibility) significantly increase when the tube feet are treated with cell-disrupting solutions, even in the absence of calcium (Fig. 1.5B; Santos et al., 2005b). In the sea stars *M. glacialis* and *A. rubens*, on the other hand, there are differences in the mechanical response to the different solutions (Fig. 1.5B; Santos et al., 2005b; Hennebert and Flammang, unpubl. obs.). A detailed TEM study has shown that, when no increase in tensile strength was observed, juxtaligamental-like cells were not lysed (Hennebert and Flammang, unpubl. obs.). When cell disruption is complete, however, a similar pattern of response is observed in both echinoids and asteroids. Following cell lysis, a stiffening factor would thus be released from the juxtaligamental-like cells in the extracellular matrix, as is the case in other echinoderm MCTs (Wilkie, 2005). MCT provides to both sea stars and sea urchins an obvious adaptative advantage. In its soft state, MCT could assist the muscles in tube foot protraction, bending and retraction; whereas in its stiff state, it could play a role in the energy-saving maintenance of position, for example during strong attachment to the substratum to resist external loads (Santos et al., 2005b). It is noteworthy that, in their stiff state, the stems of sea urchin tube feet can be as tough as the byssal threads of mussels, which are considered as one of the toughest biomaterials known (Gosline et al., 2002).

1.2.2.2 Retractor Muscle

The retractor muscle allows the flexion and the retraction of the unattached tube feet during their activities. When tube foot discs are attached, contraction of the stem retractor muscle allows sea stars and sea urchins to climb vertical surfaces or right themselves (Lawrence, 1987). It also allows these animals to clamp their body against the substratum, a behaviour that may play an important role in the echinoderm adhesion mechanism because friction generated in this way decreases the risk of dislodgement by shear forces. Finally, some sea stars use retractor muscle contraction to open the bivalve molluscs, on which they feed (Lawrence, 1987). In the sea stars *A. rubens* and *M. glacialis*, the contraction force of the tube foot retractor muscle has been measured and averages about 0.013 N (i.e., ~1.3 g) for one tube foot, with a maximum of 0.04 N (~4 g) (Hennebert and Flammang, unpubl. obs.). These data are in the range of the few asteroid pulling forces reported in the literature. Kerkut (1953) for example measured a pulling force of 0.3 g for a single tube foot of an individual of *A. rubens* climbing a vertical surface. Several workers also measured the total pull exerted by the tube feet of asteroids during prey opening, and reported forces over 30 N (Feder, 1955; Lavoie, 1956; Christensen, 1957). Although these data do not take into account the size of the asteroid and hence of its tube feet, they indicate that hundreds of tube feet are used cooperatively in the feeding behaviour. Using sea star retractor muscle active tension (about 0.05 MN/m^2; Hennebert and Flammang, unpubl. obs.) and sea urchin retractor muscle cross-sectional area (about 0.02 mm^2; Santos and Flammang, 2005), the contraction force can be extrapolated to sea urchin tube feet in which it would be about 0.001 N (i.e., ~0.1 g) for one foot.

To estimate the contribution of the retractor muscle to the tensile strength of the tube foot stem in sea stars and sea urchins, breaking forces had to be measured on an echinoderm muscle lacking dense connective tissue. This has been done on holothurian longitudinal muscles which contain only a small amount of loose connective tissue and are considered as a good experimental model to study mechanical properties of echinoderm muscles (Hill, 2004). Traction tests were applied on holothurian muscles in relaxed or contracted states (i.e. bathed in a Ca^{++}-free or in a K^{+}-rich solution, respectively; Hennebert and Flammang, unpubl. obs.). The forces needed to break tube foot retractor muscles were then calculated using the data obtained for holothurian muscles and the respective cross-sectional surface areas of both types of muscles. Comparison of the calculated retractor muscle breaking force with the total breaking force of whole tube feet shows that the retractor muscle accounts for only about 1% of the stem breaking force in sea urchins and for 18 to 25% in sea stars.

These differences can be explained by the fact that in sea urchins, the connective tissue layer is the most developed layer in the stem wall, while in sea stars the retractor muscle is the principal tissue layer (Santos et al., 2005b). It is apparent therefore that, when a tensile force is exerted on a tube foot stem, the connective tissue is the tissue layer bearing most of the load.

1.2.3 The Disc

The discs of both asteroid and echinoid tube feet consist of two superposed layers of approximately equal thickness: a proximal supporting structure bearing the tensions associated with adhesion, and a distal adhesive pad making contact with the substratum and producing the adhesive secretion that fastens the tube foot to this substratum (Fig. 1.6A,B; Santos et al., 2005a). There are, however, differences in the organization of these layers between sea star and sea urchin discs. The distal surface of the adhesive pad is a complex interface specialized for adhesion but also

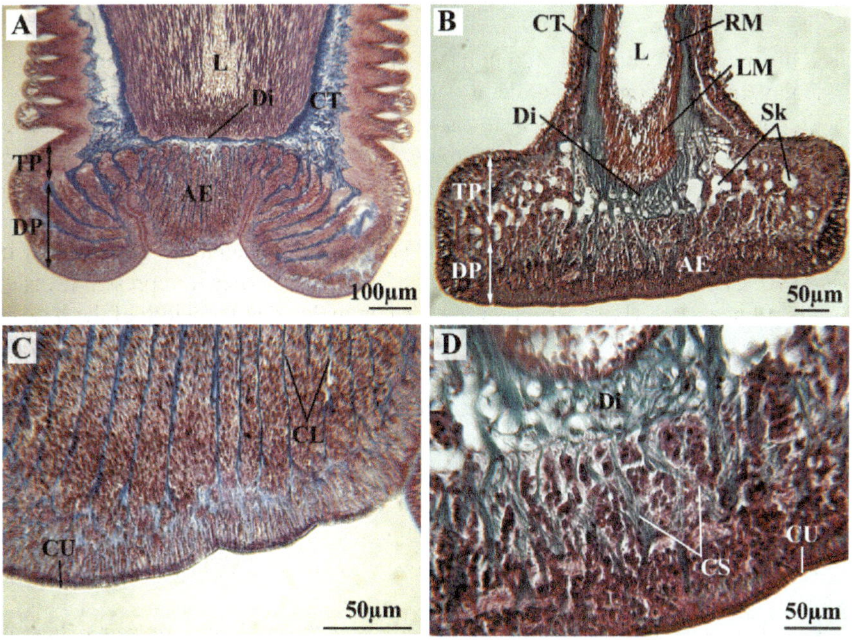

Fig. 1.6 Longitudinal histological sections through the discs of the tube feet of the asteroid *Asterias rubens* (**A**,**C**; the section goes through the margin of the disc to show the connective tissue radial lamellae) and the echinoid *Paracentrotus lividus* (**B**,**D**) (from Santos et al., 2005a). Abbreviations: AE, adhesive epidermis; CL, connective tissue radial lamellae; CS, connective tissue septa; CT, connective tissue; CU, cuticle Di, diaphragm; DP, distal pad; L, lumen; LM, levator muscle; RM, retractor muscle; Sk, skeleton; TP, terminal plate

for sensory perception. Indeed, all echinoderm tube feet have important mechano- and chemosensory abilities (Flammang, 1996). In sea stars, the rim and the centre of the disc are not clearly demarcated and the whole surface is regularly covered by small cilia and secretory pores (Fig. 1.7A). Conversely, in sea urchins, the disc

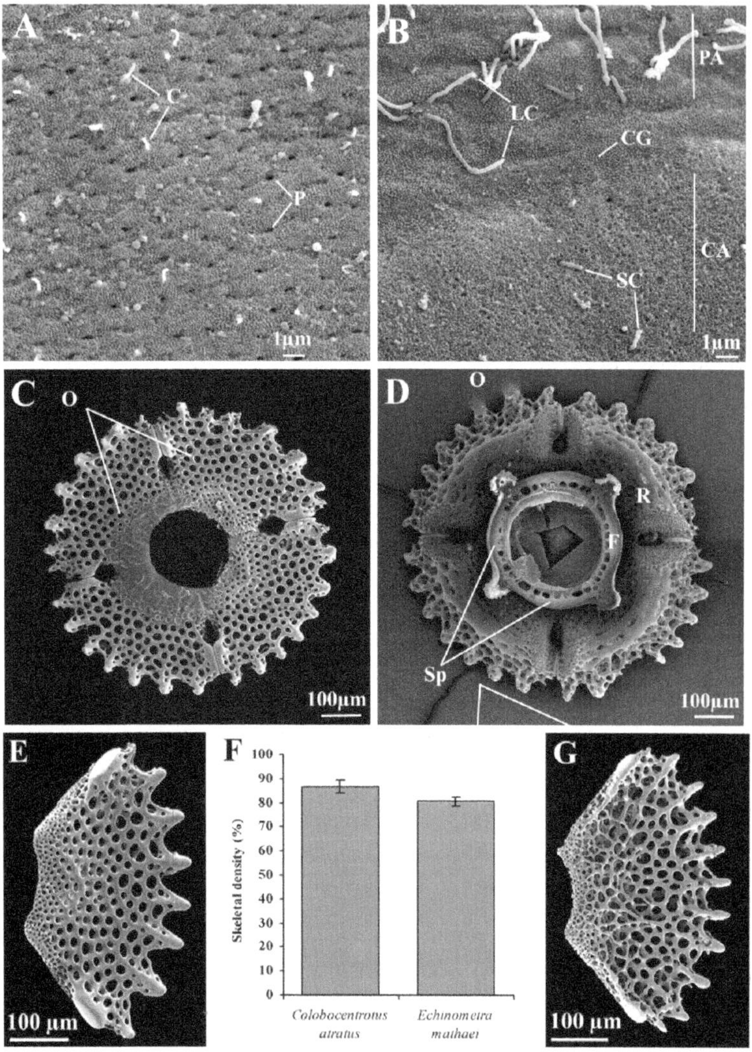

Fig. 1.7 Distal surface of the tube foot in the asteroid *Asterias rubens* (**A**) and the echinoid *Paracentrotus lividus* (**B**) (from Santos et al., 2005a). SEM images of the disc-supporting skeleton from the tube feet of the echinoid *Colobocentrotus atratus* showing its distal (**C**) and proximal (**D**) surfaces (originals). Details of the distal surface of one rosette ossicle in *C. atratus* (**E**) and in *Echinometra mathaei* (**G**) (originals), and comparison of their superficial skeletal density (**F**; mean ± SD, n=10, $p_{t-test} < 0.001$). Abbreviations: C, cilia; CA, central area; CG, circular groove; F, frame; LC, long cilia; O, ossicle; P, pore; PA, peripheral area; R, rosette; SC, short cilia; Sp, spicule

presents a distal circular groove that clearly separates a large central area with small cilia from a narrow peripheral area covered with rows of clustered cilia (Fig. 1.7B). No secretory pores are visible on the surface suggesting another mean of secretion delivery to the disc surface (see below). In both sea stars and sea urchins, the cilia are presumably involved in substratum sensing prior to attachment and in the control of adhesive secretion release (Flammang, 1996).

1.2.3.1 Supporting Structure

The supporting structure consists mostly of a circular plate of connective tissue, the so-called terminal plate that is composed of densely packed collagen fibres. The centre of the terminal plate or diaphragm is very much thinner than its margin and caps the ambulacral lumen (Fig. 1.6A, B). In both sea stars and sea urchins, numerous branching connective tissue septa (made up mostly of collagen fibers) emerge from the distal surface of the terminal plate, manoeuvring themselves between the epidermal cells of the adhesive pad (see below). The thinnest, distal branches of these septa attach apically to the support cells of the epidermis. In sea stars, these septa are arranged as well-defined radial lamellae, whereas in sea urchins they form a more irregular meshwork (Fig. 1.6C, D; Santos et al., 2005a). On its proximal side, the terminal plate is continuous with the connective tissue sheath of the stem (Fig. 1.6A, B). This structure is therefore a mechanical centerpiece of the tube foot, transmitting the tensions created at the level of the adhesive pad during disc fixation to the outer sheath of the stem connective tissue layer. The terminal plate is also the structure to which the retractor muscle anchors itself (Fig. 1.6A, B; Flammang, 1996).

In sea urchins, the terminal plate encloses a calcified skeleton made up of two superimposed structures: a distal rosette and a proximal frame. The rosette is made up of four or five large ossicles (Fig. 1.7C, D) whereas the frame consists of numerous small spicules (Fig. 1.7D). Both structures are ring-shaped and their centre is occupied by the ambulacral lumen (Nichols, 1961; Flammang and Jangoux, 1993). Investigating three sea urchin species, Santos and Flammang (2006) suggested a correlation between the superficial density of the skeleton (i.e., the relative proportion of trabeculae and pores in the stereom at the surface of the rosette) and species habitat. Species inhabiting areas with high hydrodynamic forces possess denser skeletal elements (higher proportion of trabeculae) than species from less exposed zones, which is not surprising since both the branching connective tissue septa reinforcing the adhesive pad and the longitudinal collagen fibers of the stem attach themselves to these ossicles. This correlation was corroborated by measurements made on two tropical species, *Colobocentrotus atratus* which dwells in fully-exposed areas and *Echinometra mathaei* which shelters in holes or crevices. The ossicles of the rosette in *C. atratus* have a significantly higher superficial skeletal density than those of *E. mathaei* (Fig. 1.7E-G; Mellal and Flammang, unpubl. obs.). In sea stars, on the other hand, there is usually no skeleton within the terminal plate, although ossicles have been described in the disc of some oreasterid species (Santos et al., 2005c). In the other species, it is a dense meshwork of collagen fibers that gives the terminal plate its structural stiffness (Fig. 1.6A; Smith, 1947; Flammang et al., 1994).

1.2.3.2 Adhesive Pad

The adhesive pad is composed of a thick adhesive epidermis reinforced by branching connective tissue septa (Fig. 1.6C, D). This epidermis is much thicker than the stem epidermis. As a general rule, epidermal adhesive areas of echinoderm tube feet always consist of four cell categories: support cells, sensory cells, adhesive cells of one (in echinoid tube feet) or two types (in asteroid tube feet) and de-adhesive cells. The two last cell categories are presumably involved in a duo-gland adhesive system as proposed by Hermans (1983). Externally, the epidermis is covered by a well-developed, multilayered glycocalyx, the so-called cuticle (Ameye et al., 2000).

Support cells (Fig. 1.8A, B, D) are the most abundant epidermal cells and form a supportive meshwork in which the other cell types are homogeneously distributed (Holland, 1984). Support cells are traversed by a conspicuous bundle of intermediate filaments joining their apical and basal membranes (Harris and Shaw, 1984). According to Alberts et al. (2002), these filaments may act as tension-bearing structures. At their apex, support cells bear numerous microvilli that are closely associated with the fibrous and/or granular materials constituting the cuticle (Holland, 1984). In addition to their supportive function, support cells are presumably involved in the uptake of dissolved organic material and in cuticular material synthesis (Souza Santos and Silva Sasso, 1970, 1974; Engster and Brown, 1972; Flammang et al., 1998).

Sensory cells are usually scattered singly or in small groups. They are narrow and characterized by a single short apical cilium that traverses the cuticle and protrudes into the outer medium (Fig. 1.7A, B). Basally, these cells terminate within the nerve plexus. Therefore, they are presumed to transduce environmental stimuli into electrical signals within the nerve plexus, and are generally assumed to be chemoreceptors, mechanoreceptors or photoreceptors. Their cytoplasm encloses longitudinally arranged microtubules, elongated mitochondria, and small apical vesicles (Holland, 1984; Cobb and Moore, 1986; Cobb, 1987; Flammang, 1996).

Adhesive cells are generally flask-shaped. Their enlarged cell bodies are located basally and each sends out one long apical process that reaches the distal surface of the disc (Fig. 1.8). The cytoplasm of both the cell body and the apical process is filled with large heterogeneous secretory granules (ranging from 0.3 to 1 μm in diameter according to the species considered; Flammang, 1996). Granules are usually made up of at least two materials of different texture (fibrillar or granular) and different electron density, which gives them a complex ultrastructure varying from one taxon to another (Fig. 1.8E–G; Flammang, 2006). Some authors speculate that these ultrastructural differences in the internal organisation of the adhesive secretory granules reflect the adhesive strength of the tube foot and thus are related with species habitat (Engster and Brown, 1972). However, a recent study has shown that there is no clear relationship between the tenacity of single tube feet from three sea urchin species with contrasted habitats and the variable ultrastructure of their adhesive secretory granules (Santos and Flammang, 2006). In the cell body of adhesive cells, developing secretory granules are closely associated with

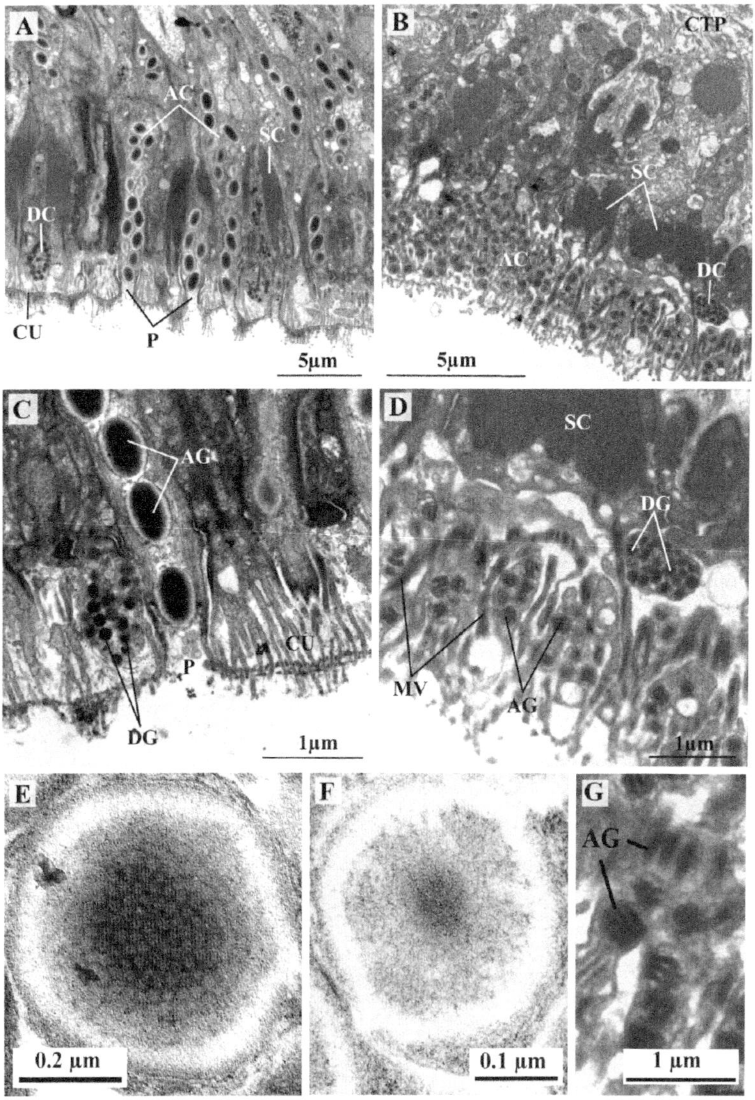

Fig. 1.8 Longitudinal TEM sections through the adhesive pad of tube foot discs of the asteroid *Marthasterias glacialis* (**A**,**C**) and the echinoid *Sphaerechinus granularis* (**B**,**D**) and detailed view of the adhesive secretory granules (type 1 [**E**] and 2 [**F**] from *M. glacialis*, and [**G**] from *S. granularis*) (originals). Abbreviations: AC, adhesive cell; AG, adhesive granule; CTP, connective tissue protrusion; CU, cuticle; DC, de-adhesive cell; DG, de-adhesive granule; MV, microvilli; P, secretory pore; SC, support cell

Golgi membranes and rough endoplasmic reticulum cisternae, suggesting that these organelles are involved in the synthesis of the granule contents (Flammang, 1996). Two modes of granule secretion can be recognized according to the morphology of the apex of the adhesive cell (McKenzie, 1988a; Flammang and Jangoux, 1992;

Flammang, 1996). In "apical duct" cells, secretory granules are extruded through a duct delimited by a ring of microvilli and opening onto the tube foot surface as a cuticular pore (Figs. 1.7A and 1.8A, C). This kind of adhesive cell occurs in asteroid, ophiuroid and crinoid tube feet, as well as in holothuroid locomotory tube feet (Flammang, 1996). In "apical tuft" cells, secretory granules are released at the tip of microvillar-like cell projections which are arranged in a tuft at the cell apex (Fig. 1.8B, D). This second kind of adhesive cell has been observed only in echinoid tube feet and holothuroid locomotory tube feet and buccal tentacles (Flammang, 1996).

De-adhesive cells are narrow and have a centrally-located nucleus. They are filled with small homogenous electron-dense secretory granules whose ultrastructure is remarkably constant from one echinoderm taxon to another (Fig. 1.8C, D). The cytoplasm of de-adhesive cells also contains numerous RER cisternae, a small Golgi apparatus and longitudinally arranged microtubules. Their basal end is tapered and penetrates the nerve plexus while their apex usually bears a short subcuticular cilium (Flammang, 1996).

A cuticle, consisting of fibrous and sometimes granular material, covers the epidermal cells of echinoderm tube foot adhesive areas (Fig. 1.8) (Holland and Nealson, 1978; McKenzie, 1988b; Ameye et al., 2000). Depending on the species, there are from three to five cuticular sublayers. The most external, present in all species, is the "fuzzy coat" and consists of numerous fine fibrils (McKenzie, 1988b; Ameye et al., 2000).

Attachment of the tube feet is a multi-step process. First, the disc adapts its distal surface to the substratum profile, as shown by SEM observation of the surface topography of tube foot discs attached to substrata with various roughnesses (Santos et al., 2005a). Both asteroid and echinoid tube feet attached to smooth substrata present flat and relatively smooth surfaces, whereas those attached to rough substrata have irregular surfaces, replicating the substratum profile (Fig. 1.9). This replication effect is permitted by the material properties of the disc adhesive pad. Indeed, tube foot discs are very soft (E-modulus of 6.0 and 8.1 kPa for the sea star *Asterias rubens* and the sea urchin *Paracentrotus lividus*, respectively) and have viscoelastic properties. Therefore, it can be hypothesized that, under slow self-imposed forces, disc material behaves viscously to adapt to substratum roughness. To the best of our knowledge, no other studies have been published on the material properties of adhesive surfaces from marine invertebrates but there are several reports on the adhesive pads of insect legs. Insects such as the grasshopper *Tettigonia viridissima* attach through a combination of an adhesive secretion and a highly deformable pad material with an elastic modulus of 27.2 kPa (Gorb et al., 2000), which is in the range of those measured for the echinoderm tube foot discs. Insect smooth attachment pads are also strikingly similar to echinoderm tube foot discs in their organization. The cuticle constituting the pad is made up of uniformly distributed fibres, orientated perpendicularly to the pad surface and branching into numerous smaller fibrils in the vicinity of the cuticle surface (Gorb et al., 2000). In both echinoderms and insects, such an organization presumably provides flexibility at two levels: (1) the local level, when preferably branched fibrils deform, and (2) the global level, when the main

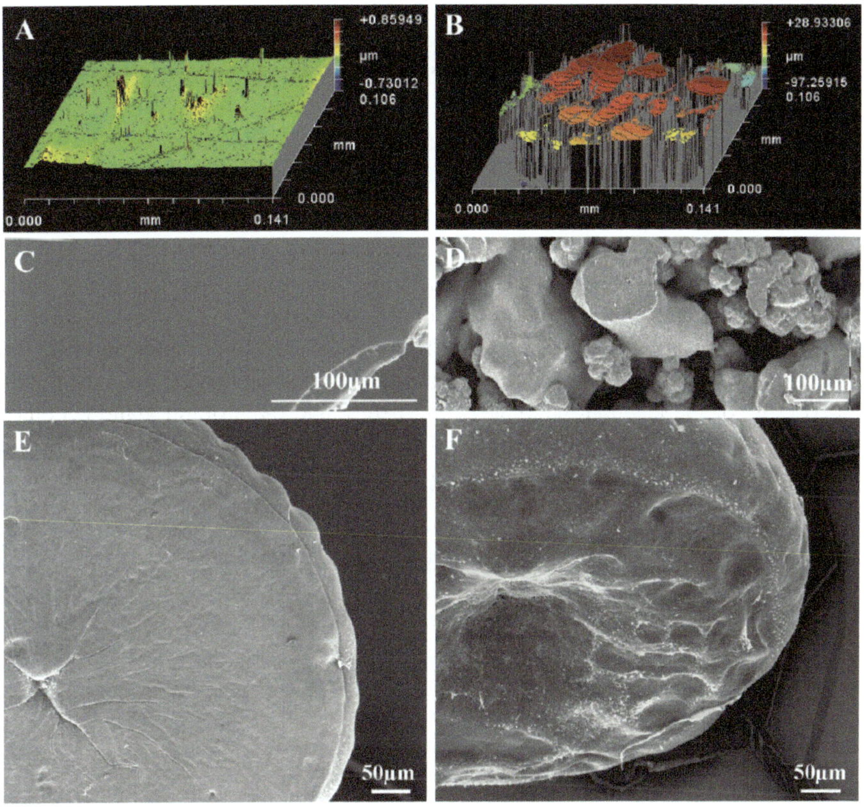

Fig. 1.9 3-D (scanning white light interferometer) and SEM images of the surface of smooth (**A**,**C**) and rough (**B**,**D**) polypropylene samples, and of the distal surfaces of tube foot discs from the echinoid *Paracentrotus lividus* attached to each of these substrata (**E**,**F**) (from Santos et al., 2005a)

fibres also deform. The first level of deformation is responsible for adapting the adhesive organ surface to the substratum micro-roughness, whereas the second one can fit it to substratum macro-roughness (Gorb et al., 2000; Santos et al., 2005a). Due to disc surface plasticity, tube feet show increased adhesion on rough substrata in comparison with their smooth counterparts because of an increase in the geometrical area of contact between the disc and the substratum (Santos et al., 2005a). Once the disc surface has been pressed against the substratum, the secretions of adhesive cells are delivered through the disc cuticle and bind the tube foot to the substratum (Fig. 1.2; Flammang et al., 1994, 1998). The adhesive is deposited as a thin film ideal for the generation of strong adhesion and it fills out only very small surface irregularities in the nanometer range (Santos et al., 2005a). Under short pulses of wave-generated forces, attached discs probably behave elastically, distributing the stress along the entire contact area, in order to avoid crack generation and thus precluding disc peeling and tube foot detachment (Santos et al., 2005a).

For detachment, secretions produced by the de-adhesive cells are released within the cuticle, as evidenced by TEM pictures of empty de-adhesive cells in tube feet that have just detached voluntarily (Flammang, 1996; Flammang et al., 2005). These secretions are not incorporated in the footprint material; they might instead function as enzymes, causing the peeling of the outermost-layer of the cuticle, the fuzzy coat (Flammang et al., 1994, 1998, 2005). After detachment, the adhesive material and part of the fuzzy coat remain on the substratum as a footprint (Fig. 1.2).

1.3 Tube Foot Adhesive

1.3.1 Ultrastructure

In all echinoderm species investigated so far, after detachment of the tube foot, the adhesive secretion usually remains firmly bound to the substratum as a footprint. The material constituting these footprints can be stained, allowing the observation of their morphology under the light microscope (Chaet, 1965; Thomas and Hermans, 1985; Flammang, 1996; Santos and Flammang, 2006). In both sea stars and sea urchins, the footprints have the same shape and the same diameter as the distal surface of the tube foot discs (Fig. 1.2). Various techniques have been used to study the fine structure of the material constituting the footprints and, whatever the method used, the adhesive material always appears as a foam-like or sponge-like material made up of a fibrillar matrix with numerous holes in it (Flammang et al., 1994, 1998; Flammang, 2006; Gorb and Flammang, unpubl. obs., Hennebert et al., 2008). This aspect has been observed in cryo-SEM (Fig. 1.10A), conventional SEM (Fig. 1.10B, C), and AFM (Fig. 1.11); and it does not differ according to whether the footprint has been fixed or not (Flammang et al., 1998; Flammang, 2006). So far, however, it has not been possible to visualize fresh footprints directly in sea water. In both asteroid and echinoid footprints, one can distinguish a very thin and homogeneous priming film covering the substratum on which the fibrillar matrix is deposited (Figs. 1.10 and 1.11). The occurrence of such a priming film has already been reported in the adhesives of mussels and barnacles (Waite et al., 2005; Kamino, 2006, respectively). The thickness of the fibrillar matrix may vary from one footprint to another but also between different areas of the same footprint (Fig. 1.10B; Flammang et al., 1994; Hennebert et al., 2008). In sea stars, the fibrils tend to form a loose meshwork with relatively large meshes, about 2 to 5 μm in diameter (Figs. 1.10A and 1.11). The walls delimiting the meshes may be quite thick (up to 1 μm) and, under the AFM, they appear as strings of little beads (Fig. 1.11; Hennebert et al., 2008). In sea urchin and sea cucumber footprints, the meshwork appears denser, with smaller meshes (<1 μm) delimited by very fine fibrils (about 50 nm in diameter) (Fig. 1.10B, C). These differences in ultrastructure could be linked to the way the adhesive secretions are delivered to the substratum, viz. through secretory pores in asteroids and at the apex of microvillar-like cell projections in echinoids (see Section 3.1.2.). Indeed, the loose meshwork of sea star

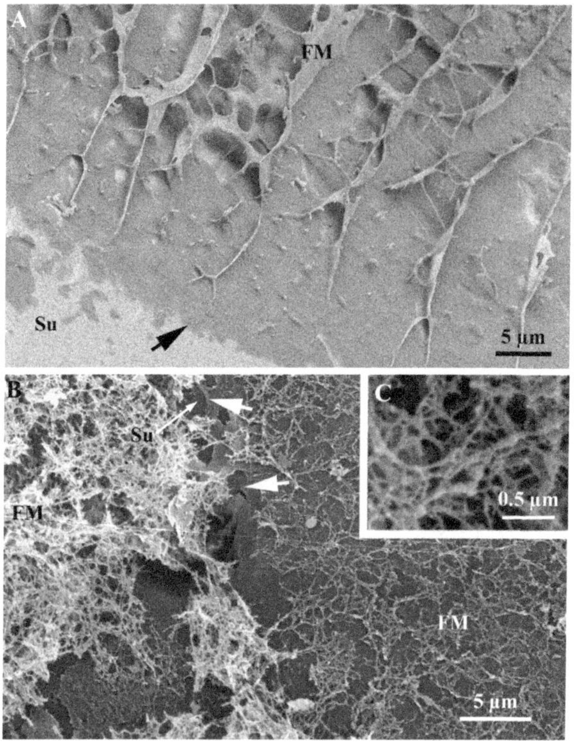

Fig. 1.10 Ultrastructure of echinoderm adhesive footprints (originals). (**A**) Cryo-SEM view of the rim of a footprint of the asteroid *Asterias rubens* deposited on a metal-coated plastic substratum. The frozen footprint was sputter-coated directly on the cooling stage of the microscope after ice sublimation. (**B**) SEM view of a footprint of the echinoid *Paracentrotus lividus* deposited on polymethyl-metacrylate showing the different thicknesses of the fibrillar matrix. (**C**) Detailed SEM view of the fibrillar matrix from a footprint of the holothuroid *Holothuria forskali* deposited on glass. Abbreviations: FM, fibrillar matrix; Su, substratum. Arrows indicate the priming film

footprints reflects approximately the distribution of the secretory pores on the tube foot disc surface (Figs. 1.7A and 1.8A), while the denser meshwork of sea urchin footprints is more reminiscent of the dense array of cell projections covering their disc surface (Figs. 1.7B and 1.8B).

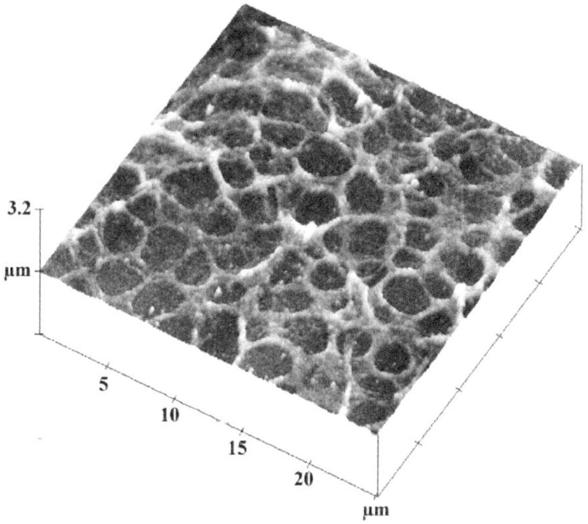

Fig. 1.11 AFM image (tapping mode) of a footprint of the asteroid *Asterias rubens* deposited on mica (Hennebert et al., 2008). The footprint has been briefly rinsed in distilled water and slightly air-dried before observation

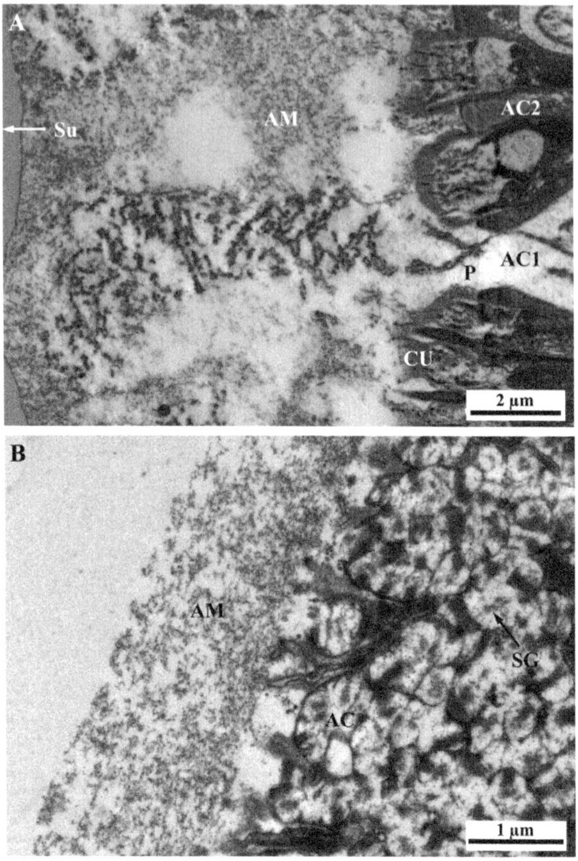

Fig. 1.12 Ultrastructure of echinoderm adhesive layer and attached tube feet (TEM originals). (**A**) View of a tube foot disc of the asteroid *Asterias rubens* attached to an epoxy resin substratum by adhesive material. (**B**) Distal surface of a tube foot disc of the echinoid *Paracentrotus lividus* with adhesive material still attached. Abbreviations: AC1, type one adhesive cell; AC2, type 2 adhesive cell; AC, adhesive cell; AM, adhesive material; CU, cuticle; P, secretory pore; SG, secretory granule; Su, substratum

Ultrastructure of the adhesive has also been investigated in TEM by dissecting and fixing tube feet while they were firmly attached to an epoxy resin substratum (Flammang et al., 1994; Hennebert et al., 2008). In sea stars, the adhesive material has the ultrastructure of a fibre-reinforced composite, which is able to fill out the very small surface irregularities of the substratum (Fig. 1.12A). The electron-dense fibres present in the adhesive material derive directly from the rods described in the secretory granules of type 1 adhesive cells (Figs. 1.8E and 1.12A). There is a continuous thin layer of homogeneous material directly covering the substratum that could correspond to the priming film observed with the other techniques. On the other hand, there are not many areas completely devoid of material within the whole adhesive layer and this suggests that the meshes observed in footprints could be filled with a loose material that collapses during drying. In sea urchins, the adhesive material appears more homogeneous (Fig. 1.12B). However, the entire adhesive layer extending from the disc surface to the substratum has not been observed because echinoid tube feet usually fall off the resin substratum during sample preparation (Hennebert and Flammang, unpubl. obs.). Based on TEM observations, the thickness of the adhesive layer ranges from 0.2 to 9 μm in *A. rubens* and, at least, from 0.3

to 2 μm in *P. lividus* (Fig. 1.12; Flammang et al., 1994; Hennebert unpubl. obs.). This is an order of magnitude higher than the mean maximum thickness reported for fresh but slightly air-dried footprints using an interference-optical profilometer which generates three-dimensional images of the footprint surface (i.e., 230 nm in *A. rubens* and 100 nm in *P. lividus*; Flammang et al., 2005). It suggests that footprints in their native state may be swollen compared to dried footprints. On the other hand, these measurements were made on different substrata (epoxy resin and glass, respectively), and it is known that substratum properties may influence the thickness and bulk properties of bioadhesives. In barnacles, it was shown that the adhesive laid on low-energy surfaces was thicker, softer (lower Young's modulus) and more hydrated than adhesive laid on high-energy surfaces (Berglin and Gatenholm, 2003; Wiegemann and Watermann, 2003; Sun et al., 2004). Similar observations have been made with adsorbed protein films made up of mussel foot protein-1 (Mefp-1; see Berglin et al., 2005, for review). This protein forms an elongated, flexible film with substantial amounts of hydrodynamically coupled water on non-polar surfaces, whereas it forms a rigidly attached adlayer with little hydrodynamically coupled water on polar surfaces. The influence of substratum properties on the thickness and ultrastructure of echinoderm footprints has not been investigated yet. However, there is indirect evidence that this influence exists. In the echinoid *P. lividus*, tube foot tenacity increases with the increase in the total surface energy of the substratum and the predominance of polar over non-polar forces in the surface [0.07 MPa on polypropylene, 0.12 MPa on polystyrene, 0.22 MPa on PMMA and 0.15 MPa on glass] (Santos and Flammang, 2006). These variations in tenacity result mostly from significant differences in the cohesiveness of the adhesive layer (Santos and Flammang, 2006), as is the case in barnacles (Sun et al., 2004). Adhesive footprints deposited by echinoid tube feet on non-polar surfaces (such as epoxy resin) would therefore be softer and maybe thicker, being more prone to cohesive failure and leading to a decreased tenacity.

1.3.2 Composition

Although the fine structure of echinoderm adhesive systems has been extensively studied, little is known on the composition of their adhesives. All the information available comes from histochemical studies on tube foot sections and from investigations on footprint biochemical composition. Results of histochemical tests performed on tube foot epidermal adhesive areas show that, in most species, the adhesive cells, which are the most prominent cells of the adhesive areas, stain with dyes specific for acid mucopolysaccharides (see Flammang, 1996, for review). However, in some species, the histochemical tests indicate the presence of both acid mucopolysaccharides and proteins in the granules of adhesive cells. As for the de-adhesive secretions, they do not stain with any classical histochemical dyes (Flammang, 1996).

At present, data on the biochemical composition of echinoderm adhesive footprints are only available for one sea star and one sea urchin species. Inorganic residues apart (40%), the footprints of the asteroid *A. rubens* are made up mainly of proteins (20.6%) although carbohydrates are also present in significant amounts (3% neutral sugars, 1.5% amino sugars and 3.5% uronic acids) (Flammang et al., 1998). The protein moiety contains slightly more polar (55%) than non-polar (45%) residues and, among polar residues, more charged (34%, of which 22% are acidic) than uncharged residues (21%) (Table 1.3). As far as the echinoid *P. lividus* is concerned, its footprints also contain a significant amount of inorganic residues (45.5%) but a much lower amount of proteins (6.4%) than those of *A. rubens*. Moreover, the amino acid composition of the protein fraction of sea urchin footprints revealed the presence of slightly more non-polar (57%) than polar residues (43%), the latter being composed of equivalent amounts of both charged (23%) and uncharged (20%, of which 12% acidic) residues (Santos et al., 2009) (Table 1.3). These results are in accordance with previous studies on footprints using various dyes, in which asteroid footprints stained for proteins and acid mucopolysaccharides (Chaet, 1965; Flammang et al., 1994) whereas echinoid footprints stained only for acid mucopolysaccharides but not for proteins (Flammang and Jangoux, 1993). Comparison of footprint amino acid composition between sea stars and sea urchins by the method of Marchalonis and Weltman (1971), in which a parameter called SΔQ is calculated by pair-wise comparison of the percentages of each amino acid constituting the protein moiety, gives a value of 114 (it is generally considered that values of SΔQ $\leq$ 100 indicate relatedness; Marchalonis and Weltman, 1971; Flammang, 2006). Common to both echinoderm adhesives, however, are the higher levels of glycine, proline, isoleucine and cysteine than for the average eukaryotic protein (Table 1.3). Cysteine residues could be involved in intermolecular disulphide

Table 1.3 Amino acid composition of the footprint material from the asteroid *Asterias rubens* and the echinoid *Paracentrotus lividus* (values in residues per thousand)

Amino acid	*Asterias rubens*[a]	*Paracentrotus lividus*[b]
ASX	118	48
THR	78	74
SER	76	86
GLX	102	74
PRO	61	68
GLY	97	147
ALA	62	98
CYS/2	32	26
VAL	67	89
MET	17	19
ILEU	45	50
LEU	61	72
TYR	27	38
PHE	38	31
LYS	56	27
HIS	21	13
ARG	41	40

[a] Flammang et al., 1998; [b] Santos et al., 2009

bonds reinforcing the cohesive strength of the adhesive (Flammang et al., 1998; Flammang, 2006). Alternatively, they may form intramolecular disulphide bonds, holding proteins in the specific shape required for interaction with their neighbours, as is the case in barnacle or periwinkle adhesives (Kamino, 2006; Smith, 2006). The richness in small side-chain amino-acids as well as in charged and polar amino acids is common to most marine adhesives characterized so far (see Flammang, 2006, for review). Small side-chain amino acids are often found in large quantities in elastomeric proteins that can withstand significant deformation without rupture, and thus probably account for the high cohesive strength of marine adhesives (Tatham and Shewry, 2000). As for charged and polar amino acids they may be involved in adhesive interactions with the substratum through hydrogen and ionic bonding, and therefore contribute to the high adhesive strength of these bioadhesives (Waite, 1987). Apparently, the de-adhesive secretion is not incorporated into the footprints (Flammang et al., 1998) and its biochemical composition remains unknown.

Although the detailed composition of tube foot adhesive is only known for two species (see above), the variability of the adhesive secretions from twenty-four echinoderm species, representing the five extant classes, has been investigated by immunohistochemistry (Table 1.4; Santos et al., 2005c; Santos and Flammang, unpubl. obs.). This was done using polyclonal antibodies raised against the footprint material from the asteroid *A. rubens*, which is mainly composed of the contents of the disc epidermis adhesive cells, but also contains some constituents of the cuticle (Flammang et al., 1998). As far as asteroids are concerned, the results are very homogeneous (Table 1.4). For the thirteen species considered, there is a very strong and reproducible immunolabelling at the level of the disc adhesive area due to the labelling of numerous granule-containing adhesive cells, the cuticle being the only other immunoreactive structure (Fig. 1.13A, B). This immunoreactivity seems to be independent of the taxon considered, of the tube foot morphotype or function, or of the species habitat (Santos et al., 2005c). It indicates that sea star adhesives are closely related, probably sharing many identical molecules or, at least, many identical epitopes on their constituents. In the other echinoderm classes, different immunoreactivity patterns occur. Crinoid digitate tube feet are strongly immunoreactive, labelling being restricted to the adhesive epidermis of the numerous papillae arranged along the tube feet (Table 1.4; Fig. 1.13C, D). Ophiuroid digitate tube feet also present a strong immunolabelling of their epidermal papillae, together with a slight non-specific labelling of the nerve plexus (Table 1.4; Fig. 1.13E, F). As for echinoids, of the five orders investigated, only members of the order Echinoida present clusters of immunolabelled adhesive cells whereas members of the other four orders do not present any labelling in the tube foot adhesive areas (Table 1.4). Common to all sea urchin species, however, is a moderate immunolabelling of the cuticle (Fig. 1.13G, H, I). In holothuroids, the antibodies do not recognize either the tube foot adhesive epidermis or the cuticle (Table 1.4; Fig. 1.13J, K). These results suggest that both crinoids and ophiuroids possess adhesive secretions sharing many similarities with the adhesive material of asteroids. On the other hand, in echinoids and holothuroids, the immunoreactivity was clearly weak or even absent indicating

Table 1.4 Variability of echinoderm tube feet in terms of morphology, function, and biochemical composition of the epidermal secretions

Species	Type of tube foot	Function of the tube feet	Immunoreactivity of the epidermal layer: Adhesive epidermis	Immunoreactivity of the epidermal layer: Non-adhesive epidermis	Cuticle	Ref.
Asteroidea						1
Paxillosida						
Astropecten aranciacus	Knob-ending	Locomotion, burrowing	++	+	+	
Astropecten polyacanthus	Knob-ending	Locomotion, burrowing	++	+	+	
Luidia savignyi	Knob-ending	Locomotion, burrowing	++	+	+	
Valvatida						
Acanthaster planci	Simple disc-ending	Locomotion, fixation	++	+	+	
Asterina gibbosa	Reinforced disc-ending	Locomotion, fixation	++	+	+	
Linckia laevigata	Simple disc-ending	Locomotion, fixation	++	+	+	
Culcita schmideliana	Simple disc-ending	Locomotion	++	+	+	
Protoreaster lincki	Simple disc-ending	Locomotion	++	+	+	
Pentaceraster mammillatus	Simple disc-ending	Locomotion	++	+	+	
Velatida						
Crossaster papposus	Reinforced disc-ending	Locomotion, fixation	++	+	+	
Spinulosida						
Echinaster sepositus	Reinforced disc-ending	Locomotion, fixation	++	+	+	
Forcipulatida						
Asterias rubens	Reinforced disc-ending	Locomotion, fixation	++	+	+	
Marthasterias glacialis	Reinforced disc-ending	Locomotion, fixation	++	+	+	
Crinoidea						2
Comatulida						
Tropiometra carinata	digitate	Feeding	++	–	–	
Ophiuroidea						2
Ophiurida						
Ophiocoma scolopendrina	digitate	Feeding	++	–	–	
Ophiomastix venosa	digitate	Feeding (?)	++	–	–	

Table 1.4 (continued)

Species	Type of tube foot	Function of the tube feet	Immunoreactivity of the epidermal layer: Adhesive epidermis	Non-adhesive epidermis	Cuticle	Ref.
Echinoidea						2
Diadematoida						
Diadema savignyi	Reinforced disc-ending	Locomotion, fixation	–	–	+	
Arbacioida						
Arbacia lixula	Reinforced disc-ending	Locomotion, fixation	–	–	+	
Temnopleuroida						
Sphaerechinus granularis	Reinforced disc-ending	Locomotion, fixation	–	–	+	
Echinoida						
Echinometra mathaei	Reinforced disc-ending	Locomotion, fixation	+	–	+	
Paracentrotus lividus	Reinforced disc-ending	Locomotion, fixation	+	–	+	
Spatangoida						
Echinocardium cordatum	Penicillate (preapical)	Burrowing	–	–	+	
	Penicillate (peribuccal)	Feeding	–	–	+	
Holothuroidea						2
Aspidochirotida						
Holothuria forskali	Reinforced disc-ending	Locomotion, fixation	–	–	–	
	Ramified (tentacles)	Feeding	–	–	–	
Apodida						
Synapta maculata	Ramified (tentacles)	Feeding	–	–	–	

–, no immunolabelling; +, weak to moderate immunolabeling; ++, strong immunolabeling.
1, Santos et al., 2005c; 2, Santos and Flammang, unpubl. data.

that there are no common epitopes between their adhesive secretions and those of *A. rubens* (Santos and Flammang, unpubl. data).

1.4 Discussion

Tube feet are interfaces between echinoderms and their environment. All of them have important sensory abilities and are also usually involved in respiratory gas exchange (Flammang, 1996). In addition, most of them possess temporary adhesive systems allowing echinoderms to interact with their environment by manipulating

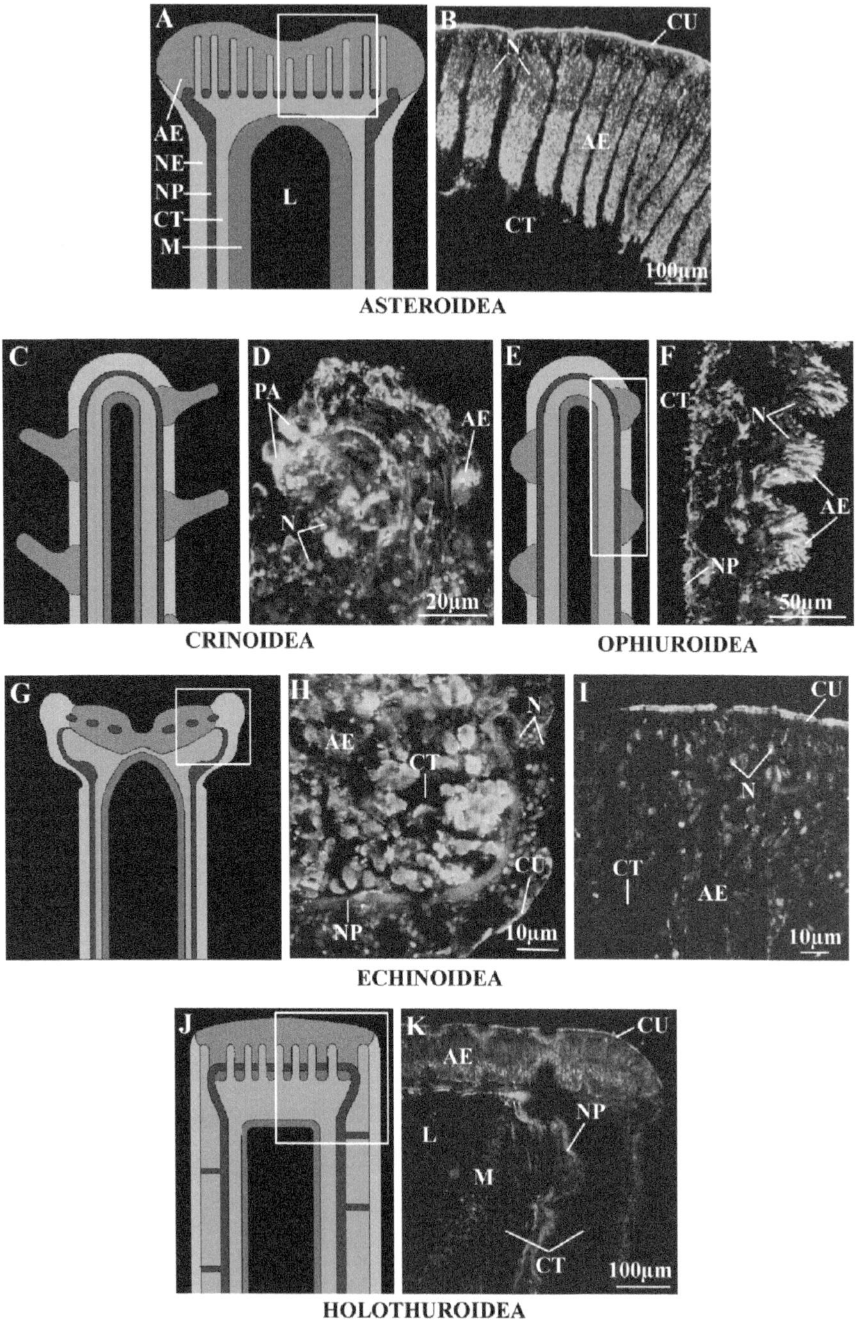

Fig. 1.13 (continued)

items for burrowing or feeding, or to cope structurally with their environment by attaching strongly to the substratum to withstand the action of waves. In the latter case, attachment involves a multitude of independent adhesive organs, the disc-ending tube feet, capable of voluntary attachment, detachment and re-attachment. Each tube foot consists of a distal disc connected to a stem forming together a functional unit. The stem acts as a tough tether that bears the tensions placed on the animal by hydrodynamic forces but is also highly mobile and flexible allowing the tube foot to perform complex movements. The disc often has an enlarged flattened surface ideal for enhancing the contact with the substratum, the ability to deform in order to replicate the surface profile, and possesses internal support structures (connective tissue and calcified skeleton) to bear the tensions associated with adhesion. Moreover, the disc epidermis possesses a duo-glandular adhesive system that produces adhesive secretions fastening the tube foot to the substratum as well as de-adhesive secretions allowing easy detachment. Although little is known about the biochemical composition of these secretions, it seems that the adhesive material is a protein-polysaccharide complex whose composition varies from one taxon to another. Asteroid tube feet seem to possess a protein-rich adhesive secretion whereas echinoid tube feet seem to rely on a carbohydrate-rich adhesive secretion. As far as the other three classes are concerned, cross-reactivity experiments using antibodies raised against a sea star adhesive material indicate a relationship between the adhesive secretions of asteroid, crinoid and ophiuroid tube feet but not, on the other hand, between those of asteroid, echinoid and holothuroid tube feet (see Section 3.2.). These observations are congruent with the phylogenetic hypothesis on the evolution of echinoderm adhesive systems (McKenzie, 1988a) according to which asteroids, crinoids and ophiuroids would share a common ancestral adhesive system, the adhesive being extruded through apical duct cells while a common echinoid/holothuroid adhesive system would have arisen later in the evolution in which the adhesive is released through apical tuft cells (see Section 3.1.2). This model, as well as our results on the adhesive composition, fit well with the most recent and commonly accepted echinoderm phylogeny (Fig. 1.14). According to these phylogenetic reconstructions, based on both morphological and molecular characters, echinoids and holothuroids form a closely related grouping, the Echinozoa, which was the last to diverge from the other classes (Wada and Satoh, 1994; Littlewood et al., 1998; David and Mooi, 1998). A drawback to this model is the moderate immunoreactivity observed in the tube feet of two of the six echinoid species studied, *Echinometra mathaei* and *P. lividus*, meaning that the adhesive in these species share some common epitopes with that of the asteroid *A. rubens*. It is noteworthy,

Fig. 1.13 Schematic representations of longitudinal sections through the tube feet of the different echinoderm classes (**A**, **C**, **E**, **G**, **J**) and immunofluorescent labelling of corresponding tube foot sections with antibodies raised against the adhesive material of the asteroid *Asterias rubens* (immunoreactive structures are labelled in green while nuclei appear in red) (**B**, **D**, **F**, **H**, **I**, **K**) (originals; see text for details). Abbreviations: AE, adhesive epidermis; CU, cuticle; CT, connective tissue; L, lumen; N, nuclei; NE, non-adhesive epidermis; NP, nerve plexus; M, mesothelium

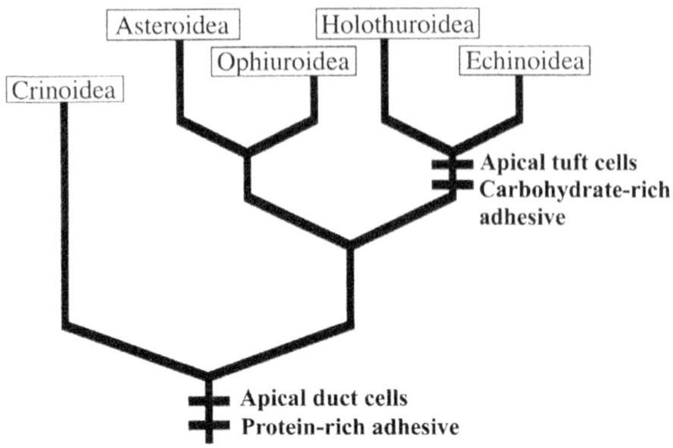

Fig. 1.14 A hypothesized evolutionary history of adhesive systems among echinoderm classes reconstructed on the most commonly accepted phylogenetic tree (tree modified after David and Mooi, 1998). The characters mapped on the tree are the type of adhesive cell (apical duct vs apical tuft) and the putative composition of the adhesive (protein-rich vs carbohydrate-rich)

however, that the two species belong to the same order and possess similar life habits: both inhabit moderately-exposed to open zones and are able to dig in the hard substrata on which they live. There is a possibility that these echinoids convergently acquired their immunoreactivity with asteroids because of common selective pressures. More studies are needed to address this question and to understand the evolution and functioning of echinoderm adhesive systems.

Acknowledgments The authors wish to acknowledge Stas Gorb for providing the opportunity to write this chapter; Paula Chicau for providing data from the Amino Acid Analysis Service at the Instituto de Tecnologia Química e Biológica, Universidade Nova de Lisboa, Oeiras, Portugal; as well as the Aquário Vasco da Gama, Dafundo, Portugal for sea urchin maintenance. R.S. benefited from a post-doctoral grant from the Foundation for Science and Technology of Portugal (FCT) (grant n° SFRH/BPD/21434/2005) and E.H. benefited from a FRIA doctoral grant (Belgium). P.F. is a Senior Research Associate for the Fund for Scientific Research of Belgium (F.R.S.-FNRS). Work supported in part by a FCT project grant (PPCDT/DG/MAR/82012/2006) and a FRFC Grant n° 2.4532.07. This study is a joint contribution of the "Centre Interuniversitaire de Biologie Marine" (CIBIM, Belgium) and the "Instituto de Tecnologia Química e Biológica" (ITQB, Portugal).

References

Alberts, B., Johnson, A., Lewis, J., Raff, M., Roberts, K., and Walter, P. (2002) *Molecular biology of the cell (fourth edition)*. New York: Garland Science.

Ameye, L., Hermann, R., Dubois, Ph., and Flammang, P. (2000) Ultrastructure of the echinoderm cuticle after fast freezing/freeze substitution and conventional chemical fixation. *Microsc Res Tech* 48: 385–393.

Berger, V.Ya., and Naumov, A.D. (1996) Influence of salinity on the ability of starfishes *Asterias rubens* L. to attach to substrate. *Biologiya Morya* 22: 99–101.

Berglin, M., and Gatenholm, P. (2003) The barnacle adhesive plaque: morphological and chemical differences as a response to substrate properties. *Colloids Surf B Biointerfaces* 28:107–117.

Berglin, M., Hedlund, J., Fant, C., Elwing, H. (2005) Use of surface-sensitive methods for the study of adsorption and cross-linking of marine bioadhesives. *J Adhesion* 81: 805–822.

Chaet, A.B. (1965) Invertebrates adhering surfaces: Secretions of the starfish, *Asterias forbesi*, and the coelenterate, *Hydra pirardi*. *Ann NY Acad Sci* 118: 921–929.
Characklis, W.G. (1981) Fouling biofilm development: a process analysis. *Biotechnol Bioeng* 23: 1923–1960.
Christensen, A.M. (1957) The feeding behaviour of the sea star *Evasterias troscheli*. *Limnol Oceanogr* 2: 180–197.
Cobb, J. (1987) Neurobiology of the Echinodermata. In: *Nervous Systems in Invertebrates*, ed. by Ah M. New York: Plenum Press, pp. 483–525.
Cobb, J., and Moore, A. (1986) Comparative studies on receptor structure in the britlestar *Ophiura ophiura*. *J Neurocytol* 15: 97–108.
David, B., and Mooi, R. (1998) Major events in the evolution of echinoderms viewed by the light of embryology. In: *Echinoderms: San Francisco*, ed. by Mooi, R., and Telford, M. Rotterdam: Balkema, pp. 21–28.
Engster, M., and Brown, S. (1972) Histology and ultrastructure of the tube foot epithelium in the phanerozonian starfish, *Astropecten*. *Tissue Cell* 4 (3): 503–518.
Feder, H.M. (1955) On the methods used by the starfish *Pisaster ochraceus* in opening three types of bivalve molluscs. *Ecology* 36: 764–767.
Flammang, P. (1996) Adhesion in echinoderms. In: *Echinoderm Studies Vol. 5*, ed. by Jangoux, M., and. Lawrence, J.M. Rotterdam: Balkema, pp. 1–60.
Flammang, P. (2006) Adhesive secretions in echinoderms: an overview. In: *Biological Adhesives*, ed. By Smith A.M. and Callow, J.A. Berlin, Heidelberg: Springer-Verlag, pp. 183–206.
Flammang, P., and Jangoux, M. (1992) Functional morphology of the locomotory podia of *Holothuria forskali* (Echinodermata, Holothuroidea). *Zoomorphology* 11: 167–178.
Flammang, P., and Jangoux, M. (1993) Functional morphology of coronal and peristomeal podia in *Sphaerechinus granularis* (Echinodermata, Echinoida). *Zoomorphology* 113: 47–60.
Flammang, P., and Walker, G. (1997) Measurement of the adhesion of the podia in the asteroid *Asterias rubens* (Echinodermata). *J Mar Biol Assoc*. UK 77:1251–1254.
Flammang, P., Demeulenaere, S., and Jangoux, M. (1994) The role of podial secretions in adhesion in two species of sea stars (Echinodermata). *Biol Bull* 187: 35–47.
Flammang, P., Santos, R., and Haesaerts, D., (2005) Echinoderm adhesive secretions: from experimental characterization to biotechnological applications. In: *Marine Molecular Biotechnology: Echinodermata*, ed. by Matranga, V. Berlin: Springer-Verlag, pp. 201–220.
Flammang, P., Van Cauwenberge, A., Alexandre, H., and Jangoux, M. (1998) A study of the temporary adhesion of the podia in the sea star *Asterias rubens* (Echinodermata, Asteroidea) through their footprints. *J Exp Biol* 201: 2383–2395.
Fletcher, M. (1994) Bacterial biofilms and biofouling. *Curr Opin Biotechnol* 5: 302–306.
Gallien, W.B. (1986) A comparison of hydrodynamic forces on two sympatric sea urchins: implications of morphology and habitat. MSc thesis, University of Hawaii, Honolulu, HI, USA.
Gorb, S., Jiao, Y., and Scherge, M. (2000) Ultrastructural architecture and mechanical properties of attachment pads in *Tettigonia viridissima* (Orthoptera, Tettigoniidae). *J Comp Physiol* A 186: 821–831.
Gosline, J., Margo, L., Carrington, E., Guerette, P., Ortlepp, C. and Savage, K. (2002) Elastic proteins : biological roles and mechanical properties. *Phil Trans R Soc Lond* 357: 121–132.
Guidetti, P., and Mori, M. (2005) Morpho-functional defences of Mediterranean sea urchins, *Paracentrotus lividus* and *Arbacia lixula*, against fish predators. *Mar Biol* 147: 797–802.
Harris, P., and Shaw, G. (1984) Intermediate filaments, microtubules and microfilaments in epidermis of sea urchin tube foot. *Cell Tiss Res* 236: 27–33.
Hennebert, E., Viville, P., Lazzaroni, R., and Flammang, P. (2008) Micro- and nanostructure of the adhesive material secreted by the tube feet of the sea star. *Asterias rubens*. *J Struct Biol* 164: 108–118.
Hermans, C. (1983) The duo-gland adhesive system. *Oceanogr Mar Biol Ann Rev* 83: 283–339.
Hill, R. B., (2004) Active state in echinoderm muscle. In: *Echinoderms: München*, ed. by Heinzeller, T., and Nebelsick, J. H. Leiden: Balkema, pp. 351–352.

Holland, N. (1984) Epidermal cells. In: *Biology of the integument. Vol. I. Invertebrates*, ed. by Bereiter-Hahn, J., Maltoltsy, A.G., and Richards K.S. Berlin: Springer-Verlag, pp. 756–774.
Holland, N., and Nealson, K. (1978) The fine structure of the echinoderm cuticle and subcuticular bacteria of echinoderms. *Acta Zool* 59: 169–185.
Kamino, K. (2006) Barnacle underwater attachment. In: *Biological adhesives*, ed. by Smith, A.M., and Callow, J.A. Berlin: Springer-Verlag, pp 145–166..
Kerkut, G.A. (1953) The forces exerted by the tube feet of the starfish during locomotion. *J Exp Biol* 30: 575–583.
Lavoie, M.E. (1956) How sea stars open bivalves. *Biol Bull* 111: 114–122.
Lawrence, J.M. (1987). *A Functional Biology of Echinoderms*. London: Croom Helm.
Leddy, H.A., and Johnson, A.S., (2000) Walking *versus* breathing: mechanical differentiation of sea urchin podia corresponds to functional specialization. *Biol Bull* 198: 88–93.
Littlewood, D.T.J., Smith, A.B., Clough, K.A., and Emson, R.H. (1998) Five classes of echinoderm and one school of though. In: *Echinoderms: San Francisco*, ed. by Mooi, R., and Telford, M. Rotterdam: Balkema, pp. 47–49.
Marchalonis, J.J., and Weltman, J.K. (1971) Relatedness among proteins: a new method of estimation and its application to immunoglobins. *Comp Biochem Physiol* 38B: 609–625.
Märkel, K., and Titschack, H. (1965) Das Festhaltevermögen von Seeigeln und die Reißfestigkeit ihrer Ambulacralfüßchen. *Sond Zeit Naturw* 10:268.
McKenzie, J.D. (1988a) The ultrastructure of tube foot epidermal cells and secretions: Their relationship to the duo-glandular hypothesis and the phylogeny of the echinoderm classes. In: *Echinoderm Phylogeny and Evolutionary Biology*, ed. by Paul, C.R.C., and Smith, A.B. Oxford: Clarendon Press, pp. 287–298.
McKenzie, J.D. (1988b) Ultrastructure of the tentacles of the apodous holothurian *Leptosynapta* spp. (Holothuroidea: Echinodermata) with special reference to the epidermis and surface coats. *Cell Tiss Res* 251: 387–397.
Nichols, D. (1961) A comparative histological study of the tube-feet of two regular Echinoids. Q. *J Microsc Sci* 102: 157–180.
Nichols, D. (1966) Functional morphology of the water vascular system. In: *Physiology of Echinodermata*, ed. by Boolootian, R.A. New York: Interscience Publishers, pp. 219–244.
Paine, V.L. (1926) Adhesion of the tube feet in starfishes. *J Exp Zool* 45: 361–366.
Santos, R., and Flammang, P. (2005) Morphometry and mechanical design of tube foot stems in sea urchins: a comparative study. *J Exp Mar Biol Ecol* 315: 211–223.
Santos, R., and Flammang, P. (2006) Morphology and tenacity of the tube foot disc of three common European sea urchin species: a comparative study. *Biofouling* 22: 187–200.
Santos, R., and Flammang, P. (2007) Intra- and interspecific variation of the attachment strength in sea urchins. *Mar Ecol Prog Ser* 332: 129–142.
Santos, R., da Costa, G., Franco, C., Gomes-Alves, P., Flammang, P., and Coelho, A.V. (2009) First insights into the biochemistry of tube foot adhesive from the sea urchin *Paracentrotus lividus* (Echinoidea, Echinodermata). *Mar Biotechnol*, DOI 10.1007/s10126-009-9182-5.
Santos, R., Gorb, S., Jamar, V., and Flammang, P. (2005a) Adhesion of echinoderm tube feet to rough surfaces. *J Exp Biol* 208: 2555–2567.
Santos, R., Haesaerts, D., Jangoux, M., and Flammang, P. (2005b). The tube feet of sea urchins and sea stars contain functionally different mutable collagenous tissues. *J Exp Biol* 208: 2277–2288.
Santos, R., Haesaerts, D., Jangoux, M., and Flammang, P. (2005c) Comparative histological and immunohistochemical study of sea star tube feet (Echinodermata, Asteroidea). *J Morphol* 263: 259–269.
Shadwick, R.E. (1992). Soft composites. In: *Biomechanics. Materials. A practical approach*, ed. By J.F.V. Vincent., Oxford: Oxford University Press pp. 133–164.
Sharp, D.T., and Gray, I.E. (1962) Studies on factors affecting the local distribution of two sea urchins, *Arbacia punctulata* and *Lytechinus variegatus*. *Ecology* 43: 309–313.
Siddon, C.E., and Witman J.D. (2003) Influence of chronic, low-level hydrodynamic forces on subtidal community structure. *Mar Ecol Prog Ser* 261: 91–110.
Smith, A.B. (1978) A functional classification of the coronal pores of echinoids. *Palaeontology* 21: 759–789.

Smith, A.M. (2006) The biochemistry and mechanics of gastropod adhesive gels. In: *Biological adhesives*, ed. by Smith, A.M., and Callow, J.A. Berlin: Springer-Verlag, pp 167–182.

Smith, A.M., and Callow, J.A. (2006) *Biological adhesives*. Berlin: Springer-Verlag.

Smith, J.E. (1947) The activities of the tube feet of *Asterias rubens* L. I. The mechanics of movement and of posture. *Q J Microsc Sci* 88: 1–14.

Souza Santos, H., and Silva Sasso, W. (1970) Ultrastructural and histochemical studies on the epithelium revestment layer in the tube feet of the starfish *Asterina stellifera*. *J Morph* 130: 287–296.

Souza Santos, H., and Silva Sasso, W. (1974) Ultrastructural and histochemical observations of the external epithelium of echinoderm tube feet. *Acta Anat* 88: 22–33.

Sun, Y., Guo, S., Walker, G.C., Kavanagh, C.J., and Swain, G.W. (2004) Surface elastic modulus of barnacle adhesive and release characteristics from silicone surface. *Biofouling* 20: 279–289.

Szulgit, G.K., and Shadwick, R.E. (2000) Dynamic mechanical characterization of a mutable collagenous tissue: Response of sea cucumber dermis to cell lysis and dermal extracts. *J Exp Biol* 203: 1539–1550.

Tatham, A.S., Shewry, P.R. (2000) Elastomeric proteins: biological roles, structures and mechanisms. *Trends Biochem Sci* 25: 567–571.

Thomas, L.A., and Hermans, C.O. (1985) Adhesive interactions between the tube feet of a starfish, *Leptasterias hexactis*, and substrata. *Biol Bull* 169: 675–688.

Trotter, J. A., and Koob, T. J. (1995) Evidence that calcium-dependent cellular processes are involved in the stiffening response of holothurian dermis and that dermal cells contain an organic stiffening factor. *J Exp Biol* 198: 1951–1961.

Trotter, J. A., Tipper, J., Lyons-Levy, G., Chino, K., Heuer, A. H., Liu, Z., Mrksich, M., Hodneland, C., Dillmore, W. S., Koob, T. J., Koob-Emunds, M. M., Kadler, K. and Holmes, D. (2000) Towards a fibrous composite with dynamically controlled stiffness. Lessons from echinoderms. *Biochem Soc Trans* 28: 357–362.

Tyler, S. (1988) The role of function in determination of homology and convergence – examples from invertebrates adhesive organs. *Fortsch Zool* 36: 331–347.

Vincent, J. (1990). *Structural Biomaterials*. Princeton: Princeton University Press.

Vogel, S. (2003) *Comparative Biomechanics – Life's physical world*. Princeton: Princeton University Press.

Wada, H., and Satoh, N. (1994) Phylogenetic relationships among extant classes of echinoderms, as inferred from sequences of 18S rDNA, coincide with relationships deduced from the fossil record. *J Mol Evol* 38: 41–49.

Waite, J.H. (1983) Adhesion in byssally attached bivalves. *Biol Rev* 58: 209–231.

Waite, J.H. (1987) Nature's underwater adhesive specialist. *Int J Adhes Adhes* 7: 9–14.

Waite, J.H. (2002) Adhesion à la moule. *Integr Comp Biol* 42: 1172–1180.

Waite, J.H., Andersen, N.H., Jewhurst, S., and Sun, C. (2005) Mussel adhesion: Finding the tricks worth mimicking. *J Adhesion* 81: 297–317.

Walker, G. (1987) Marine organisms and their adhesion. In: *Synthetic Adhesives and Sealants*, ed. by Wake, W.C. Chichester: John Wiley & Sons, pp. 112–135.

Whittington, I.D., and Cribb, B.W. (2001) Adhesive secretions in the Platyhelminthes. *Adv Parasitol* 48: 101–224.

Wiegemann, M., and Watermann B. (2003) Peculiarities of barnacle adhesive cured on non-stick surfaces. *J Adhesion Sci Technol* 17: 1957–1977.

Wilkie, I.C. (1996) Mutable collagenous tissues: extracellular matrix as mechano-effector. In : *Echinoderm Studies*, ed. by Jangoux, M., and Lawrence, J.M., Rotterdam: Balkema, pp. 61–102.

Wilkie, I.C. (2005) Mutable collagenous tissue: overview and biotechnological perspective. In: *Marine molecular biotechnology: Echinodermata*, ed. by Matranga, V. Berlin: Springer-Verlag, pp. 221–250.

Wood, R.L., and Cavey, M.J. (1981) Ultrastructure of the coelomic lining in the podium of the starfish *Stylasterias forreri*. *Cell Tissue Res* 218: 449–473.

Yamasaki, S., Nashimoto, K., Yamamoto, K., Hiraishi, T. (1993) Fluid forces on short-spined sea urchin and northern sea urchin. *Nippon Suisan Gakkaishi* 59: 1139–1146.

Chapter 2
Mechanisms and Principles Underlying Temporary Adhesion, Surface Exploration and Settlement Site Selection by Barnacle Cyprids: A Short Review

Nicholas Aldred and Anthony S. Clare

2.1 Background

The diminutive stature of many barnacle species can lead to their being easily overlooked or ignored by the casual observer of rocky intertidal shores, so it is surprising to discover that a multi-billion dollar industry (Yebra et al., 2004) exists solely to prevent the settlement and growth of fouling organisms on man-made marine structures (Callow and Callow, 2002). The systematic study and taxonomy of acorn barnacles (Cirripedia; Thoracica) owes its initiation and many of its fundamental observations to Charles Darwin who, after his research classifying species aboard the *Beagle*, selected a Chilean barnacle (that he named *Cryptophialus*) to form the basis of much of his research into evolution by natural selection (Darwin, 1854). From that point on, research into barnacles never abated.

As Darwin was completing his work, the maritime shipping industry was in a period of transition. Wooden hulled ships, that were generally copper plated to prevent the settlement of barnacles, were being replaced by cast iron hulled ships. At this time the concern over biofouling accumulation shifted from primarily corrosion-related issues towards the significant reduction in propulsion efficiency due to fouling, and its resulting impact on fuel costs. This is still considered to be the primary deleterious effect of marine fouling, increasing hydrodynamic drag across the surface of a ship's hull, reducing propulsion efficiency and increasing fuel consumption by up to 30% (Brady, 2001). It was not until the mid 1960s that an effective method was developed to prevent the settlement of biofouling organisms.

Biocidal paints containing tributyltin (TBT) provided some respite for the maritime industry (Yebra et al., 2004), with an estimated 70–80% (Yebra et al., 2006) of the world's ocean-going fleet using some form of TBT self polishing copolymer

N. Aldred (✉)
Newcastle University, School of Marine Science and Technology, Newcastle upon Tyne, UK
e-mail: nicholas.aldred@ncl.ac.uk

S.N. Gorb (ed.), *Functional Surfaces in Biology*, Vol. 2,
DOI 10.1007/978-1-4020-6695-5_3, © Springer Science+Business Media B.V. 2009

during the 1990s. Unacceptable environmental consequences brought an end to the use of TBT-based coatings, however, and the biofouling problem re-emerged (Evans et al., 2000).

As a result, work in the field of anti-biofouling has diversified significantly over the last decade, with research into synthetic biocides being matched by novel strategies such as natural product antifoulants (e.g. Hellio et al., 2001), surface textures that reduce the settlement of larvae/spores (Carman et al., 2006) and further study into the fundamental characteristics of surfaces, such as elastic modulus (Chaudhury et al., 2005) and free energy or charge that are now known to modulate adhesion in fouling organisms.

Barnacles are a particular concern since their large size and hard, calcareous nature increases hydrodynamic drag on vessels; significantly increasing fuel costs and the necessity for costly cleaning procedures (Brady and Singer, 2000; Christie and Dalley, 1987). Research into their settlement and adhesion is fundamental to combating biofouling. There remains, however, a significant bias towards studying adult representatives of these species, rather than their larvae. This may partly reflect difficulties in experimentation or larval culture, however, larval settlement is considered by many to be the key stage at which biofouling prevention could be most effective. From a more fundamental perspective, the barnacle cypris larva also represents the only example, to the authors' knowledge, of an aquatic organism putatively using adhesion mechanisms similar to those of terrestrial invertebrates (e.g. flies) and vertebrates (e.g. geckos). Unlike its terrestrial counterparts, it is able to adhere effectively underwater. For this reason alone, the cyprid (Fig. 2.1) is surely worthy of extensive study.

The following review will discuss the importance of the cypris larva to barnacle ecology and, hopefully, highlight the need for further research into their interactions with surfaces; specifically with regard to adhesion. No definitive mechanism has

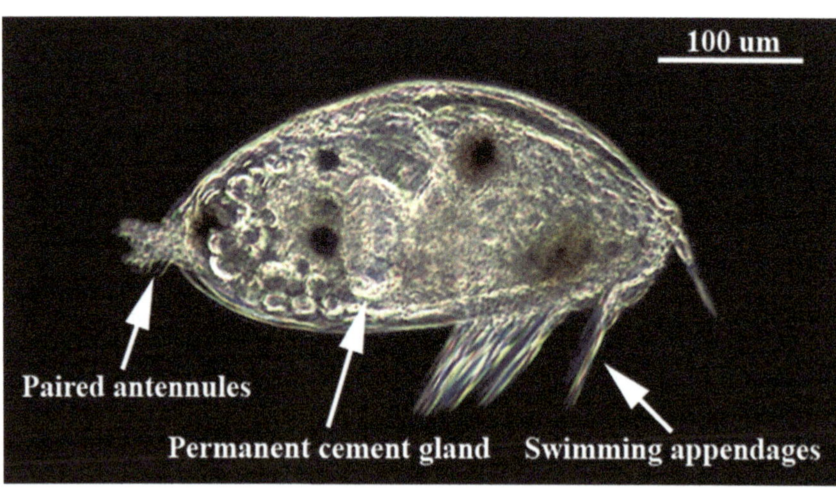

Fig. 2.1 A light micrograph of a *Balanus amphitrite* cypris larva, imaged using dark field optics

ever been described regarding cyprid adhesion, so drawing firm conclusions from the literature is a challenging task. However, novel arguments based on analogous systems are proposed in this review and it is hoped that they will inspire further interdisciplinary study of the system.

2.2 Larval Release and Settlement

The majority of sessile barnacles, including balanid (Sessilia) and lepadid (Pedunculata) barnacles, are simultaneous hermaphrodites, possessing both male and female reproductive systems. They therefore retain the ability to self fertilise should isolation prohibit cross fertilisation; although cross-fertilisation is generally regarded as the rule (Anderson, 1994). The necessity to cross fertilise in order to maintain the genetic health of offspring and the population is the basis of gregarious settlement in sessile barnacles, since potential mates must be within reach of the acting male's extensible penis. Settlement adjacent to at least one potential mate is, therefore, of importance for settling larvae that also select surfaces based on a range of biological and physical factors.

A generalised barnacle life cycle involves 6 planktonic nauplius stages, all except the first of which are planktotrophic (i.e. feeding), a non-feeding cyprid stage and the adult (see Fig. 2.2).

The broadcast release of larvae into the plankton may seem irreconcilable with the clearly gregarious habit of these organisms. How do the microplanktonic and relatively slow-swimming larvae of barnacles return to the adult population and settle? Answers to these fundamental questions began to arise in the 1950s with a spate of research into barnacle biology, much of which was performed under the guidance of the late Prof. Dennis Crisp at the former Natural Environment Research Council (NERC) Institute for Marine Invertebrate Biology (Menai Bridge, Anglesey, N. Wales). The conventional wisdom suggested that, if cypris larvae had inherent preferences for certain settlement substrata, then clustered settlement would automatically result by selective rejection of unfavourable surfaces – assisted by the local hydrology transporting larvae to certain areas. Alternatively, or concurrently, cyprids could settle uniformly on appropriate surfaces and subsequent mortality by physical and/or biological factors, such as sedimentation and grazing, could produce the apparent clustering of adults characteristic of gregarious settlement. It was shown, however, that cyprids can be highly discriminating in their choice of settlement sites (Yule and Walker, 1985), which they explore prior to settling using their paired antennules (Figs. 2.1 and 2.5) for bi-pedal walking. The antennules bear attachment discs that effect reversible adhesion underwater as well as an array of sensory processes, specialised for surface discrimination. Cyprids are known to discriminate on the basis of:

Texture: In general, invertebrate larvae have a predilection for rough as opposed to smooth surfaces for settlement (Crisp, 1974). This preference, it is suggested, could result from the requirement of favourable surfaces for the 'lock and key'

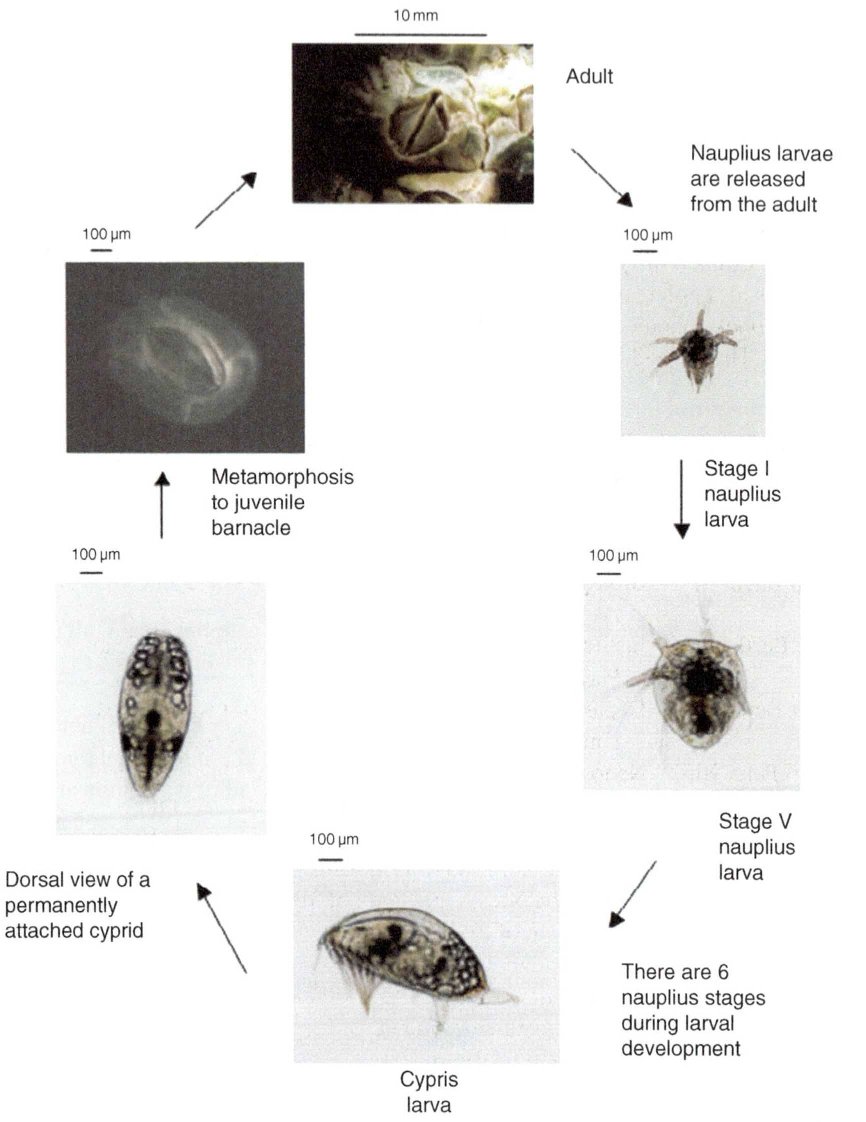

Fig. 2.2 A typical balanid barnacle life history (credit, Dr. M. Kirby, Newcastle University)

method of adhesion by the sessile adult. Textured surfaces usually facilitate stronger adhesion than smooth ones when this strategy is employed since roughened surfaces would allow greater interaction between the adult adhesive and the surface. *Semibalanus balanoides* cyprids are reported to settle preferentially on resin tiles with a surface texture in the range of 0–0.5 mm (Hills and Thomason, 1998), which is similar to their body size of ~1 mm. Wethey (1986) on the other hand, concluded that larger scale surface contour was the most dominant, positive influence on settlement of *S. balanoides*. So for this species at least, both surface texture and contour appear to be important variables in stimulating cyprid settlement. In contrast, there have been few reports of a negative influence of surface topography on barnacle settlement, although Barnes and Powell (1950) reported that glass fibres of 8 μm × 1 mm were inhibitory to *Balanus crenatus* cyprid settlement. Crisp and Barnes (1954) established that *S. balanoides* cyprids rejected surfaces with sharp ridges and grooves and, in a more detailed investigation, Lemire and Bourget (1996) found that settlement of *Balanus* sp. cyprids was lower on surfaces with V-shaped grooves with a lateral dimension of 1 mm, compared to surfaces bearing grooves with lateral dimensions of 10 and 100 mm; though settlement was still higher than on flat surfaces. In contrast to the aforementioned species, *Balanus improvisus*, an important temperate fouling species, prefers smooth over roughened surfaces for settlement. The scale of microtexture that effects settlement inhibition in *B. improvisus* has been investigated in both laboratory and field experiments with moulded surfaces (Berntsson et al., 2000, 2004). This revealed a narrow range of surface roughness that was strongly inhibitory to settlement of *B. improvisus*; namely a roughness height within the range 30–45 μm, an average roughness of 5–10 μm and a roughness width of 150–200 μm (Berntsson et al., 2000). The inhibitory action appeared to be effected by a behavioural rejection of the microtextured surfaces; cyprids explored ribbed surfaces less and engaged less in so-called close exploratory behaviour, indicating strongly that these surfaces were less attractive to the cyprids (Berntsson et al., 2004). Recently, microtextured polydimethylsiloxane elastomer (PDMSe) surfaces with topographies of ~20 μm have been shown to reduce the settlement of *Ulva* spores by up to 85%. The design of these surfaces was inspired by the skin of fast moving sharks (Carman et al., 2006) and their topography also renders them superhydrophobic (Marmur, 2006).

Larval age: Larval age is known to affect barnacle settlement (Rittschof et al., 1984) with older cyprids generally settling at a higher rate, and with less discrimination, than younger cyprids. This occurs as lipid energy reserves, accumulated as a feeding nauplius, are gradually consumed (Satuito et al., 1996). This area of larval ecology has, however, received comparatively little attention in the literature since historically the majority of cyprids used in assays have been wild and thus of indeterminate age (Crisp and Meadows, 1963). More recently, laboratory rearing of cyprids has presented the opportunity to study this area further and it has been demonstrated empirically, in *B. amphitrite* at least, that older cyprids settle more rapidly and in larger numbers than younger cyprids. During surface inspection, in their planktonic stage, barnacle cyprids may explore multiple surfaces and return to the water column many times before committing to permanent settlement.

Flow: The hydrodynamic characteristics of settlement locations are of great importance to *S. balanoides* cyprids during surface exploration. Crisp (1955) elegantly demonstrated this using a number of techniques to test the relative effects of illumination and shear stress on the swimming and attachment of cyprids in laminar flow. Mullineaux and Butman (1991) refined Crisp's study and demonstrated that settlement in *B. amphitrite* was consistently negatively correlated to shear stress in a hydrodynamic flume. Advection of cyprids to the settlement surfaces increased larval supply in certain areas of their apparatus, with cyprids subsequently migrating to their preferred locations. Indeed, this preference for slow moving/static water has resulted in the rise in popularity of *B. amphitrite* for laboratory assays (Rittschof et al., 1984), and provided further evidence that site selection by cyprids is not a passive process, but the result of a highly complex sequence of behavioural adaptations. Eckman et al. (1990) showed that detachment from surfaces during exploration is governed primarily by instantaneous increases in flow, rather than the mean maximal flow after slow ramping. It is not known whether this release is voluntary on behalf of the cyprids, or a manifestation of their temporary attachment mechanism.

Surface chemistry: A reasonable hypothesis might be that cyprids (and indeed other marine invertebrate larvae) preferentially select surfaces to which they attach more strongly both pre- and post-metamorphosis. A complimentary area of study concerns release of these organisms from different surfaces, based on their chemistry. This forms the basis of the current philosophy regarding fouling-release coatings (Almeida et al., 2007).

The physicochemical phenomenon, commonly referred to as 'surface free energy' (SFE), has been the basis for many studies into cyprid settlement and adhesion. It is hypothesised that high SFE would afford stronger attachment to exploring cyprids and, consequently, increase the attractiveness of surfaces. Ironically SFE has never been specifically tested in this regard. Yule and Walker (1984) and Crisp et al. (1984) investigated the adhesion of cyprids onto diverse surfaces such as beeswax, polytetrafluoroethylene (PTFE), glass, Perspex, slate and Tufnol and their findings suggested that temporary adhesion of *S. balanoides* was significantly greater on high energy surfaces, especially those with a large polar component to the SFE. Importantly, however, these surfaces also differed in a range of other characteristics including chemical functionality, rugosity and material properties such as Young's modulus. This variability, as in many other studies, renders any conclusion drawn on the basis of SFE equivocal.

Gerhart et al. (1992) and Rittschof and Costlow (1989) also determined that *B. amphitrite* settled in significantly greater numbers on 'high SFE' surfaces, such as muffled glass, than on 'low SFE' surfaces such as glass modified by silanisation chemistry. Silanisation resulted in surface functional groups, such as $R—CH_3$ and $R—NH_3$. The surface energies, or reactivity, of these materials, as well as those used by Crisp et al. (1984) , andYule and Walker (1984), were, by convention, characterised using an analogous surface property – the advancing water contact angle (θ_{AW}). $\cos\theta_{AW}$ is often used to infer SFE and, while contact angles from a range of diverse probe liquids can be used to accurately infer SFE (in dynes.cm^{-1}) from the critical surface tension of a solid, water contact angle is enormously variable depending on surface texture (Marmur, 2006), contamination and hysteresis. The terms 'wettability', based on $\cos\theta_{AW}$, and surface energy as a finite energy in Joules should not, therefore, be used interchangeably. Wettability is an observable manifestation of SFE, and is the property most often measured and referred to in the literature as SFE. Modulating the SFE of a smooth surface can only be achieved by altering the surface chemistry. Vastly different surface chemistries can manifest identical θ_{AW}, so the specific surface chemistry becomes a covariate and is not adequately covered by the umbrella term, SFE.

Recent work by Aldred et al. (2006) has demonstrated the potential for error associated with drawing assumptions about a single surface characteristic from such diverse materials. In Aldred et al. (2006), mussels showed increased spreading of their byssal plaque on hydrophilic, self-assembled monolayers (SAMs) of ω-terminated alkanethiolates, as opposed to on hydrophobic SAMs – as was predicted by Crisp et al. (1984). This observation is contrary to the predictions of Young's theory (Young, 1805), and further demonstrates the importance of surface chemistry on adhesion, rather than simply a balance of thermodynamic forces. Even using accurately characterised surfaces such as SAMs, the chemistry and reactivity of the surface functional groups, by definition, change to manifest different wettabilities. Specific surface chemistry, rather than wettability per se, may, therefore, have a greater effect on settlement. Assays with cyprids (Aldred, unpublished) suggested this to be the case. *B. amphitrite* cyprids settled in significantly higher numbers on hydrophilic —OH (θ_{AW} =< 20°) terminated SAMs rather than hydrophobic —CH_3 SAMs ($\theta_{AW} = 113°$). The rate of cyprid settlement was clearly based on more than wettability, since significantly more cyprids also settled on trimethylamine (—Nme3) terminated SAMs, compared to carboxylic acid (—COOH) SAMs, both of which had $\theta_{AW} < 20°$ (Fig. 2.3).

Surface chemistry clearly dictates the attractiveness of surfaces to cyprids. Whether it does so via the modulation of surface wettability (Lindner, 1992; Callow and Fletcher, 1994), or by more direct effects of surface chemistry on bonding is a matter of debate. Furthermore, the cypris larvae of *B. improvisus* prefer to settle

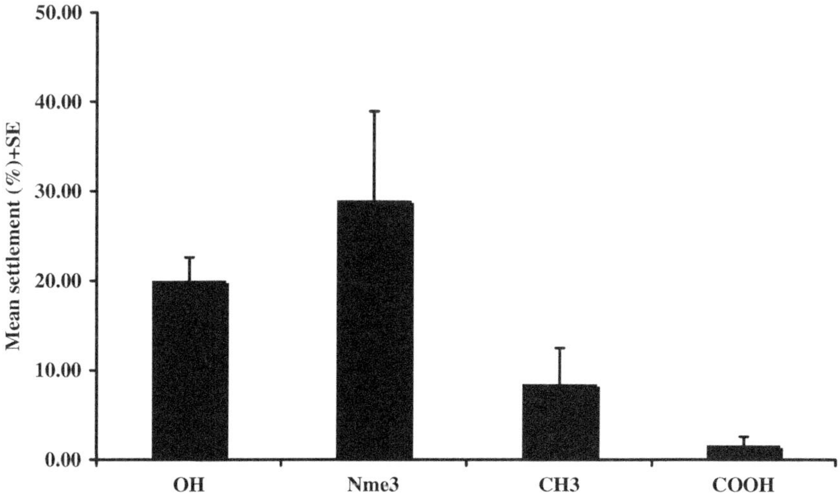

Fig. 2.3 *B. amphitrite* cyprid settlement after 24 h on SAMs with different chemical terminations: —OH (θ_{AW} =< 20°), —Nme3 (θ_{AW} =< 20°), —CH_3 ($\theta_{AW} = 113°$) and —COOH (θ_{AW} =< 20°)

on relatively hydrophobic surfaces (Dahlström et al., 2004), suggesting that either attachment tenacity is not an important consideration for cyprids during exploration and surface selection, or that underlying physicochemical factors, other than wettability, affect the attachment tenacity of exploring and settling cyprids. The latter seems a reasonable hypothesis, and the effects of differing surface chemistry will consequently be considered during the following discussion of the thermodynamics of cyprid temporary adhesion.

Adult Pheromones: In addition to texture, hydrodynamics and chemistry, the literature points to many other factors that potentially influence the 'choice' of settlement site, including e.g. surface colour (Yule and Walker, 1984) and biofilm (Wieczorek and Todd, 1998). A combination of these factors could, it seemed, restrict populations of barnacles to certain locations, thus approximating gregariousness. However, it was the discovery of inductive pheromonal effects in barnacles, and the degree to which adult conspecifics affect the behaviour of cyprids, that elucidated the system – but also raised many questions. The cyprid settlement cue, found to be present in adult barnacles, was the most intriguing and, as a result, the most comprehensively studied settlement cue for cyprids during the 1960s–80s. Indeed, it seemed to finally explain the mechanism for gregariousness in barnacles.

Knight-Jones and Stevenson (1950) were the first to observe that a water soluble protein, termed 'arthropodin' after Fraenkel and Rudall (1940), was important in the gregarious settlement of barnacles. The integument of adult barnacles (and extract thereof) was known to induce conspecific settlement of these organisms (Knight-Jones, 1953; Crisp and Meadows, 1962, 1963; Larman et al., 1982; Yule and Crisp, 1983) and a comprehensive review of this field can be found in Clare and Matsumura (2000). This settlement inducer is now termed the settlement-inducing protein complex (SIPC) (Matsumura et al., 1998a, b), a large glycoprotein which has recently been cloned and sequenced (Dreanno et al., 2006a). In fact, the presence of adult barnacles is thought to be of overriding importance to exploring cyprids, increasing the attractiveness of surfaces that would, otherwise, be rejected (Prendergast et al. 2009: G Prendergast regarding *S. balanoides*).

One question was clear, however: If conspecific pheromones are of paramount importance to settling cyprids and they will, unless 'desperate', reject all surfaces except those inhabited by adults, how then do barnacles colonise new areas? Yule and Walker (1985; see also Yule and Crisp, 1983) provided an answer to this paradox with the discovery that proteinaceous 'footprints' deposited by exploring cyprids effect a positive settlement response in subsequently exploring larvae.

Antennular secretion as a settlement cue: The footprints are of a glycoproteinaceous antennular secretion that is deposited onto surfaces during exploration by cyprids (Fig. 2.4). Although not conspicuous after staining on all surfaces, an invisible trace seemed to be sufficient to induce an increased settlement response in competent cyprids (Clare et al. 1994: N. Aldred). It was hypothesised that cyprids explore more extensively on surfaces that present attractive physical characteristics, as discussed previously, and therefore deposit more footprints. This increases the attractiveness of the surface in a positive feedback mechanism and results, ultimately, in gregarious settlement of cyprids without the necessity of an encounter

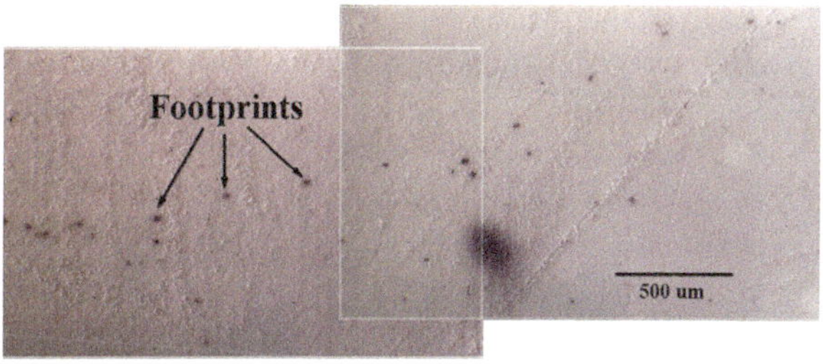

Fig. 2.4 *B. amphitrite* footprints, stained on nitrocellulose membrane using an antibody specific to a 76 kDa subunit of the SIPC

with a conspecific adult. This finding answered important questions regarding the mechanisms behind gregarious colonisation, by barnacles, of conspecific-free surfaces, but raised many more concerning cyprid temporary adhesion.

The mechanism by which cyprids detect this surface-bound protein remains unclear. Contact stimulation of sensory processes on the 3rd and 4th antennular segments (Fig. 2.5), as proposed by Crisp and Meadows (1963), would seem probable since the antennular secretion is not detected in solution. Barnes (1970), however, preferred the suggestion by Knight-Jones (1953) that the antennular secretion of the exploring cyprids may contain enzymes that are capable of breaking down the protein cue into small perceptible units. Presumably (assuming that the secretion contributes to adhesion) this secreted enzyme would also digest the glycoprotein maintaining the cyprid's attachment to the surface, making this theory unlikely. Indeed, enzymatic attack of the cyprid antennular secretion was suggested as a possible antifouling method by Pettitt et al. (2004).

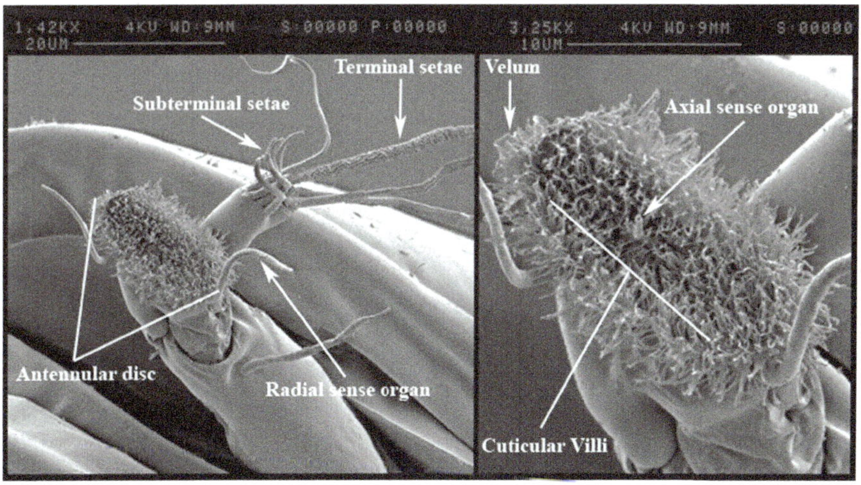

Fig. 2.5 A scanning electron micrograph of the antennular structure of *B. amphitrite,* including an enlargement of the adhesive disc (right hand side) and associated sensory structures (left hand side)

There is now strong evidence that the footprint protein and the SIPC are related, thereby explaining the pheromonal activity of the former. Matsumura et al. (1998b), using a polyclonal antibody raised against a 76 kDa subunit of the SIPC, showed that footprint protein reacted with the antibody. More recent evidence of a close relationship has been obtained using antibodies raised against peptides near the C- and N-terminals of the SIPC (Dreanno et al., 2006b). Moreover, the SIPC, a cuticular protein, and footprint protein are both likely to be epidermal in origin (Dreanno et al., 2006b, c).

2.3 Temporary Adhesion

It is clearly erroneous to suggest that one stage in the life history of an organism is more or less important than any other, however, it is clear that the cypris larva fulfils a special role. The successful location of an appropriate settlement site predetermines both the success of the adult barnacle *and* that of subsequent generations. Cyprids, therefore, must be particularly discerning during their exploration of surfaces. So critical is this exploratory phase that 'fatal errors', invoked by permanent attachment to a surface deleterious to the organism, are thought to be the major driving force behind the evolution of these organisms (Holm, 1990). For the assessment of, and temporary attachment to, potential settlement sites, cyprids have developed an array of sensory and 'adhesive' adaptations that have been referred to previously.

2.3.1 Morphological Adaptations

From casual observation, the most noticeable change that takes place during metamorphosis from the nauplius to the cyprid is the change in body morphology to a more streamlined aerofoil shape. Streamlining, in combination with their small size, allows cyprids to effectively exploit the fluid boundary layers surrounding immersed objects and dramatically reduces the degree of shear and, therefore, drag that they experience (Crisp, 1955; Eckman et al., 1990). This reduces the tenacity that is necessary for cyprids to remain attached during surface exploration; however a significant force is still required for attachment.

The majority of morphological changes that take place during the transition from the 6th nauplius stage to the cyprid facilitate temporary attachment and successful settlement site selection. The modification that is of primary importance in this capacity is the development of the cyprid antennular structures (see Figs. 2.1 and 2.5).

The antennules and their associated structures, are undoubtedly the primary sensory tools of barnacle cyprids (Clare and Matsumura, 2000), and the 3rd and 4th segments of the antennules bear an array of setae that, in the case of *S. balanoides* and *B. amphitrite*, have been studied in minute detail (Nott and Foster, 1969; Blomsterberg et al., 2004; Bielecki et al. 2009). The fourth segment especially is almost entirely filled with neurone dendrites (Barnes, 1970).

Some of the structures described by Nott and Foster (1969) are presented for *B. amphitrite* in Fig. 2.5 (see also Clare et al., 1994). The 3rd antennular segment bears an attachment disc with encircling velum. The attachment disc (3rd segment) surface is covered with micro-scale cuticular villi. The 3rd segment also possesses radial sensory setae, a central, 'axial', sense organ and numerous pores for the secretion of a glycoproteinaceous 'adhesive' originating from specialised hypodermal glands within the 2nd segment. It is suggested that this substance, deposited as footprints during exploration, is secreted onto the surface of the disc to facilitate temporary attachment to surfaces. Also present on the surface of the antennular disc is the cement duct opening, through which another proteinaceous adhesive is expressed on permanent settlement (Walker, 1971). This permanent adhesive, or cement, is secreted directly from paired glands within the body of the cyprid and cures within 2 h in *B. amphitrite* (Phang et al., 2006), effectively anchoring the cyprid until the onset of adult adhesive production (Okano et al., 1996; Ödling et al., 2006). This can take as long as 40 days in the case of *S. balanoides* (Yule and Walker, 1987).

The 4th antennular segment bears an array of sensory setae, the function of which is unknown. However, behavioural observations by Clare et al. (1994) support their sensory function. On contact with surfaces, or when immersed in a solution containing stimulatory substances, the 4th segment is flicked at a rate that has been shown to increase with concentration of the stimulant.

2.3.2 Contemporary Theories

It is, perhaps, fair to say that the majority of fundamental observations on cyprid settlement behaviour were completed before studies of cyprid adhesion had really begun. The ability of cyprids to attach and explore underwater was originally recorded and explained in some detail by Darwin (1854), and subsequently by Visscher (1928) and Doochin (1951). It is the descriptions of Crisp (1955, 1976), however, that still stand today (Jonsson, 2005). Crisp described wide searching, close searching and inspection behaviours, suggesting that exploration becomes gradually more refined on more favourable surfaces, whereas unfavourable surfaces are rejected during wide exploration.

Physiological and behavioural studies of exploring cyprids (Lagersson and Høeg, 2002; Marechal et al., 2004) have provided useful information pertaining to temporary adhesion on a macro-scale. However, suggestions of how cyprids may attach to surfaces during exploration, based on empirical evidence, were not made until the early 1980s. Work at Menai Bridge (Yule and Crisp, 1983; Walker and Yule, 1984; Yule and Walker, 1985) demonstrated the strength of *S. balanoides* temporary adhesion to be of the order of 0.068–0.076 MNm^{-2} on smooth glass; precluding the earlier suggestion (Lindner, 1984 that their attachment may be due to suction developed by the encircling cuticular velum (Fig. 2.5). This conclusion was supported by A. B. Yule's observation that cyprids can explore on the edge of fine surfaces, where the antennular disc is not in complete contact with the surface. The suction hypothesis was already considered unlikely by Barnes (1970) who claimed that the musculature of the antennule would not allow for the creation of a vacuum beneath the adhesive disc. Further, tenacity was shown to vary with surface composition, which would not be the case if the mechanism relied purely on suction. Temporary attachment on high-energy surfaces was shown to be significantly stronger than attachment to low energy surfaces, and these data compare favourably to more contemporary published studies (Kesel et al., 2003) in other organisms. Although the cyprid's modified antennules have long been known to facilitate its temporary attachment, quite how cyprids generate this great tenacity underwater remains poorly understood.

Under certain circumstances, deposits of the antennular secretion are left as 'footprints' on explored surfaces and can easily be stained using conventional protein reagents, or by immunoblotting (see Fig. 2.4 and Clare et al., 1994; Matsumura et al., 1998b). Assuming that the secretion acts as a visco-elastic adhesive, Yule and Walker (1987) acknowledged that this mechanism raises many questions regarding adhesive production and detachment from surfaces. It is our opinion, taking into account the structure of the antennular discs, the behaviour of cyprids during exploration and the inconsistent deposition of footprints of this adhesive onto diverse surfaces, that the antennular secretion only provides part of the adhesive force generated by the attachment disc (Phang et al., 2008).

2.4 Theoretical Principles of Cyprid Temporary Adhesion

It is surprising, given the research effort on barnacle settlement, that no definitive mechanism has ever been described to adequately explain the means by which cyprids explore surfaces underwater. This is even more surprising when it is considered that the temporary attachment of larvae is a fundamental 'first step', necessary before an adult fouling population can become established. It would seem logical that this initial exploratory phase would be a target for fouling prevention. Perhaps the relative ease with which quantities of the adult adhesive can be collected for study (Cheung et al., 1977) explains why larval adhesion mechanisms have been neglected to such a degree. In fact, the current depth of knowledge regarding the temporary and permanent adhesion/adhesives of cypris larvae is limited to the literature presented here; none of which has been able to make anything more than a speculative description of the underlying mechanics of cyprid adhesion. This clearly has more to do with the technical difficulties associated with carrying out micro-adhesion studies underwater than a lack of interest. However, there is potentially a great deal that can be learnt from similar studies carried out in a dry state (e.g. Gorb et al., 2000). Although it presents a daunting technical challenge to researchers, adhesion underwater is theoretically similar to adhesion in air. Both the ambient media are fluids, albeit with one considerably more viscous and having markedly different electrostatic properties than the other. Importantly, basic principles remain the same.

Many organisms use temporary adhesion on dry surfaces routinely. For example, certain insects (Arthropoda; Uniramia) and spiders (Arthropoda; Chelicerata) can move vertically up a wall, or inverted across a ceiling. Geckos are also one of nature's great climbers. In fact, the mechanism underlying this ability may be more relevant to cyprid temporary adhesion than has previously been recognised. The so-called 'dry' adhesion system used by geckos has never been seriously proposed as an explanation for the cyprid attachment mechanism, possibly under the misapprehension that the system cannot work hydrated. A review of the dry adhesion mechanisms used by terrestrial organisms may, therefore, provide a valuable insight into the temporary adhesion of cyprids.

2.4.1 Adhesion by Contact-Splitting and Van der Waals Forces

It has been almost 40 years since the detailed morphological study by Walley (1968) on *S. balanoides* cyprids. Since then, although similarities between the antennular structures of cyprids and the tarsal structures of flies (see Niederegger et al., 2002) have been noted (Walker et al., 1985), there has been no direct comparison of how these apparatuses may function under water.

Non-charged/non-polar intermolecular forces, such as Liftshitz-van der Waals (LW) and London dispersion forces are long range (up to 100 Å), but relatively weak forces, relying upon the establishment of weak temporary dipoles within molecules. The probability that electrons surrounding atomic nuclei within molecules are equally distributed at all times is very low, so temporary dipoles establish and are in a constant state of flux. Areas of different electron density attract each other and so, over short distances at least, these interactions can contribute strongly to adhesion. It is this type of force that contributes to non-polar SFE and is the only intermolecular force present in hydrocarbons such as methane (CH_4). Importantly, for these weak forces to operate, contact between two surfaces must be intimate, so it is necessary for one or both surfaces to be able to deform in order to account for micro-scale surface roughness.

The Johnson–Kendall–Roberts (JKR) model (Johnson et al., 1971) of contact mechanics demonstrates that, paradoxically, the splitting of a single contact into multiple smaller contacts provides enhanced adhesion strength, based on increased contact area with the surface (a summary of biologically inspired contact-splitting technology is provided by Spolenak et al. (2005)). Contact splitting is now accepted as a key principle of adhesion and has gained much support with respect to biological adhesion systems. Principal evidence for this theory is provided by the observation that the degree of contact splitting, or the 'hairiness' of attachment organs, is scaled to the opposing forces acting on the attachment. So for small, light organisms such as flies and beetles, micron scale terminal setae are sufficient to provide intimate contact and allow enough intermolecular interactions to occur to support the animal. In larger organisms such as the gecko (*Gecko gecko*), many more sub-micron setae are necessary to increase the contact area proportionally and ensure strong adhesion to the surface (Arzt et al., 2003).

The antennular disc morphology suggests that adhesion may be imparted by contact-splitting in barnacle cyprids. Cyprids are small (*B. amphitrite* = ~500μm, *S. balanoides* = ~1000 μm length) and their mass is also supported by the surrounding seawater. In addition to gravity, however, they must also contend with enormous external forces in the wave-beaten inter-tidal environment (Denny and Gaines, 1990). Although cyprids are hydrodynamically shaped and protected from hydrodynamic forces to a degree by the boundary layer overlying immersed surfaces, the high density and sub-micron terminations of *B. amphitrite* and *S. balanoides* cyprid cuticular villi (Fig. 2.5) suggest the potential and presumably the requirement, for considerable tenacity. When compared to flies, for example, the contact area: mass ratio of cyprids is sufficiently high to suggest that their mass should be much greater than it actually is. This over-compensation may have evolved in response to the huge forces present in the marine environment that terrestrial organisms do not have to contend with but, importantly, could also have arisen due to interference and tenacity reduction by water. The electron micrographs presented by Moyse et al. (1995) demonstrate considerable interspecies variation both in overall antennule morphology and in the size and density of the cuticular villi. As the effects of water on adhesion are constant for all species, it would seem that this diversity is driven either by the different physical properties of the favoured settlement sites of different species, and/or by the physical conditions of the settlement environment – i.e. wave action, currents etc. In either case, the implication is that the cuticular villi have at least a part to play in temporary adhesion. For example, cyprids of *Lepas australis* are considerably larger than those of another stalked barnacle *Pollicipes pollicipes* and have proportionally larger antennular attachment discs. Interestingly though, the cuticular villi of *L. australis* are considerably *smaller* and more densely packed than those of *P. pollicipes* (Moyse et al., 1995). This, for the reasons discussed above, would certainly be predicted by JKR contact splitting theory. *Megatrema anglica* has even more specialised 3rd antennular segments, presumably suited to its coralline habitat. *Capitulum mitella*, an inhabitant of shaded intertidal rocky crevices where hydrodynamic shear is likely to be high, had the highest density of villi (10.2 villi μm^{-2}) of the species examined. This relationship between villus density and potential tenacity was noted by Moyse et al. (1995) although they, as Nott and Foster (1969) before them, suggested only that the extra villi may serve to better retain temporary adhesive. Although this is undoubtedly important, as will be discussed in the following section, it is difficult to envisage how a greater volume of adhesive would directly increase tenacity.

2.4.2 Capillarity and Stèfan Adhesion

Huber et al. (2005) used Atomic Force Microscopy (AFM) to demonstrate that adhesion of gecko spatulae is stronger on high- than low-energy surfaces, and far weaker when totally immersed underwater than when in air. Geckos, especially, are almost totally reliant upon weak intermolecular/electrostatic forces for adhesion, whereas insects have been shown to use a different adhesion method in addition to contact splitting. Flies leave deposits of an oily substance on surfaces that they have attached to (Walker et al., 1985; Langer et al., 2004) and this, it is believed, is used to exploit capillary or viscous forces and maximise adhesion. Interestingly, increasing the relative humidity when geckos are attaching in air also significantly increases tenacity (Huber et al., 2005), suggesting some contribution of capillary forces to over-all adhesion strength.

Capillarity, and adhesion resulting from its effects, arises from the interaction of adhesion, cohesion and surface tension within fluids and by pressure differences in a fluid joint as described by the Laplace-Kelvin equation. For fluids, such as water between two smooth surfaces, capillarity functions when the *adhesive* force between the fluid (usually a liquid) and the solid is greater than the *cohesive* force within the liquid. Under these circumstances the liquid will 'run' between the surfaces, drawing them together as it spreads. For this reason, two polished steel plates are drawn together by paraffin oil (Budgett, 1911) but are not if liquid mercury is the fluid in the joint. This theory alone does not, however, provide a complete explanation for how cyprids reversibly attach to surfaces.

In water, according to the Laplace-Kelvin theory, capillary forces would be compromised if water were trapped at the interface between the antennular secretion and the surface. When the liquids in the adhesive joint (water) and the surrounding medium (also water) met and merged, the difference in tension inside and outside the joint would equilibrate and adhesion would be lost. If this situation occurs, the glycoproteinaceous temporary adhesive of the cyprid must make an effective water displacer in order to contact immersed, hydrophilic surfaces. There may, however, be a contribution by Stèfan adhesion (Stèfan, 1874), or resistance to viscous flow (Smith, 2002). Indeed, the Stèfan adhesion theory predicts very large attachment forces for materials like the cyprid temporary adhesive (Grenon and Walker, 1981). However, the assumptions for this mechanism are not entirely met, as the antennular disc is flexible rather than rigid and undulating rather than smooth, and true Stèfan adhesion would also assume that the fluid within the adhesive joint is also present outside the joint. If the temporary adhesive truly functions as a kind of viscous adhesive joint, then it is likely that a combination of these forces is involved. More questions are raised, however: If the antennular secretion is a viscose adhesive, why are the extensive cuticular villi on the attachment disc required? It has been suggested that they may serve to retain the secretion on the adhesive disc, but it is perhaps more likely that they serve a dual role, as in flies, providing electrostatic adhesion with a fluid joint aspect. Flies, however, never have to attach underwater. In this scenario, barnacle cyprid temporary adhesion, without the antennular secretion, could be considered analogous to a gecko attempting to attach underwater. Overall tenacity relying on LV interactions would be dramatically reduced in an aqueous medium, and these forces alone would almost certainly be insufficient to allow a cyprid to remain firmly attached. The advantage that cyprids have over the hypothetical marine gecko, is that, like flies, cyprids can introduce a viscous fluid into the adhesive joint, increasing capillary forces and strengthening tenacity; just as increasing humidity does for geckos in air. In this way, the antennular secretion can be considered an 'adhesive'. Cyprids may have evolved an immiscible and highly viscous glycoproteinaceous secretion, derived from modified epidermal cells (Dreanno et al. 2006b, c), to achieve surface dewetting and temporary adhesion under water. Interestingly, the secreted oils that serve this function in air for some insects have been shown to be very similar in composition to cuticlar lipids (Kosaki and Yamaoka 1996), thus demonstrating a similar adaptation of body tissue for this specialised purpose.

A case, albeit speculative, has now been made for cyprids exploiting contact splitting and the theory of dry adhesion in a similar way to flies and geckos, but also enhancing this mechanism with the use of a viscous secretion to bridge the adhesive joint and increase tenacity through capillary forces. Even this, however, may not be a complete explanation for the use of an antennular secretion during surface exploration. Huber et al.'s work (2005) with geckos suggests that, while LV forces are important to both gecko and fly adhesion, adhesion is weakened when the system is immersed in water. It was also mentioned above that a possible reason for the large degree of contact splitting on cyprid antennular discs could be due, similarly, to the dampening effect of water on electrostatic adhesion, resulting in reduced tenacity. This phenomenon is known as water's dielectric effect, and could suggest another use for the antennular secretion.

2.4.3 Dielectrics and Adhesion

The specific property of water that causes this reduction in electrostatic adhesion is its high dielectric constant (ε). The dielectric constant is a measure of the extent to which a substance concentrates electrostatic lines of flux. Thus, it describes the storage of electricity by a medium and, therefore, the insulative properties of a substance. Materials with high dielectric constants (termed 'dielectrics') are often used as electrical insulators e.g., mica, hydrocarbon plastics and metal oxides (ceramics). Pure water has a high dielectric constant of ε=80 and is considered a good insulator. Dielectrics have the effect of increasing the relative distance between charged surfaces. That is to say, in a vacuum (which, by definition has ε=1), two charged surfaces will be attracted 80 times more strongly than they would be in an aqueous medium. This clearly has serious implications for organisms trying to adhere electrostatically to immersed surfaces, as was alluded to above, and is sufficient to greatly diminish the interaction energies of charge-charge, non-covalent bonds (Zhao et al., 2006) when the reaction occurs in an aqueous medium. It would then be advantageous for cyprids to reduce the dielectric constant of the interface between the antennular attachment disc and the substratum to increase the strength of the adhesive joint. Indeed, evidence exists to suggest that the glycoprotein secretion of barnacle cyprids could perform this function.

Proteins can, in fact, circumvent this issue by restricting ligand-receptor reactions to highly hydrophobic areas of their conformation, thus displacing water, lowering ε and increasing the interaction energy with surfaces (Chen et al., 2002). Fürth (1923) originally demonstrated, using gelatin and albumen proteins, that solutions of these proteins tend to have dielectric constants far lower than that of pure water by an amount increasing in proportion with the protein concentration. Interestingly, ε of the whole proteins did not seem to be dependant on ε of single constituent amino acids. Glycocoll, for example, has a higher dielectric constant than water (ε=90) (Hedestrand, 1928). Sugars were also shown to reduce ε in aqueous solutions (Malmberg and Maryott, 1950) suggesting that glycoproteins, consisting of both peptide and carbohydrate moieties, could be well suited to increasing intermolecular attraction between the attachment disc and the substratum. Providing that proteins are non-polar and have no permanent electrical moment, then this phenomenon is provided for by Debye's theory – relating the dielectric constant of a substance to its molecular dipole moment (Onsager, 1936). Non-polar proteins within the temporary adhesive would also be beneficial, as they would allow for effective dewetting of the antennular structure's adhesive surface. This would, theoretically, make initial surface contact more successful and enhance adhesion through successful hydrophobic interactions as predicted by the DLVO theory for colloidal adhesion (Derjaguin and Landau, 1941; Verwey and Overbeek, 1948).

Finally, it is important to consider that there are two adhesive interfaces during the temporary adhesion of cyprids. The interface between the antennular secretion, or cuticular villi, and the surface is the obvious one. However, adhesion between the antennular disc and the antennular secretion must be at least as strong. In fact, it would make good sense for the cyprid to bond more strongly to its 'adhesive' than the adhesive does to the surface to reduce the necessity for adhesive production, therefore conserving valuable energy. With this in mind, the cuticular villi could serve *yet another* function, increasing the effective surface area of the antennular disc and facilitating retention of the antennular secretion as discussed previously. Of course, footprint adhesive needs to be deposited to serve as a settlement cue to other cyprids. However, only a trace of the glycoprotein on a surface is sufficient to effect settlement and the majority probably remains on the antennular disc (depending on the surface).

Low-polarity glycoproteins may, therefore, provide an ideal solution to the problems encountered by cyprids exploring surfaces under water. Using these secretions would, theoretically, exclude water and accentuate 'dry' adhesion as explained above, while simultaneously providing a capillary/Stèfan type adhesion mechanism and enhancing the effect of favourable hydrophobic interactions. If the antennular secretion is not permanently present on the antennular disc, or if it is removed and deposited onto a surface for any reason, then the spreading of more glycoprotein would, once again, be aided by the morphology of the disc surface which, by its nature, would likely be superhydrophobic (Marmur, 2006).

2.5 Conclusions

Given that novel technologies are already being sought to replace the current generation of (primarily copper-based) biocidal marine coatings, there is a clear need to improve our understanding of under-water adhesive mechanisms. Presently, the low modulus (Swain and Schultz, 1996), low surface energy (Brady and Singer, 2000) approach is based upon removal of established adult barnacles, and little consideration seems to be given to the effects of these properties on cyprid temporary adhesion. This is unfortunate since, first, deterring settlement at this stage would seem a far more logical approach to the problem – prevention always being better than cure – and secondly, from the previous discussion, it seems that many of the properties currently included in fouling-release coatings to facilitate hard fouling release could also interfere with cyprid temporary adhesion. Radical modification of the current technologies may not be necessary to develop present coatings into surfaces to which cyprids find it very difficult to attach. For example, modulation of wettability could, theoretically, inhibit cyprid temporary attachment in one of two ways. The development of superhydrophobic surfaces has been by far the most scrutinized avenue of study, based on the conventional theory that wet adhesion should be more difficult on hydrophobic surfaces. In addition an adhesive joint, once established, would be assumed to be weaker if only weak non-polar bonding was responsible for its maintenance. Unfortunately, extension of basic wetting (Young's) theory suggests, in fact, that an adhesive would readily spread on immersed hydrophobic surfaces, providing that the adhesive is not as polar as water. This is likely to be the case for cyprids, especially if the antennular secretion is non-polar, so cyprids should still be able to attach very effectively to hydrophobic surfaces (see Callow et al., 2005 and Aldred et al., 2006 for further discussion). Alternatively, superhydrophilic surfaces could attract water so strongly that cyprids would find it difficult to displace the water and thus make intimate contact with the surface. Currently, it seems that cyprids will attach to surfaces of any wettability, although serious modulation of underlying chemical functionalities is only in its infancy.

The theoretical argument presented here for cyprid temporary adhesion can be experimentally tested. AFM studies of footprints on glass (Fig. 2.6) have begun to clarify the mechanism of temporary adhesion (Phang et al., 2008) although much more work is needed. As predicted, it appears that the adhesion strategy of cyprids is complex and relies neither entirely on a viscous mechanism, facilitated by the antennular secretion, or on the pseudo-'dry' adhesion system proposed above but perhaps requires a combination of the two. Micro-mechanical studies, such as AFM, are likely to yield invaluable information regarding the processes of adhesion at

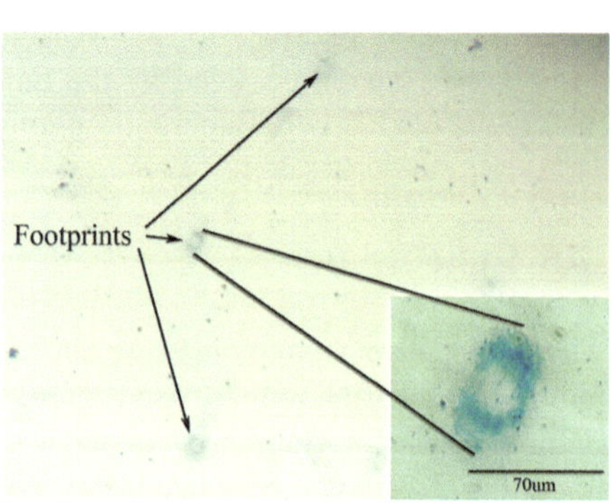

Fig. 2.6 *S. balanoides* footprints on glass could be used for AFM studies. Staining using Coomassie Brilliant Blue protein dye reagent allows visualisation

biological interfaces in the future. Difficulties with acquisition of material samples and the manipulation of these small planktonic organisms will persist, but the temporary adhesive strategy of barnacle cyprids is likely to occupy researchers in this field for many years to come.

Acknowledgments We are grateful for support from the US Office of Naval Research (N00014-05-1-0767 and N00014-08-1-1240 to ASC) for studies of cyprid temporary adhesion. We would also like to express our appreciation to Dr. L.K. Ista and Prof. G.P. Lopez at the University of New Mexico, USA for allowing us to use unpublished data from our collaborative work.

References

Aldred, N., Ista, L. K., Callow, M. E., Callow, J. A. Lopez, G. P., and Clare, A. S. (2006). Mussel (*Mytilus edulis*) byssus deposition in response to variations in surface wettability. *J. R. Soc. Interface* 3: 37–43.

Almeida, E., Diamantino, T. C., and de Sousa, O. (2007). Marine paints: The particular case of antifouling paints. *Prog. Org. Coat.* 59: 2–20.

Anderson, D. T. (1994). *Barnacles: Structure, function, development and evolution.* Chapman and Hall, London.

Arzt, E., Gorb, S., and Spolenak, R. (2003). From micro to nano contacts in biological attachment devices. *Proc. Natl. Acad. Sci. USA* 100: 10603–10606.

Barnes, H. (1970). A review of some factors affecting settlement and adhesion of some common barnacles. In:*Adhesion in Biological Systems.* Ed. R. S. Manly. Academic Press, New York and London, pp. 89–111.

Barnes, H., and Powell, H. T. (1950). Some observations on the effect of fibrous glass surfaces upon the settlement of certain sedentary marine organisms. *J. Mar. Biol. Assoc. UK* 29: 299–302.

Berntsson, K. M., Jonsson, P. R., Larsson, A. I., and Holdt, S. (2004). Rejection of unsuitable substrata as a potential driver of aggregated settlement in the barnacle *Balanus improvisus*. *Mar. Ecol. Prog. Ser.* 275: 199–210.

Berntsson, K. M., Jonsson, P. R., Lejhall, M., and Gatenholm, P. (2000). Analysis of behavioural rejection of micro-textured surfaces and implications for recruitment by the barnacle *Balanus improvisus*. *J. Exp. Mar. Biol. Ecol.* 25: 59–83.

Bielecki, J., Chan, B.K.K., Hoeg, J. T., and Sari, A., (2009). Antennular sensory organs in cyprids of balanomorphan cirripedes: standardizing terminology using Megabalanus rosa. *Biofouling*25:203-214.

Blomsterberg, M., Høeg, J. T., Jeffries, W. B., and Lagersson, N. C., (2004). Antennulary sensory organs in cyprids of *Octolasmis angulata* and three species of *Lepas* (Crustacea: Thecostraca: Cirripedia: Thoracica): a scanning electron microscopy study. *J. Morphol.* 200: 141–153.

Brady, R. F. (2001). A fracture mechanical analysis of fouling release from non-toxic antifouling coatings. *Prog. Organ. Coat.* 43: 188–192.

Brady, R. F., and Singer, I. L. (2000). Mechanical factors favoring release from fouling release coatings. *Biofouling* 15: 73–81.

Budgett, H. M. (1911). The adherence of flat surfaces. *Proc. R. Soc. Lond. A* 86: 25–35.

Callow, J. A., Callow, M. E., Ista, L. K., Lopez, G., and Chadhury, M. K. (2005). The influence of surface energy on the wetting behaviour of the spore adhesive of the marine alga *Ulva linza* (synonym *Enteromorpha linza*). *J. R. Soc. Interface* 2: 319–325.

Callow, M. E., and Callow, J. A. (2002). Marine biofouling: a sticky problem. *Biologist* 49: 10–14.

Callow, M. E., and Fletcher, R. L. (1994). The influence of low surface energy materials on bioadhesion – a review. *Int. Biodeter. Biodeg.* 34: 333–348.

Carman, M.L., Estes, T.G., Feinberg, A.W., Schumacher, J.F., Wilkerson, W., Wilson, L.H., Callow, M.E., Callow, J.A., and Brennan, A.B. (2006). Engineered antifouling microtopographies – correlating wettability with cell attachment. *Biofouling* 22: 11–21.

Chaudhury, M. K., Finlay, J. A., Chung, J. Y., Callow, M. E., and Callow, J. A. (2005). The influence of elastic modulus and thickness on the release of the soft-fouling green alga *Ulva linza* (syn. *Enteromorpha linza*) from poly(dimethylsiloxane) (PDMS) model networks. *Biofouling* 21: 41–48.

Chen, B., Piletsky, S., and Turner, A. P. F. (2002). Molecular recognition: Design of "keys". *Combinat. Chem. High. Throughput Screen.* 5: 409–427.

Cheung, P. J., Ruggieri, G. D., and Nigrelli, R. F. (1977). A new method for obtaining barnacle cement in the liquid state for polymerisation studies. *Mar. Biol.* 43:157–163.

Christie, A. O., and Dalley, R. (1987). Adhesion in barnacles. In: *Barnacle Biology*. Crustacean Issues, 5 Ed. A. J. Southward, AA Balkema. Rotterdam, The Netherlands, pp. 419–433.

Clare, A. S., and Matsumura, K. (2000). Nature and perception of barnacle settlement pheromones. *Biofouling* 15: 57–71.

Clare, A. S., and Nott, J. A. (1994). Scanning electron microscopy of the fourth antennular segment of *Balanus amphitrite amphitrite*. *J. Mar. Biol. Assoc. UK* 74: 967–970.

Clare, A. S., Freet, R. K., and Mclary, M. J. (1994). On the antennular secretion of the cyprid of *Balanus amphitrite amphitrite*, and its role as a settlement pheromone. *J. Mar. Biol. Assoc. UK* 74: 243–250.

Crisp D.J., Walker, G., Young, G.A., and Yule, A.B. (1984). Adhesion and substrate choice in mussels and barnacles. *J. Coll. Int. Sci.* 104: 41–50.

Crisp DJ (1976) Settlement responses in marine organisms. In: *Adaptation to Environment: Essays on the Physiology of Marine Animals*. Ed. R. C. Newell., Butterworths, London, pp. 83–124.

Crisp, D. J. (1955). The behaviour of barnacle cyprids in relation to water movement over a surface. *J. Exp. Biol.* 32: 569–590.

Crisp, D. J. (1974). Factors influencing the settlement of marine invertebrate larva. In: *Chemoreception in Marine Organisms*. Ed. P. T. Grant and A. M. Mackie. Academic Press, New York, 177–265.

Crisp, D. J., and Barnes, H. (1954). The orientation and distribution of barnacles at settlement with particular reference to surface contour. *J. Anim. Ecol.* 23: 142–162.

Crisp, D. J., and Meadows, P. S. (1962). The chemical basis of gregariousness in cirripedes. *Proc. Roy. Soc. Lond. B.* 156: 500–520.

Crisp, D. J., and Meadows, P. S. (1963). Adsorbed layers: the stimulus to settle in barnacles. *Proc. Roy. Soc. Lond. B.* 158: 364–387.

Dahlström, M., Jonsson, H. Jonsson, P. R., and Elwing, H. (2004). Surface wettability as a determinant in the settlement of the barnacle *Balanus improvisus* (Darwin). *J. Exp. Mar. Biol. Ecol.* 305: 223–232.

Darwin, C. (1854). A monograph on the subclass Cirripedia, with figures of all the species. *Ray. Soc. Publ. London.* 684pp.

Denny, M. W., and Gaines, S. D. (1990). On the prediction of maximal intertidal wave forces. *Limnol. Oceanogr.* 35(1): 1–15.

Derjaguin, B. V., and Landau, L. (1941). Theory of the stability of strongly charged lyophobic sols and of the adhesion of strongly charged particles in solutions of electrolytes. *Acta Phys. Chim. URSS.* 14: 633–662.

Doochin, H. D. (1951). The morphology of *Balanus improvisus* Darwin and *Balanus amphitrite niveus* Darwin during initial attachment and metamorphosis. *Bull. Mar. Sci. Gulf. Caribb.* 1: 15–39.

Dreanno, C., Kirby, R. R., and Clare, A. S. (2006b). Smelly feet are not always a bad thing: The relationship between cyprid footprint protein and the barnacle settlement pheromone. *Biol. Lett.* 2: 423–425.

Dreanno, C., Kirby, R. R., and Clare, A. S. (2006c). Locating the barnacle settlement pheromone: spatial and ontogenetic expression of the settlement-inducing protein complex of *Balanus amphitrite*. *Proc. R. Soc. B.* 273: 2721–2728.

Dreanno, C., Matsumura, K., Dohmae, N., Takio, K., Hirota, H., Kirby, R. R., and Clare, A. S. (2006a). A novel alpha 2-macroglobulin-like protein is the cue to gregarious settlement of the barnacle, *Balanus amphitrite*. *Proc. Natl. Acad. Sci.USA* 103: 14396–14401.

Eckman, J. E., Savidge, W. B., and Gross, T. F. (1990). Relationship between duration of cyprid attachment and drag forces associated with detachment of *Balanus amphitrite* cyprids. *Mar. Biol.* 107: 111–118.

Evans, S. M, Birchenough, A. C., and Brancato, M. S. (2000). The TBT ban: out of the frying pan into the fire? *Mar. Poll. Bull.* 40: 204–211.

Fraenkel, G., and Rudall, K. M. (1940). A Study of the physical and chemical properties of the insect cuticle. *Proc. Roy. Soc. Lond. B* 129: 1–35.

Fürth, R. (1923). The dielectrical constants of some fluid solutions and their significance according to the dipal theory of Debye-With some supplements of dielectrical constants of some biologically and technologically interesting material. *Ann. Physik,* 70: 63–63.

Gerhart, D. J., Rittschof, D., Hooper, I. R., Einsenman, K., Meyer, A. E., Baier, R. E., and Young, C. (1992). Rapid and inexpensive quantification of the combined polar components of surface wettability: Application to biofouling. *Biofouling* 5: 251–259.

Gorb, S., Jiao, Y., and Scherge, M. (2000). Ultrastructural architecture and mechanical properties of attachment pads in *Tettigonia viridisima* (Orthoptera Tettigoniidae). *J. Comp. Physiol. A.* 186: 821–831.

Grenon, J. F., and Walker, G. (1981). The tenacity of the limpet, *Patella vulgata* L.: An experimental approach. *J. Exp. Mar. Biol. Ecol.* 54: 277–308.

Hedestrand, G. (1928). Uber die Dielektrizitaitskonstanten wüsseriger Losungen einiger Aminosauren. *Zeitschr. Physik. Chemie* 135: 36–48.

Hellio, C., De La Broise, D., Dufosse, L., Le Gal, Y., and Bourgougnon, N. (2001). Inhibition of marine bacteria by extracts of macroalgae: potential use for environmentally friendly antifouling paints. *Mar. Environ. Res.* 52: 231–47.

Hills, J. M., and Thomason, J. T. (1998). The effect of scales of surface roughness on the settlement of barnacle (*Semibalanus balanoides*) cyprids.*Biofouling* 12: 57–69.

Holm, E. R. (1990). Attachment behaviour in the barnacle *Balanus amphitrite amphitrite* (Darwin): Genetic and environmental effects. *J. Exp. Mar. Biol. Ecol.* 135(2): 85–98.

Huber, G., Mantz, H., Spolenak, R., Mecke, K., Jacobs, K., Gorb, S. N., and Arzt, E. (2005). Evidence for capillarity contributions to gecko adhesion from single spatula nanomechanical measurements. *Proc. Natl. Acad. Sci. USA* 102: 16293–16295.

Johnson, K. L., Kendall, K., and Roberts, A. D. (1971). Surface energy and the contact of elastic solids. *Proc. R. Soc. Lond. A.* 324: 301–313.

Jonsson, P. R. (2005). A classic hydrodynamic analysis of larval settlement. *J. Exp. Biol.* 208: 3431–3432.

Kesel, A. B., Martin, A., and Seidl, T. (2003). Adhesion measurements on the attachment devices of the jumping spider *Evarcha arcuata*. *J. Exp. Biol.* 206: 2733–2738.

Knight-Jones, E. W. (1953). Laboratory experiments on gregariousness during settlement in *Balanus balanoides* and other barnacles. *J. Exp. Biol.* 30: 584–598.

Knight-Jones, E. W., and Stevenson, J. P. (1950). Gregariousness during settlement in the barnacle *Elminius modestus* Darwin. *J. Mar. Biol. Assoc. UK* 29:291–297.

Kosaki, A., and Yamaoka, R. (1996). Chemical composition of footprints and cuticular lipids of three species of lady beetles. *Jpn. J. Appl. Entomol. Zool.* 40: 47–53.

Lagersson N. C., and Høeg, J. T. (2002). Settlement behaviour and antennulary biomechanics in cypris larvae of *Balanus amphitrite* (Crustacea: Thecostraca: Cirripedia). *Mar. Biol.* 141: 513–526.

Langer, M. G., Ruppersberg, J. P., and Gorb, S. (2004). Adhesion forces measured at the level of a terminal plate of the fly's seta. *Proc. R. Soc. Lond. B* 271: 2209–2215.

Larman, V. N., Gabbott, P. A., and East, J. (1982). Physico-chemical properties of the settlement factor proteins from the barnacle *Balanus balanoides*. *Comp. Biochem. Physiol.* 72(b): 329–338.

Lemire, M., and Bourget, E. (1996). Substratum heterogeneity and complexity influence microhabitat selection of *Balanus* sp. and *Tubulariacrocea* larvae. *Mar Ecol. Prog. Ser.* 135: 77–87.

Lindner, E. (1984). The attachment of macrofouling invertebrates. In J. D. Costlow and R. C. Tipper (eds.). Marine Biodeterioration: An Interdisciplinary Study: 183–202. London: Spon.

Lindner, E. (1992). A low surface energy approach in the control of marine biofouling. *Biofouling* 6: 193–205.

Malmberg, C. G., and Maryott, A. A. (1950). Dielectric constants of aqueous solutions of dextrose and sucrose. *J. Res. Natl. Bureau. Stds.* 48(4):299–303.

Marechal, J-P., Hellio, C., Sebire, M., and Clare, A. S. (2004). Settlement behaviour of marine invertebrate larvae measured by Ethovision 3.0.*Biofouling* 20: 211–217.

Marmur, A. (2006). Super-hydrophobicity fundamentals: implications to biofouling prevention. *Biofouling* 22: 107–115.

Matsumura, K., Nagano, M., and Fusetani, N. (1998a). Purification of a larval settlement-inducing protein complex (SIPC) of the barnacle, *Balanus amphitrite. J. Exp. Zool.* 281: 12–20.

Matsumura, K., Nagano, M., Kato-Yoshinaga, Y., Yamazaki, M., Clare, A. S., and Fusetani, N. (1998b). Immunological studies on the settlement-inducing protein complex (SIPC) of the barnacle *Balanus amphitrite* and its possible involvement in larva-larva interactions. *Proc. Roy. Soc. Lond. B.* 265: 1825–1830.

Moyse, J., Høeg, J. T., Jensen, P. G., and Al-Yahya, H. A. H. (1995). Attachment organs in cypris larvae: Using scanning electron microscopy. In: *New Frontiers in Barnacle Evolution.* Crustacean issues 10. Ed. F. R. Schram and J. T. Høeg, AA. Balkema publishers, Rotterdam, Netherlands, pp. 153–177.

Mullineaux, L. S., and Butman, C. A. (1991). Initial contact, exploration and attachment of barnacle (*Balanus amphitrite*) cyprids settling in flow. *Mar. Biol.* 110: 93–103.

Niederegger, S., Gorb, S., and Jiao, Y. (2002). Contact behaviour of tenent setae in attachment pads of the blowfly *Calliphora vicina* (Diptera, Calliphoridae) *J. Comp. Physiol. A.* 187: 961–970.

Nott, J. A., and Foster, B. A. (1969). On the structure of the antennular attachment organ of the cypris larva of *Balanus balanoides. Phil. Trans. R. Soc. Lond.* B 256: 115–134.

Ödling, K., Albertsson, C., Russell, J. T., and Mårtensson, L. G. E. (2006). An in vivo study of exocytosis of cement proteins from barnacle *Balanus improvisus* (D.) cyprid larva. *J. Exp. Biol.* 209: 956–964.

Okano, K., Shimizu, K., Satuito, C. G., and Fusetani, N. (1996). Visualisation of cement exocytosis in the cypris cement gland of the barnacle *Megabalanus rosa. J. Exp. Biol.* 199: 2131–2137.

Onsager, L. (1936). Electric moments of molecules in liquids. *J. Am. Chem. Soc.* 58: 1486 –1493.

Pettitt, M. E., Henry, S. L., Callow, M. E., Callow, J. A., and Clare, A. S. (2004). Mode of action of commercial enzymes on the settlement and adhesion processes used by cypris larvae of barnacles (*Balanus amphitrite*), spores of the green alga *Ulva linza*, and the diatom *Navicula perminuta. Biofouling* 20: 299–311.

Phang, I. Y., Aldred, N. A., Clare, A. S., and Vancso, G. J. (2008). Towards a nanomechanical basis for temporary adhesion in barnacle cyprids (*Semibalanus balanoides*). *J. R. Soc. Interface* 5: 397–401. Available Online doi:10.1098/rsif.2007.1209.

Phang, I. Y., Aldred, N. A., Clare, A. S., Callow, J. A., and Vancso, G. J. (2006). An in situ study of the nanomechanical properties of barnacle (*Balanus amphitrite*) cyprid cement using atomic force microscopy (AFM). *Biofouling* 22: 245–250.

Prendergast, G., Zurn, C., Bers, A. V., Head, R., Hansson, L., and Thomason, J. (2009). The relative magnitude of the effects of biological and physical settlement cues for cypris larvae of the acorn barnacle, Semibalanus balanoides L. *Biofouling* 25: 35–44.

Rittschof, D., and Costlow, J. D. (1989). Bryozoan and barnacle settlement in relation to initial surface wettability: A comparison of laboratory and field studies. In: *Topics in Marine Biology.* ed. J. D. Ross.*Scient. Mar.* 53: 411–416.

Rittschof, D., Branscomb, E. S., and Costlow, J. D. (1984). Settlement and behavior in relation to flow and surface in larval barnacles, *Balanus amphitrite* Darwin. *J. Exp. Mar. Biol. Ecol.* 82: 131–146.

Satuito, C. G., Shimizu, K., Natoyama, K., Yamazaki, M., and Fusetani, N. (1996). Age-related settlement success by cyprids of the barnacle *Balanus amphitrite*, with special reference to consumption of cyprid storage protein. *Mar. Biol.* 127: 125–130.

Smith, A. M. (2002). The structure and function of adhesive gels from invertebrates. *Integr. Comp. Biol.* 42(6): 1164–1171.

Spolenak, R., Gorb, S., and Arzt, E. (2005). Adhesion design maps for bio-inspired attachment systems. *Acta Biomater.* 1: 5–13.

Stèfan, J. (1874). Sitzungsbericht Akadem. *Wiss. Wien II* 68: 325.

Swain, G. W., and Schultz, M. P. (1996). The testing and evaluation of non toxic antifouling coatings. *Biofouling* 10: 187–197.

Verwey, E. J., and Overbeek J. T. G. (1948). *Theory of the Stability of Lyophobic Colloids*. Elsevier, Amsterdam.

Visscher, J. P. (1928). Reaction of the cyprid larvae of barnacles at the time of attachment. *Biol. Bull.* 54: 327–335.

Walker, G., and Yule, A. B. (1984). Temporary adhesion of the barnacle cyprid: the existence of an antennular adhesive secretion. *J. Mar. Biol. Assoc. UK* 64: 679–686.

Walker, G., Yule, A. B., and Ratcliffe, J. (1985). The adhesive organ of the blowfly, *Calliphora vomitoria*: a functional approach (Diptera: Calliphoridae). *J. Zool. Lond.* 205: 297–307.

Walley, L. J. (1968). Studies on the larval structure and metamorphosis of*Balanus balanoides* (L.). *Phil. Trans. Roy. Soc. Lond.* 256: 237–279.

Wethey, D. S. (1986). Ranking of settlement cues by barnacle larvae: Influence of surface contour. *Bull. Mar. Sci.* 39: 393–400.

Wieczorek, S. K., and Todd, C. D. (1998). Inhibition and facilitation of settlement of epifuanal marine invertebrate larvae by microbial biofilm cues. *Biofouling* 12: 81–118.

Yebra, D. M., Søren, K., and Dam-Johansen, K. (2004). Antifouling technology – past, present and future steps towards efficient and environmentally friendly antifoulings. *Prog. Organ. Coat.* 50: 75–104.

Yebra, D. M., Søren, K., Weinell, C. E., and Dam-Johansen, K. (2006). Presence and effects of marine microbial biofilms on biocide-based antifouling paints. *Biofouling* 22: 33–41.

Young, T. (1805). On the cohesion of fluids. *Phil. Trans. R. Soc. Lond.* A84. 95: 65–87.

Yule, A. B., and Crisp, D. J. (1983). Adhesion of cypris larvae of the barnacle, *Balanus balanoides*, to clean and arthropodin-treated surfaces. *J. Mar. Biol. Assoc. UK* 63(2): 261–271.

Yule, A. B., and Walker, G. (1984). The temporary adhesion of barnacle cyprids: effects of some differing surface characteristics. *J. Mar. Biol. Assoc. UK*64:429–439.

Yule, A. B., and Walker, G. (1985). Settlement of *Balanus balanoides*: The effect of cyprid antennular secretion. *J. Mar. Biol. Assoc. UK* 65: 707–712.

Yule, A. B., and Walker, G. (1987). Adhesion in barnacles. In: *Barnacle Biology*. Crustacean Issues, 5. Ed. A.J. Southward, AA Balkema. Rotterdam, The Netherlands, pp. 389–423.

Zhao, H., Robertson, N. B., Jewhurst, S. A., and Waite, J. H. (2006). probing the adhesive footprints of *Mytilus californianus* byssus. *J. Biol. Chem.* 281: 11090–11096.

Chapter 3
Alternative Tasks of the Insect Arolium with Special Reference to Hymenoptera

Dmytro Gladun, Stanislav N. Gorb and Leonid I. Frantsevich

3.1 Introduction

Apterigous terrestrial insects walk over relatively flat surfaces the radii of curvature of which are not less in size than their bodies or legs. They grasp at the surface with claws of several extremities, clasping the foothold between opposite legs. Without any sticky pads, machilids run extremely fast on stones and rocks of an arbitrary inclination. Hook-like interlocking mechanisms of much smaller dimension also are found in secondary legs of caterpillars (Nielsen and Common, 1991; Hasenfuss, 1999).

Plants may resist phytophagous insects using hairy or slippery waxed surfaces to prevent grasping (Stork, 1980; Gannon et al., 1994; Eigenbrode et al., 1996; Gorb and Gorb, 2002). The insect offensive strategy was to evolve smooth flexible sticky pads or sticky hair fields at the ventral side of tarsomers, in the joint region between tarsus and tibia, and below or between the claws (Beutel and Gorb, 2001; Gorb and Beutel, 2001). The adhesive contact between the leg and the substrate is mostly important for a rather short period of time of the stance phase and must be released during the swing phase (Fig. 3.1); these cycles are repeated infinitely during locomotion. Therefore, adhesive tarsal structures must fulfill several tasks, sometimes opposite to each other and reversibly executed in repeated step cycles.

1. The sticky surface of the tarsus must be ready to immediately contact the smooth substrate and, at the same time, should prevent an undesired contact formation with the rough ground.
2. Adhesion must be strong enough to hold the insect on the inclined and even on the overhanging substrate and, at the same time, low enough to allow detachment of the swinging leg. In other words, adhesion must be reversible.
3. It is desirable that the adhesive fluid spreads over the entire attachment area, but is spared to some extent at detachment for repeated usage.

D. Gladun (✉)
Department of Ethology and Social Biology of Insects, Schmalhausen Institute of Zoology, Kiev, Ukraine
e-mail: gladun.dmytro@gmail.com

S.N. Gorb (ed.), *Functional Surfaces in Biology*, Vol. 2,
DOI 10.1007/978-1-4020-6695-5_4,

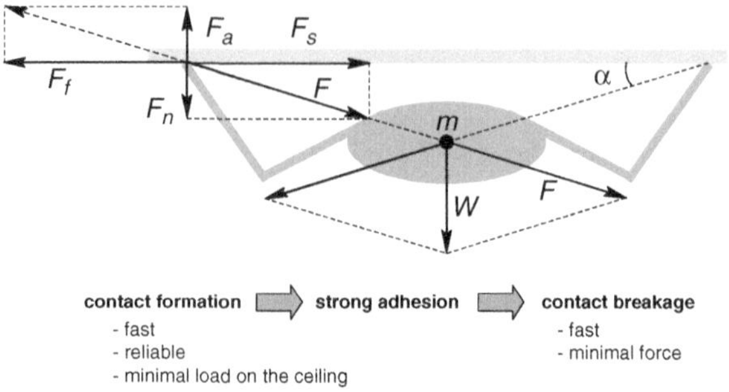

Fig. 3.1 Ceiling situation. If the mass m of the animal is at the center of gravity, then its weight is $W = m\ g$, and the angle of the ground reaction is α, the normal pull off force component is F_n and the shear force component is F_s. Thus, the weight of the insect walking on the ceiling should be counterbalanced by two forces: friction (F_f), which prevents the leg from sliding along the substrate, and adhesion (F_a), which prevents the leg from separating from the substrate. In order to walk on the ceiling, it is not sufficient just to generate strong adhesive bond between the leg and substrate. Two additional problems have to be solved: (1) contact formation must be fast, reliable and performed with minimal load on the ceiling; (2) contact must be released in a fast manner and with a minimal applied load

4. The adhesive surface of the pad must be sticky enough to generate sufficient adhesive force for supporting the body weight on the ceiling, and should be secured from adhering to itself within pad folds.

In addition, performance of these reversible tasks must be under extreme simple control. These requirements are solved in various ways in representatives of different insect lineages. For example, flies (Diptera: Brachycera), a majority of adult beetles (Coleoptera: Chrysomelidae, Coccinellidae, Cerambycidae, etc.), earwigs (Dermaptera) possess adhesive structures subdivided into little hairs having strong geometrical anisotropy. Such geometry is responsible for contact formation during the pad sliding towards body and for the contact breakage during the pad sliding in the opposite direction (Niederegger et al., 2002; Niederegger and Gorb, 2003).

Adhesive tarsal structures of the majority of Hymenoptera are of the smooth type and represented by the plantulae at the ventral side of the tarsomeres and by arolia located between the claws. In some taxa, both types of structures are present at the same tarsus. While plantulae are common in the much earlier hymenopterans ("Symphyta"), they have disappeared in the representatives of the higher lineages of the order (Parasitica, Aculeata). Hymenopteran arolium is an attachment organ of the highest complexity among insects from the structural and mechanical points of view. It reveals a great structural and functional diversity within the order. Hymenopteran arolia are able to fulfill a number of reversible tasks mentioned above.

We present here the functional morphology of the hymenopteran arolium, studied by the authors in different taxa of the order. Additionally, literature data on the hymenopteran arolium and on sticky pads in representatives of other insect orders are reviewed.

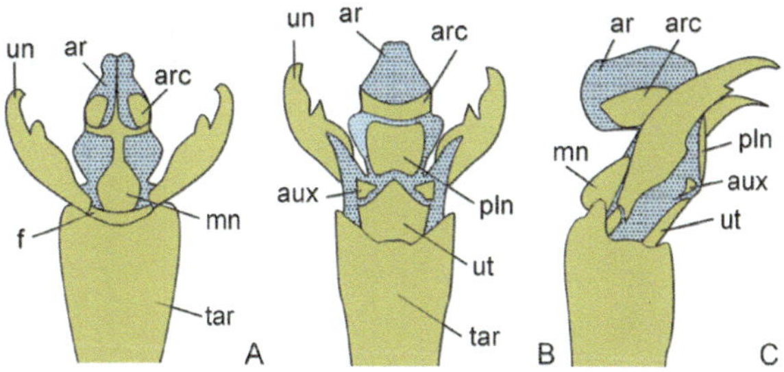

Fig. 3.2 Terminology of arolium structures (*Apis mellifera*). **A**. Dorsal aspect. **B**. Ventral aspect. **C**. Lateral aspect. ar, arolium; arc, arcus; aux, auxiliae; f, marginal flange of the terminal tarsomere; mn, manubrium; pln, planta; tar, tarsal segments; un, claw; ut, unguitractor plate

Insects were collected in the field. In addition, we used dry specimens from collections of the Schmalhausen Institute of Zoology (Kiev, Ukraine). Arolia of dry specimens were macerated in 5% lactic acid for 24 h, washed in water and then unfolded with the aid of slight pressure applied to the tarsus by forceps. Other methods of collection, fixation, dissection, preparation, microscopy and video recordings have been described in our previous publications (Frantsevich and Gorb, 2002, 2004). For our comparative studies, we have extensively used whole mounts, because they showed a degree of sclerotization and aided in exposing internal sclerites, unseen in most SEM samples.

We use names of pretarsal sclerites proposed by Dashman (1953) and Snodgrass (1956), with slight changes (Fig. 3.2). Pressure and tenacity are expressed in mN/mm^2. Force is sometimes related to body weight and expressed in body weight units (BWU).

3.2 Versatility of the Tarsus

In insect evolution, development of sticky tarsal and pretarsal structures has been preceded by two important preadaptations: development of the multisegmented tarsal chain and the ability of this chain to bend. The tarsus in hymenopterans, as well as in other terrestrial insects is adapted for attachment to the ground. It is equipped with structures attaching to the substrate by the use of either friction or adhesion: spines, bristles, hair tufts (Fig. 3.3C), or sticky pads (euplantulae). The tarsus often terminates with a special complex of attachment structures, the pretarsus, which includes several grasping, friction-enhancing or adhesive devices: the claws, the hairy planta, the sticky arolium between the claws or, in some insects, sticky hairy pads (pulvilli) under the claws (see review by Beutel and Gorb, 2001). The pretarsal structures are driven by a single muscle (retractor unguis, MRU), the long apodeme of which runs through the whole length of the tarsus to its insertion on the unguitractor plate.

The tarsus in Hymenoptera consists of five segments (tarsomeres). The tarsomeres articulate with the tibia and with each other by monocondylar joints. The condyles are skew outgrowths originating from folds of the cuticle. Paired condyles of neighboring joints come into contact with their fairly smooth external surfaces

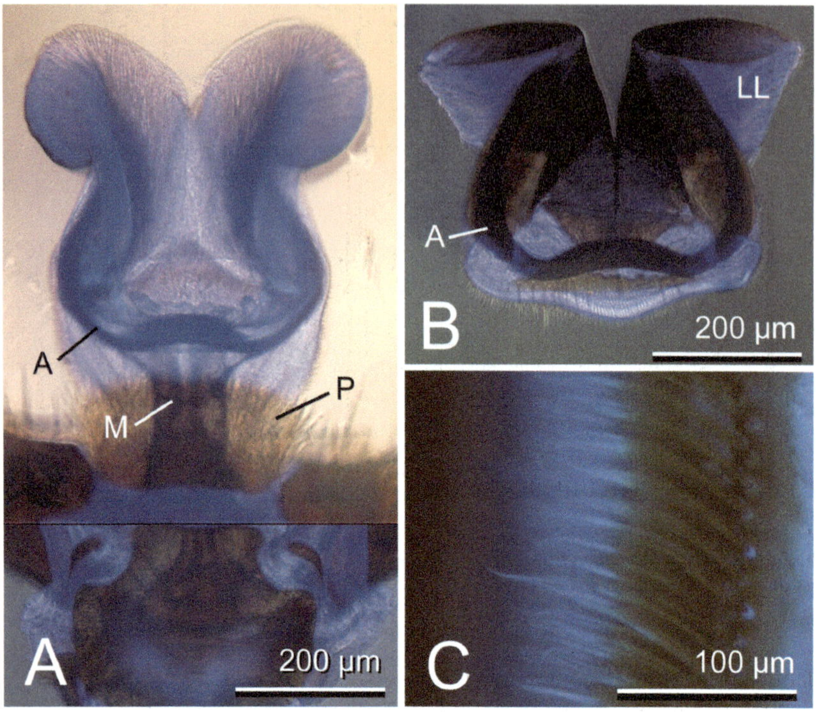

Fig. 3.3 Resilin-bearing structures in the hornet pretarsus. **A**. Arolium, ventral aspect. The distal part of the pretarsus with arolium and its proximal part with unguitractor are shown in the *upper and lower panels*, respectively. **B**. Arolium, dorsal aspect. **C**. Setae located on the ventral side of tarsomeres, lateral aspect. A, arcus; LL, lateral lobe; M, manubrium; P, planta

(Fig. 3.4A). Each proximal condyle bears a shallow concave socket, the distal one rests in the latter with a round convex head. The articulation of such a ball-and-socket type allows three rotational degrees of freedom. However, rich mechanical versatility is not supported by corresponding versatility of the muscular system: there are no intrinsic muscles between tarsomeres. Three extrinsic muscles are located in the tibia. They move the first tarsomere together with the entire tarsus as a whole: up and down, forward and backward.

Zygentoma bristle-tails are the most primitive of recent hexapods with a multisegmented tarsus. They are smart runners, able to climb on the underside of a rock (Manton, 1972). Their dominant attachment devices are claws. Evidently, the rough substrate is clenched between opposite legs, which provide horizontal forces of grasping or stretching (Zielinska, 2001). The claws are efficient on a rough substrate: the leaf beetle, *Chrysolina polita*, holding onto a cloth, resisted to an imposed lift of 50 BWU (Stork, 1980); the hornet, *Vespa crabro*, on sand paper resisted up to 24 BWU (Frantsevich and Gorb, 2004).

Why is a chain of intermediate tarsomeres between the tibial and pretarsal attachment devices advantageous for contact formation during locomotion? In our

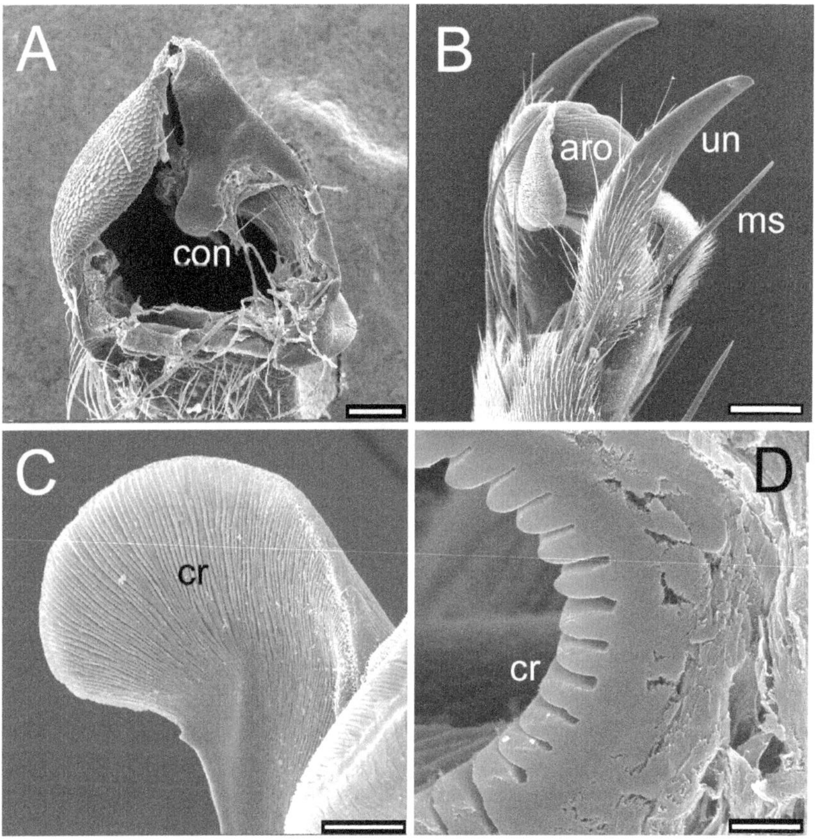

Fig. 3.4 Structures of the tarsus and arolium in the hornet, *Vespa crabro*. **A**. Condylus (con) at the proximal butt of the first tarsomere. **B**. Folded arolium in profile. **C**. Cristae at the dorsal side of the arolium. **D**. Transverse fracture across the cristae (cr). aro, arolium; ms, macroseta; un, unguis. Scale bars: A, C – 100 μm, D – 5 μm, B – 200 μm

opinion, the main reason for a flexible tarsomere chain is to compensate for the great turning of the leg relative to its end point during the stance phase of the step. Judging by our observations on walking insects, the tibia rotates, relative to the substrate, by about 40–70° or more (Frantsevich and Cruse, 1997). The elastic chain turns and twists passively during the stance, leaving claws in contact with the substrate. Insect walking posture with spread legs provides the animal with an automatic compensation of overturning disturbances due to change of leg position relative to the center of gravity. This response is mediated via the mechanical, dynamic feedback (*preflex*) without participation of the neural *reflexes* (Kubow and Full, 1999).

Flexibility and elasticity of the tarsomere chain is provided by two mechanisms:

1. resilin-bearing tendons surrounding the tibio-tarsal and intertarsomere joints (Gorb, 1996; Frazier et al., 1999; Neff et al., 2000; Niederegger and Gorb, 2003;

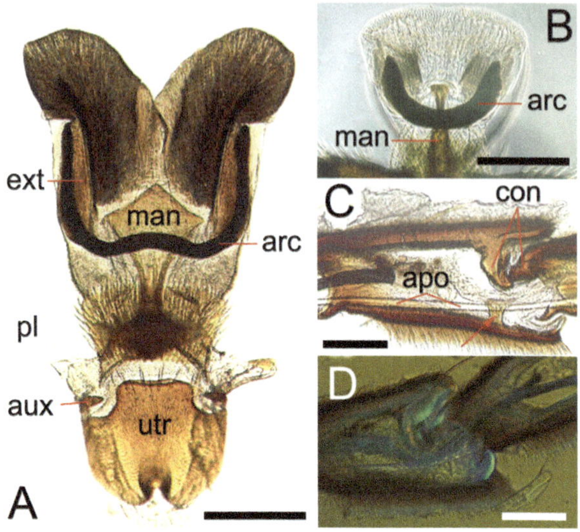

Fig. 3.5 Pretarsus and tarsomeres of hymenopteran legs. **A**. Arolium in *Vespa crabro* (Vespidae). **B**. Arolium in *Apis mellifera* (Apidae). **C**. Intertarsomere articulation in *V. crabro*. **D**. Same, fluorescence of resilin-bearing tendons. apo, apodeme; arc, arcus; aux, auxilia; con, condyli of neighbouring tarsomeres; ext, external area; man, manubrium; pl, planta; utr, unguitractor. Scale bars – 200 μm

Endlein and Federle, 2007) (Fig. 3.5D); the tendon stretched from the base of the unguitractor plate to the ventral edge of the terminal tarsomere in hornets (Frantsevich and Gorb, 2004) (Figs. 3.3 and 3.5D), bugs and beetles (Weber, 1933); and restraining tendons running from the MRU apodeme to the walls of the tibia in thrips (Heming, 1971).

2. contact between condyles supplemented by friction-generating surfaces. Even at a small translation of structures within the intertarsomere joint on retraction of the MRU apodeme, the condyles would be compressed together. The stiffness of the joint to passive rotation is enhanced several times by this action (Frantsevich and Gorb, 2004).

Stiffness of the tarsomere chain allows many higher insects to keep their legs resting in stance not only on the end of the tibia and on the tarsal segments (Fig. 3.6A, plantigrade mode of walking according to Dashman (1953)), but on terminal tarsomeres and pretarsi (Fig. 3.6B, C), as it was previously observed in the firebrat, *Thermobia domestica* (own photographs, unpublished), cockroaches (Frazier et al., 1999), tiger beetles (Nachtigall, 1996), flies (Niederegger et al., 2002; Niederegger and Gorb, 2003), and wasps (Frantsevich and Gorb, 2004). Tiny insects with reduced claws even walk on the tips of their arolia, as thrips do (Heming, 1971), or on the tips of the funnel-like pretarsal plantae, as probably some chalcidoid wasps do (Fig. 3.7A) (Gladun and Gumovsky, 2006).

The flexible multisegmented chain with the MRU apodeme inside allows the tarsus of the Pterygota to bend around the narrow ground and for grasping and climbing on plants (Gladun and Gorb, 2007). The tarsus bends due to eccentricity of the MRU apodeme relative to intertarsomere joints. The condyles at both sides of the joint emerge from the dorsal sides of neighboring tarsomeres. The apodeme lies at the ventral side of tarsomeres, being fixed to the bottom with "slings of tissue",

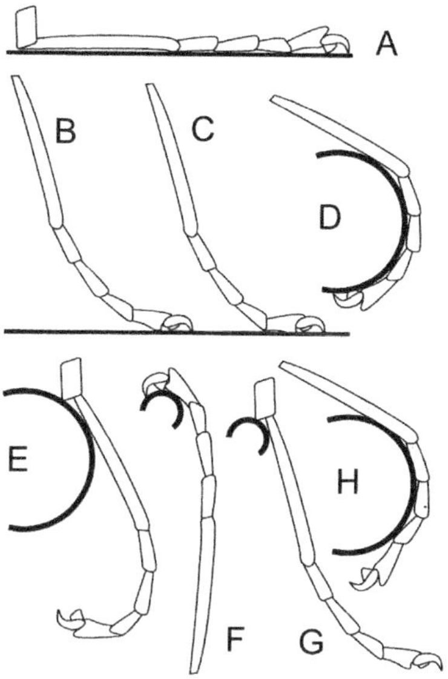

Fig. 3.6 Diagrams of leg placement methods on a *flat ground* (**A**, **B**, **C**), on a relatively *thick rod* (**D**, **E**, **H**), and on a relatively *thin rod* (**F**, **G**). From Gladun and Gorb (2007)

found in the locust, *Schistocerca gregaria* (Kendall, 1970), and invaginated cuticle forks, found in the hornet, *Vespa crabro* (Fig. 3.5C).

This eccentric mechanism of bending is quite opposite to the anti-bending mechanism situated at the more proximal level of the same MRU apodeme: this apodeme crosses the femoro-tibial articulation directly through the very axis of joint rotation (Radnikov and Bässler, 1991). Flexion and extension of the tibia do not affect the state of the tarsus and pretarsus.

Let us consider the simplest model of the curved chain, an arc α of a circular sector (Fig. 3.8A). Inside the sector, there is an internal arc, modeling the curved tendon. The distance between two arcs is h. The length of the internal arc is lesser than that of the external one by d. Bending angle of the arc (in radians)

$$\alpha = d/h \tag{1}$$

depends on the distance h between the joint and the apodeme and on the apodeme retraction d, but does not depend on the run-time curvature of the chain (Fig. 3.8A). Total maximal bending angle of a tarsomere chain has been evaluated as 130° in the stick insect, *Carausius morosus* (Radnikov and Bässler, 1991), 107° in the hornet, *Vespa crabro* (Frantsevich and Gorb, 2004).

The ability of such enormous bending is not observed during the insect walking on a flat surface, when the tarsus is almost straight or even overextended. Bending is expressed to its full extent when the tarsus embraces a thin stem, the diameter of which is comparable to the length of the tarsus. Such a grasping mode has been illustrated in the locust, *Locusta migratoria*, grasping a vertical stem (Hassenstein and Hustert, 1999). The same mode is depicted in Fig. 3.6D. It is not strange that stick-insects, orthopterans, as well as primitive hymenopterans (phytophagous sawflies

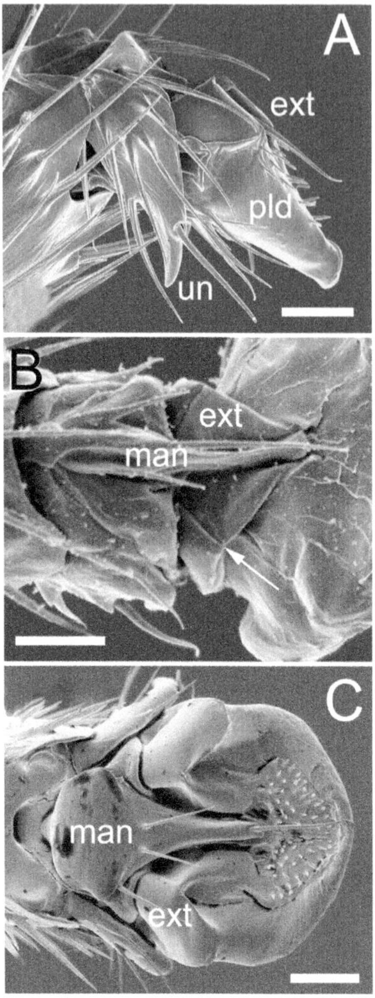

Fig. 3.7 Arolium in parasitic wasps. **A**. *Sympiesis* sp. (Eulophidae), lateral aspect, arolium folded. **B**. *Trichogramma semblidis* (Trichogrammatidae), dorsal aspect. **C**. *Perilampus* sp. (Perilampidae), dorsal aspect, arolium is partly spread. Ext, extender; man, manubrium; pld, distal planta; un, claw. *Arrow* indicates a crease between an anterior margin of the extensor area. Scale bars: A – 10 μm, B – 5 μm, C – 20 μm

"Symphyta") possess sticky pads (*euplantulae*) on their tarsomeres in addition to the pretarsal sticky pad (*arolium*) (Beutel and Gorb, 2001; Schulmeister, 2003). It is stranger that higher hymenopterans, such as parasitoid wasps, wasps, ants, and bees, lack the sticky euplantulae.

If the stem diameter is less than the tarsus length, even a strongly bent chain will not fit the stem curvature. During walking on very thin stems or rods, insects clasp them between the proximal tarsomeres of opposite legs (Fig. 3.6E, G, H). The honeybee, *Apis mellifera*, with its short tarsi and modified proximal tarsomeres, may use its arolia for attachment to extra thin rods (Gladun and Gorb, 2007).

On relaxation of the MRU during the swing phase, the straight shape of the tarsus is recovered due to the spring action of elastic tendons/ligaments located in intertarsomere joints. Both the unguitractor and MRU apodeme return distad upon

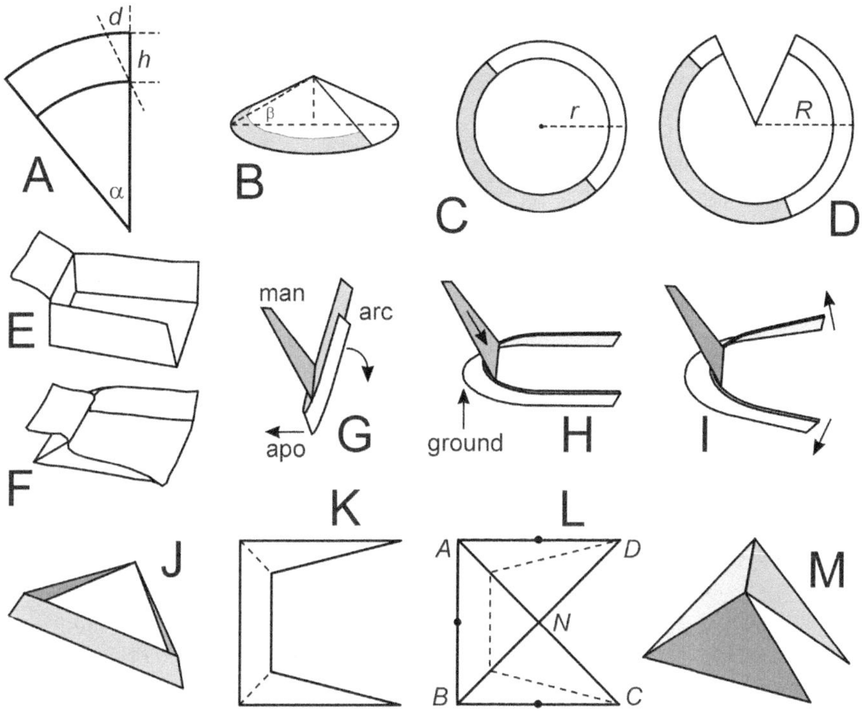

Fig. 3.8 Simple geometrical interpretations of tarsus and pretarsus kinematics. **A**. Two arcs in a sector, simulating arching of the tarsus. **B**. Cone with the *shaded belt*, the arcus. **C**. Projection of the cone onto its base. **D**. Flat development of the cone. **E**, **F**. Scoop model of Snodgrass (1956). **G**, **H**, **I**. Interaction between the manubrium and the arcus (Federle et al., 2001). **J**. Finite model of the arcus with two folds. **K**. Flat development of the former model. **L**. Complement of this development to three triangles. **M**. Pyramid made of three triangles, partly opened

such an action of previously expanded tendons. The passive recoil is not necessarily immediate, but rather quick: the recoil by 30–40° of the released terminal tarsomere in the cockroach, *Periplaneta americana*, lasted 12–15 ms in the intact leg, but only 6 ms in a preparation with the apodeme severed off the viscous MRU (Frazier et al., 1999). Such prompt, purely mechanical responses without neural control are called *preflexes* (Brown and Loeb, 1999; Full and Koditschek, 1999).

3.3 Sclerites of the Arolium

The arolium is an everted cuticular sac situated between the claws. Part of the integument of the arolium is soft and compliant, some parts are sclerotized. Aroliar sclerites are mechanical elements participating in unrolling (spreading) and rolling (folding) of the arolium. Sclerites of the hymenopteran arolium are, probably, the

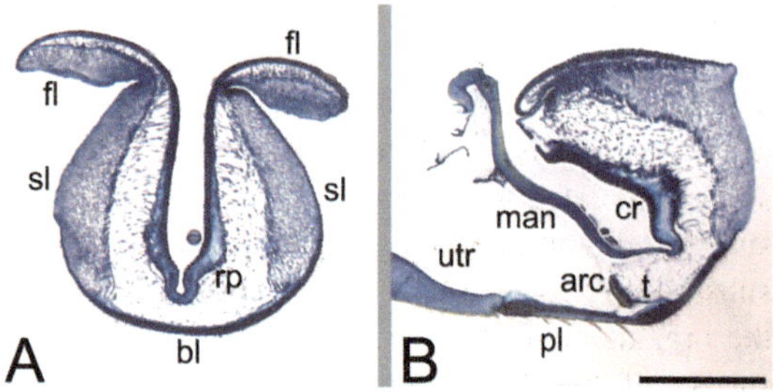

Fig. 3.9 Semi-thin sections of the folded arolium in the hornet, *Vespa crabro*. **A.** Frontal plane. **B.** Sagittal plane. arc, arcus; bl, bottom lobe; cr, system of ridges; fl, flap lobe; man, manubrium; pl, planta; rp, resilin pillow; sl, side lobe; t, trace of invagination; utr, unguitractor. Scale bar – 200 μm.

most distinct and elaborated among insects (Fig. 3.5A). Only arolia of caddis flies (Trichoptera) can compete with them.

The manubrium is the medial dorsal sclerite, movably articulated with a socket in the dorsal rim of the terminal tarsomere, between two claw condyles. It reaches to the middle of the arolium. Its longitudinal profile is straight (Federle et al., 2001) or convex ventrad (Fig. 3.9B). Shapes of hymenopteran manubria were illustrated in some figures by Domenichini (1994). Variability of manubria is illustrated in Fig. 3.10. Based on the comparative study of arolium sclerites in representatives of different lineages of Hymenoptera, it can be concluded that the width of the base of the manubrium is related to the distance between claw condyles. Without discussion on the shape of the manubrium as a character for phylogenetical reconstructions, we can suggest that the shape of the manubrium is not critical to it functions. It should be just long enough to cover the arcus. For example, the manubrium in the caddis fly, *Limnophilus* sp., is straight (Fig. 3.15C). It is found to be reduced in *Typhia* (Typhiidae) (Basibuyuk et al., 2000). Homology of the manubrium with the medial dorsal groove of the arolium in Tipulomorpha (Diptera) and Mecoptera (Röder, 1986) is dubious, because sclerotization of this groove was not observed by us in transparent whole mounts.

Cuticular dorsal plates, acting as extenders of the arolium, are paired lateral, smooth, sclerotized plates at the dorsal side of the hornet arolium (Frantsevich and Gorb, 2002) (Fig. 3.5A). They were discovered and named for the first time in phloeothripids (Heming, 1972), where their possible participation in arolium spreading has been emphasized. In the honey-bee, the extenders are broader than in the wasp; remarkably broad extenders were found in a sawfly (Fig. 3.15). In chalcidoid parasitic wasps, the extenders are also broad and divided by a crease into medial and lateral lobes, both which can fold along the crease (Fig. 3.7B, C) (Gladun and Gumovsky, 2006). Broad extenders, connected with claw bases, cover the arolium in the cryne-fly, *Tipula hortulana* (Diptera) (Fig. 3.15A). Like the cover spine of a book, they hold the narrow shape of the folded arolium, compressed from the sides (Röder, 1986). Less conspicuous is a slight sclerotization at both sides of the medial dorsal groove in the scorpion fly, *Panorpa communis* (Mecoptera) (Fig. 3.15B).

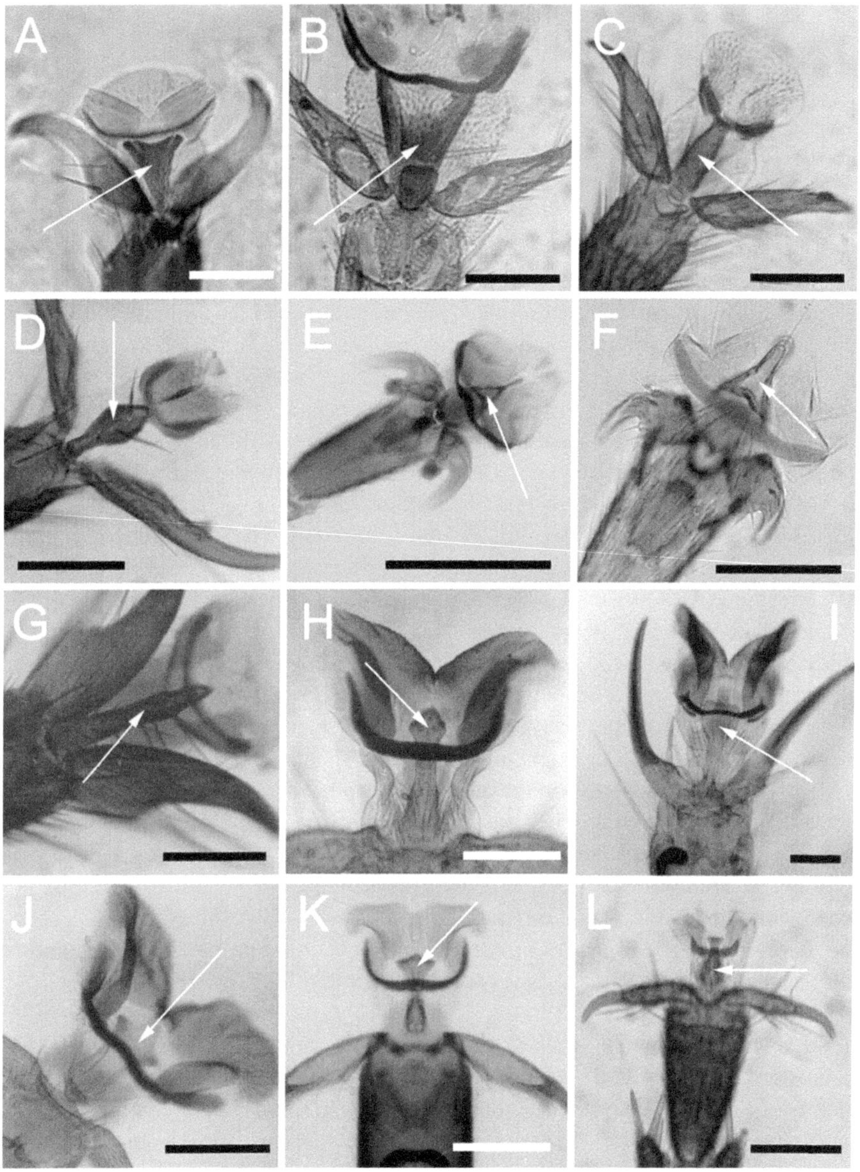

Fig. 3.10 Diversity of manubria (indicated by *arrows*) in hymenopterans: inverted triangular (**A**), bifurcated (**B**), straight (**C**, **G**), tapered (**E**, **F**, **L**), dilated (**D**), and spearlike (**H**–**K**). **A**. *Xyela julii* (Xyelidae). **B**. *Diprion similis* (Diprionidae). **C**. *Hartigia linearis* (Cephidae). **D**. *Sirex noctilio* (Siricidae). **E**. *Ormyrus* sp. (Ormyridae). **F**. *Torymus* sp. (Torymidae). **G**. *Pimpla* sp. (Ichneumonidae). **H**. *Philanthus coronatus* (Sphecidae). **I**. *Bembix rostrata* (Sphecidae). **J**. *Crabro peltarius* (Sphecidae). **K**. *Eumenes coarctatus* (Eumenidae). **L**. *Formica cunicularia* (Formicidae). Scale bars: A, F – 50 μm; B, C, E, L – 100 μm; D, G–K – 200 μm

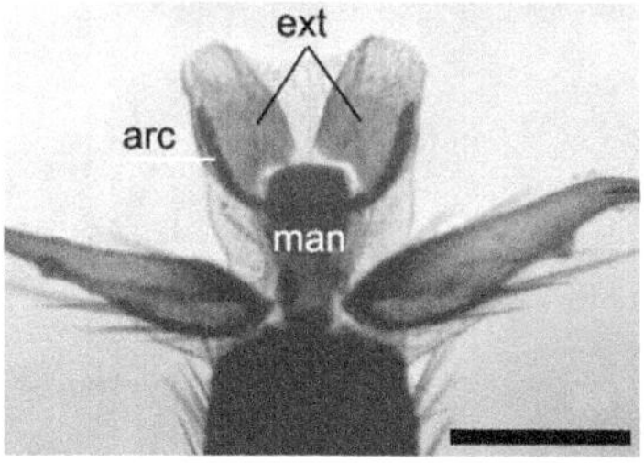

Fig. 3.11 Pretarsus in *Coenolida reticulata* (Pamphiliidae) with broad extenders (ext). Arc, arcus; man, manubrium. Scale bar – 200 μm

The arcus is internal sclerite situated inside the lumen of the arolium. It is a sort of an apophysis representing a bilayered ingrowth of the integument. The narrow trace of invagination distad to the sclerotized part of the planta is seen in the sagittal section of the arolium (Fig. 3.9B; see also Federle et al., 2001; Jarau et al., 2005). One can discern a transverse, basal part of the arcus and longitudinal side branches. In plane, the arcus of Hymenoptera is thin and U-shaped (in the honey-bee, Fig. 3.5B; in the ant, *Oecophila smaragdina*, Federle et al., 2001) or lyriform (in sawflies, Fig. 3.10E, I, J; in the hornet, Fig. 3.5A). It is possible that the arcus may be U-shaped in the folded arolium, but becomes lyriform (as it was considered to be in the honey bee by Federle et al., 2001), when side branches are spread sidewards.

In profile, the arcus is approximately a belt cut out of the cone (Fig. 3.8B). Its apex points dorsad in the spread arolium. We shall regard the base of the cone as the plane of the arcus. The arcus was discussed to be a cuticular spring, controlling the shape of the arolium (Snodgrass, 1956). Side branches of the arcus are directly supported by the adjacent extenders, preventing spontaneous coiling.

The arcus is present also in the arolium of representatives of Mecoptera (Röder, 1986), Lepidoptera, and Trichoptera (Federle et al., 2001). Short and seemingly primitive in representatives of the former two orders (Fig. 3.15B), it is well elaborated and even lyriform in the last one (Fig. 3.15C). Mechanics of spreading and folding actions of the arolium in these orders have not as yet been studied.

The dorsal side of the arolium in a hornet is sculptured in parallel longitudinal ridges, obviously sclerotized as it is seen in fractured SEM preparations (Fig. 3.4C, D). It is possible to pull the fresh arolium out of the terminal tarsomere with claws. If the dorsal integument is then separated, the arolium immediately rolls up, with its outer surface outwards. This behavior is similar to coiling of the excised arcus. Evidently, the dorsal side of the arolium is also a prestrained structure with spring-like properties. It is not clear, whether the elastic energy is stored in the dorsal system of ridges or in the resilin pillow beneath it (Figs. 3.3 and 3.9). A similar dorsal system of ridges was previously illustrated in *Paravespula germanica* erroneously referred to as the bottom surface of the arolium (Beutel and Gorb, 2001). The dorsal side of the arolium in the ant, *Oecophila smaragdina*, is granulated (Federle et al., 2004). The ventral side in the hymenopteran arolium shows no clear signs of tension after the excision described above.

We will just briefly mention other pretarsal sclerites in representatives of Hymenoptera. The unguitractor plate is shield-like, sometimes with a narrow distal process. Its side edges rest on two skew lateral apophysi of the distal side walls of the terminal tarsomere. The unguitractor slides inward and upward on contraction of the MRU, then turns down again, as has been filmed through the window in the tarsomere wall (Frantsevich and Gorb, 2002).

Auxilia or auxiliary sclerites are small. Claws are simple or double, usually longer than the rest of the pretarsus. In most Chalcidoidea, the claws are short and hardly reach the substrate during the stance phase. The planta consists of two parts: the more rigid proximal part and the thin flexible distal part (Fig. 3.9). In Chalcidoidea, the sides of the distal part are elevated, when the arolium is folded. They form a funnel which tapers distad into the narrow butt (Fig. 3.7A). The tip of the planta presumably represents a kind of sole for the walking chalcid. However, the gait of such tiny insects has not been experimentally investigated.

The dorsal part of the internal space in the terminal tarsomere is occupied by the aroliar, or tarsal gland, as previously demonstrated in the honey bee (Conde-Beutel et al., 1989; Federle et al., 2001), in the bee, *Melipona seminigra* (Jarau et al., 2005), and in ants *Oecophila smaragdina* (Federle et al., 2004), *Amblyopone reclinata* (Billen et al., 2005), *Pachycondyla* spp. (Orivel et al., 2001). The same gland resides in the tibia in adult thrips (Heming, 1971). Its lumen is connected with the arolium lumen in bees and ants. In an ant, *Amblyopone reclinata*, an additional footprint gland has been found in the proximal part of the terminal tarsomere (Billen et al., 2005).

3.4 Contact with the Ground

The arolium is elevated between the claws during walking on rough ground. The bottom surface stays almost at a right angle to the ventral pretarsal sclerites. Both the unguitractor and the proximal planta are positioned, as is demonstrated in medial sagittal semi-thin sections (Fig. 3.9B). Thus the ventral sticky surface of the arolium is spared unnecessary contact with the ground. The sides of the arolium are also folded so that we discern three or even five lobes in transverse sections (Fig. 3.9A). In SEM preparations, the arolium often appears folded into three lobes, as demonstrated in the honey bee (Conde-Beutel et al., 1989) and in the ant *Oecophila smaragdina* (Federle et al., 2004), into five lobes in the hornet *Vespa crabro* (Frantsevich and Gorb, 2002), in the ant *Pachycondyla goeldii* (Ponerinae) (Orivel et al., 2001), and, remarkably, in the caddis fly, *Limnophilus* sp. (personal observations, unpublished). Side lobes are elevated above the medial lobe. The claws on rough ground need only a short excursion until engaging with the ground, then the unguitractor plate is arrested by its inward excursion, and it no longer disturbs the elevated position of the arolium.

On smooth ground, the claws find no support and slip sidewards, dragged by further retraction of the unguitractor plate. Evidently, an everted sac, fixed at its dorsal edge and pulled inward by its ventral edge, would bend down. Arolii are bent down upon unrestrained retraction of the unguitractor plate and claws in all insects with arolii, even if they do not possess special external and internal sclerites.

We observed this forced turn in the grasshopper, *Chothippus apricarius*; scorpion fly, *Panorpa communis*; crane fly, *Tipula juncea*, and butterfly, *Argynnis lathonia*. Bending of the arolium, together with the claws, was observed in cockroaches (Roth and Willis, 1952; Frazier et al., 1999) and in fulgoroid cicadas Dictiyopharidae (Emeljanov, 1982). Similar bending of pulvilli (sticky hairy pads, derivatives of the auxillae) together with the claws was observed in the blow fly, *Calliphora vicina* (Niederegger et al., 2002).

The turn of the hornet arolium at an angle of about 90°, exerted by the micromanipulator clasping the MRU apodeme, was fairly reversible (Frantsevich and Gorb, 2002). The same result was obtained for the honey bee (Federle et al., 2001). Both teams confirmed that the arolium turns, but did not spread without contacting the ground or from pressure applied from below onto the planta. Under physiological conditions, the bent arolium touches the ground and spreads.

One must take into account that, after contacting the ground the arolium is arrested at the bottom. Further retraction of the apodeme would cause further bending of the tarsomere chain. It means that the terminal tarsomere would rise above the substrate (this movement was also discovered in the fly's tarsus by Niederegger and Gorb (2003)). This movement will cause some extension of the manubrium relative to the tarsomere.

3.5 Spreading and Folding of the Arolium

3.5.1 Mechanics Versus Hydraulics

Spreading of attachment devices during contact to the ground results in an increase of the contact area between the pretarsal attachment devices and the substrate. Spreading and folding of the arolium together with the claws was observed in cockroaches (Roth and Willis, 1952; Frazier et al., 1999). Similar bending of pulvilli together with the claws was observed in the blowfly, *Calliphora vicina* (Niederegger et al., 2002). Our observations with the hornet demonstrate that insects with clipped claws are able to walk upside down on a smooth ceiling without attachment being disturbed.

The mechanism of spreading and folding of the arolium has been studied experimentally in representatives of thrips and hymenopterans. Other groups with arolii that obviously are able to fold and spread their attachment organs are still neglected (crane flies, scorpion flies, caddis flies). Also the information about pretarsus mechanics in groups with arolii of apparently constant shape (orthopteroids, cicadellids, and butterflies) is almost completely lacking in the literature. However, even among these last groups of insects, the presence of the specialized spreading mechanisms can not be excluded. Crossland et al. (2005) noticed that arolii in alate termites "dramatically inflate" upon attachment to glass. Based on the taxa studied, there are two hypotheses about the mechanism of spreading and folding of the arolium: the hydraulic and the mechanical one. But, both hypotheses have some weak points. The hydraulic mechanism hypothesis is hindered by the lack of evidence about obvious pumps and valves. The mechanical mechanism hypothesis lacks information about the distinct joints between aroliar sclerites that are part of the mechanism. The sclerites are separated by flexible areas of the articulation membrane or resilin-bearing elastic regions (Fig. 3.3). However, both hypotheses also have some strong evidence supporting them.

The mechanical hypothesis, in spite of the structural complexity of the arolium, explains its principle of operation in the an utmost simple way. The single muscle, MRU, retracts the unguitractor plate with all sclerites and soft parts of the pretarsus. This drag bends the claws and tarsomeres within the single degree of freedom and expands the arolium. The reverse folding is passive, due to the stored elastic energy in resilin-bearing structures of the arolium (Fig. 3.3A, B).

In general, hydraulic mechanisms may be effective for fast movements of appendages, e.g. legs in salticid spiders (Foelix, 1982) or in the mask of a dragonfly nymph (Tanaka and Hisada, 1980). The arolium has the ability to evert itself under applied internal pressure. This turning inside out can be readily achieved upon applying pressure to the body in thrips (Heming, 1972), to the terminal tarsomere of a hornet (the method used by the authors for the preparation of whole mounts), to the whole tarsus using an external pump in both the honey-bee and ant, *Oecophila smaragdina* (Conde-Beutel et al., 1989; Federle et al., 2001), upon CO_2 anesthesia in the honey bee (Federle et al., 2004), and upon immersion into the hypotonic solution in the hornet (Frantsevich and Gorb, 2002). However, all these experiments remain not without doubt as to whether the methods mentioned above are comparable with physiological conditions. Similar behavior of attachment structures in non-physiological conditions have been previously observed elsewhere: the eversible tibial euplantula expands after ether vapor treatment of an aphid (Lees and Hardie, 1988). It is possible to pump out any inverted structure that is soft enough. For example, it has been previously shown that by boiling three terminal antennal segments in the beetle, *Lethrus apterus* (Scarabaeidae, Geotrupinae), which are normally compressed one into the next, the proximal flagellomere can be everted (Frantsevich et al., 1977). After this procedure, previously hidden surfaces of the flagellomeres, covered with hidden sensory hairs, were exposed.

Jordan (1888) suggested that spreading of the pad in Thysanoptera was caused by blood pressure. Later the mechanism of interaction of the compressed fluid with the pretarsal sclerites in adult thrips was proposed (Heming, 1971): (1) spoon-like claws hold the arolium from its sides; (2) on the active pull of the unguitractor, claws move laterally and release the arolium; (3) the last one everts due to internal pressure produced inside the trunk. The same traction stretches the restraining tendons. Their passive recoil returns the apodeme of the relaxed MRU back. The recoil, consequently, is strong enough to compress the claws together with the arolium against pressure of the blood. No explanation was proposed as to why the middle part of the expanded arolium retracted only upon recoil.

The pressure required to spread the arolium in the weaver ant, *Oecophyla smaragdina*, has been experimentally evaluated as 15–25 mN/mm^2 (kPa), only 15–25% of the normal atmospheric pressure. It seems very convincing, that the punctured arolium was not capable of unfolding solely upon traction of the MRU apodeme (Federle et al., 2001). The authors regarded the arolium gland as a reservoir for the hydraulic fluid. Hypothetically, when the arolium bends ventrad, the corner between the planta and the unguitractor plate is invaginated, the internal volume of the terminal tarsomere diminishes, and the arolium gland becomes compressed, which causes an increase of pressure transmitted to the arolium. We may add that the unguitractor plate, protruding inside the terminal tarsomere, may compress the gland from below. If the tarsus, severed from the body, is in contact with the smooth substrate, the arolium rolls up passively (Federle and Endlein, 2004). Contribution of pretarsal sclerites, such as the planta, arcus, and manubrium, to spreading and folding of the arolium has not been discussed for the weaver ant.

The weaver ant constructs its nest of leaves bound together with silk threads. This specific behavior requires the ant's ability to hold the leaf with its arolia for hours. It remains therefore unclear, whether the gland, once compressed, would be able to keep the internal pressure for hours. It is possible to unroll the arolium, in the separated tarsus by pulling the arolium, in contact with the substrate, in a proximal direction (Federle et al., 2001). If the spreading mechanism is the same in this case, we may suppose that pulling the arolium out of the tarsus causes a decrement of pressure in the arolium relative to the terminal tarsomere.

Conde-Beutel et al. (1989) proposed another mechanism which had to produce a decrement of pressure inside the arolium. The arcus, expanding to the sides, may lead to an increase of the interior volume of the arolium, causing the transient filling of the arolium with hemolymph. No one has measured the volume of the arolium in the two natural states, folded and spread. Also, the rapidity of the transient filling by a viscous hemolymph or by secretion of the aroliar gland through a narrow orifice into the arolium lumen remains unknown. Lees and Hardie (1988), investigating the evertable tibial euplantula in aphids, suggested that the source of hydraulic pressure was the pulsatile organ at the base of the tibia.

Experimental incisions at both sides of the arolium, at the base of the manubrium, or experimental removal of extenders in the hornet, most probably result in damaging the hydraulic mechanism. However, these operations did not prevent partial spreading of the arolium upon retraction of the MRU apodeme (Frantsevich and Gorb, 2002).

3.5.2 Scoop Model

Snodgrass (1956) regarded the tri-lobed folded arolium of the honey bee as an analog of a paper scoop made up of a bottom and three walls, with a handle on the middle wall (Fig. 3.8E, F). The walls of the arolium are situated between the manubrium (a handle) and the planta (the bottom), supported from below by the substrate. The manubrium flexes ventrad and inward upon MRU retraction. Very demonstratively, pressing the scoop handle down, the middle wall turns downwards, side walls move laterally and downwards, and thus the scoop flattens: "*a downward pressure on the base of the scoop spreads its sides widely apart*". However, observing the scoop, one may notice that its side walls move, not as flat rectangles, but are curved. Deformation is distributed along the side walls. Snodgrass did not indicate which parts of the arolium correspond to walls and corner articulations of the scoop.

Snodgrass claimed that the elastic arcus may fold the side lobes of the arolium upwards upon releasing of the pressure. The role of the arcus in spreading was estimated in subsequent models.

3.5.3 Role of the Arcus

In a honey-bee as well as in a hornet, experimental incisions amidst or at the sides of the arcus were fatal for spreading (Federle et al., 2001; Frantsevich and Gorb, 2002). The arcus, as a key link in the spreading mechanism, attracted the attention of the cited authors because of its unusual elastic properties, specific position between the external sclerites (the manubrium and the planta), and peculiar geometry.

Elastic property resides to a different degree of sclerotization in two cuticular layers of the bilayered arcus (Federle et al., 2001). These layers are distinctly seen in sagittal sections (Fig. 3.9B). Frantsevich and Gorb (2002) have experimentally demonstrated that the arcus, in situ, was prestrained: when excised free from the arolium, it coiled. Upon immersion in the concentrated alkaline solution, it coiled even more, probably because of stronger contraction of the anterior cuticular layer in the solution.

Kinematic models of the arolium turning upon MRU retraction are similar in both papers mentioned above (Federle et al., 2001; Frantsevich and Gorb, 2002). In the folded state of the arolium, the manubrium points toward the acute angle to the plane of the arcus, if one looks at the arolium from above and outside. When the arolium turns down, this angle widens to a right or even blunt angle. Thus, the arcus is gripped between the manubrium and the planta supported by the substrate. It is now essential that now the manubrium applies pressure onto the conical band of the arcus from above. The manubrium "twists" (Conde-Beutel et al., 1989) or "pushes" (Federle et al., 2001) the arcus, causing it to straighten. Federle et al. (2001) guessed that due to the peculiar the geometry of the arcus, it may translate the vertical movement into a horizontal movement; however, no model was proposed for such a translation.

Indeed, at the first approximation, the model is rather simple and more definite, than Snodgrass' scoop. The arcus in the folded arolium is not flat but coiled, like a belt cut out of the cone, as we noted above. The cone may be developed into a flat sector (Fig. 3.8B–D). The radius R of the development is larger than the radius r of the cone base:

$$r = R \cdot sin\ \beta, \tag{3.1}$$

where β is the angle between the generatrix of the cone and the base plane. The arcs γ of the base and its development Γ relate inversely to their radii:

$$\gamma/\Gamma = R/r. \tag{3.2}$$

The horizontal span of the involution is larger than the diameter of the cone base. On the contrary, the altitude of the cone tends toward zero upon development. Thus, under pressure on the cone, applied along the altitude, the vertical compression translates into horizontal expansion.

The cone model may be further simplified into a Π-shaped figure cut out of a pyramid (Fig. 3.8J–M). In the pyramid model, relations between the applied vertical force and extending moments down the ribs become elementary (Frantsevich and Gorb, 2002). We presented the flat unfolding of the pyramid as a part of the square deprived of one quarter. Folding of the side triangles down, the diagonals produce a regular pyramid of three sides and a regular triangle at the base. Pressing the apex down, one may partly unfold the pyramid. We defined the side triangle's angle of rotation about the common rib in the middle as ρ , the inclination angle between the plane of the middle triangle ABN and the substrate plane ABCD as β, and the angle between the side branches AD and BC resting on the substrate as θ. After simple transformations, we can derive:

$$tg\beta = \frac{\sqrt{2}}{1+\cos\rho}, \tag{3.3}$$

$$\sin\left(\frac{\theta}{2}\right) = \frac{\cos\rho}{\sqrt{2}}. \tag{3.4}$$

If the force F is applied downwards at the apex N, the force component, normal to the face ABN, rotates this face downward about the axis AB. The rotational moment M down AB is directly proportional to $cos\beta$. At the equilibrium, M is balanced by two moments MA, MB, complanar with and aimed at $\pm 135^\circ$ to M, each equal $M\sqrt{}2$.

Expansion of the arcus upon the vertically applied force has been confirmed by mechanical modeling (Frantsevich and Gorb, 2002). The model shows that the lyriform shape of the arcus base is more efficient than the U-shape, if the arcus is compressed between the manubrium and the substrate, because the arm of the vertical force and its moment down the base is larger. The real arcus is more complicated than its model of three discrete elements. Deformations occur not only in two folds, but occupy areas of the sclerite. Applied forces exert propagating strain.

3.5.4 Extenders

Only larvae of thrips have normal claws and, hence, the mechanical device of unfolding, when the claws slip apart on the smooth substrate. The arolium, folded as a cone, is hidden inside an indentation in the unguitractor plate. Representatives of

Phlaeothripidae (e.g. *Haplothrips verbasci*) have a symmetrical pair of sclerotized triangular plates, named extenders, at the proximal part of the dorsal side of the arolium. The extenders are joined to the claw bases. The divided claws pull the lateral sides of the extenders ventrad and unfold them like a book. Other groups of thrips lack extenders (Heming, 1972). Strongly sclerotized dorsolateral plates are present at the sides of the arolium in lantern-flies Fulgoroidea (Auchenorrhyncha), but their participation in evertion of the sticky terminal face of the arolium has not been previously studied (Fennah, 1945; Emeljanov, 1987; Frantsevich et al., 2008). We suppose that pronation of claws, slipping sidewards, also stretches the base of the arolium sidewards.

Presumably, sclerotized side plates in crane-flies from the genus *Tipula* are also extenders. Their action is illustrated by a simple foldable model of two structures, each opening like a book: the diverging claws and the flattening extenders (Fig. 3.12A, B).

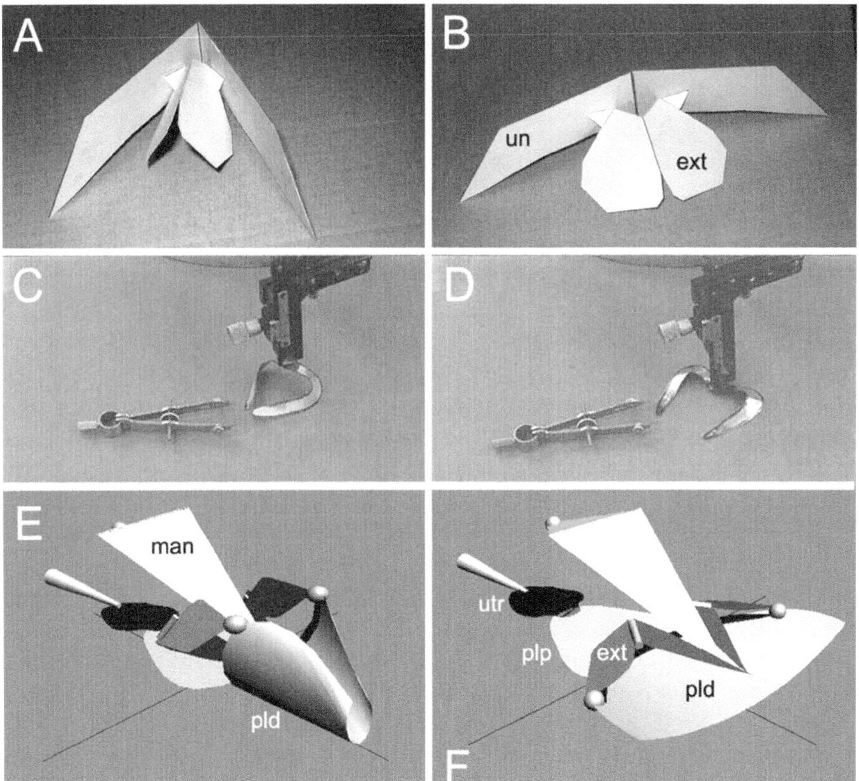

Fig. 3.12 Models of the unfolding mechanism of the arolium. **A**, **B**. Cardboard model of claws and extenders in the crane fly, *Tipula*. **C**, **D**. Tin model of the arcus compressed by a driver. **E**, **F**. 3D graphic simulation of extension in chalcids. ext, extenders; man, manubrium; pld, distal parts of the planta; plp, proximal distal parts of the planta; un, claw; utr, unguitractor

Extenders are also present in hymenopterans. However, excision of the hornet's dorsal plate did not prevent the arolium from expanding (Frantsevich and Gorb, 2002). A more active role of the extenders is supposed in chalcidoid wasps. Each extender is divided into a medial and lateral lobe by a crease. In the folded arolium, the flexible part of the planta is rolled up; each plate is folded along the crease beside the manubrium. On retraction of the unguitractor plate, the manubrium flexes ventrad, the extenders beside the manubrium unfold laterally, like an opening book. They simultaneously unroll the funnel-like elastic part of the planta and pull apart the side branches of the arcus. In addition, the manubrium presses on the base of the arcus, facilitating divergence of the side branches (Gladun and Gumovsky, 2006). This hypothesis was confirmed by 3D mechanical simulation of the unfolding mechanism (Fig. 3.12E, F).

Reverse movement of the spread arolium is operated simply by relaxation of the MRU, but executed passively by manifold elastic structures strained during expansion: the restraining tendons, the arcus, the manubrium, cristate dorsal surface of the arolium, the resilin formations beneath it (Fig. 3.3A, B), and, lastly, by the composite structure at the bottom of the arolium. Also the influence of a possible decrease of interim pressure in the terminal tarsomere on progression of the unguitractor plate may not be completely excluded. Emeljanov (1982) suggested that relaxation of the pretarsus might be caused by pumping of the hemolymph inside the tarsus.

3.6 Attachment and Detachment

Two main alternative hypotheses have been proposed to explain how the arolium holds onto a smooth surface: adhesion and the suction cup effect. Despite the fact that the last idea was already criticized 120 years ago (Dahl, 1884) and later by Slifer (1950) plus refuted by experiments under a vacuum (Roth and Willis, 1952), it has still been mentioned by recent authors (Heming, 1972; Conde-Beutel et al., 1989; Orivel et al., 2001). Based on our own experiments and data of previous authors, we reject the suction cup hypothesis because of the contradiction between the pressure decrement in the arolium, necessary for action of the suction cup, and the hydraulic pressure increment, necessary for unfolding.

The nature of animal attachment pads, adherence to the surface have several other explanations, such as electrostatic force (Maderson, 1964), cohesive force and surface tension of the adhesive secretion (Stork, 1983; Dixon et al., 1990) or an adsorbed water layer (Homann, 1957; Huber et al., 2005b), and molecular adhesion (Stork, 1980; Autumn et al., 2000; Gorb, 2001; Autumn et al., 2002; Langer et al., 2004). Probably, several mechanisms cooperate to various extents in different animals. Their contribution may also depend on the surface energy of smooth surfaces (Hiller, 1968; Autumn et al., 2002

In all cases, tight contact between the surface of an attachment device and the substrate is necessary (Gorb, 2005). The contact is improved by shearing proximal displacement of the leg at the beginning of the stance phase, firstly noticed in flies

(*Sarcophaga, Calliphora*) and the bug, *Rhodnius prolixus*, both using hairy sticky pads on their pretarsi and tibiae respectively (Wigglesworth, 1987). This proximal sliding smears an oil in the thinnest layer; a layer so thin that it does not flow and has the strong cohesion (Kendall, 2001). The initial sliding was confirmed in flies by high speed videorecordings (Niederegger and Gorb, 2003). Additionally, shear forces, applied to the pads, result in an increase of the contact area with the substrate (Niederegger et al., 2002).

The real contact area between such macrostructures as arolia and the smooth glass surface was visualized by previous authors using the effect of complete internal reflection (Roth and Willis, 1952). The measured area was compared to the body weight, in order to evaluate tenacity (adhesion force per unit contact area) as a measure for stickiness of an attachment organ. The lower limit of such an estimate was obtained by the use of the entire arolium area, assuming that the contact area was maximal. Different cockroach species hold themselves on a glass ceiling (without additional weights) with a tenecity 13–20 mN/mm^2. Assuming the tripod gait, this force must be doubled while walking on a ceiling. This force range comprises 10–20% of the normal atmospheric pressure (101.3 mN/mm^2). However, cockroaches were able to hold on, in a vacuum (pressure 20–40 mN/mm^2), with a tenecity of 6–7 mN/mm^2. Thus, in any case, the suction cup effect would not provide a stable hold (Roth and Willis, 1952). One tarsus in the aphid, *Aphis fabae*, could hold 3-fold body weight (Dixon et al., 1990). In the hornet, *Vespa crabro*, walking with four legs in stance under a polystyrene ceiling, the arolii hold 6 mN/mm^2. As the insect could walk upside down with a load equal to its body weight, this value must also be doubled (Frantsevich and Gorb, unpublished).

In the bush cricket, *Tettigonia viridissima*, the smooth sticky pads, euplantulae, have a tenacity of 1.7–2.2 mN/mm^2 under normal load on the smooth substrate (Jiao et al., 2000). Lateral tenacity in the smooth pulvilli of the bug, *Coreus marginatus*, was about 100 mN/mm^2 (Gorb and Gorb, 2004). The leaf beetle's, *Chrysolina polita*, hold on the smooth surface with the aid of sticky hairs was 50 times larger the its own weight (Stork, 1980), which (assuming the mass of the beetle to be approximately 20 mg) should be in the range of 1000 mg = 1 g = 10 mN. The relationship between tenacity and lateral tenacity has been demonstrated for several species of flies and beetles (Gorb et al., 2002). On the rough surfaces, attachment forces are strongly reduced (Peressadko and Gorb, 2004b; Gorb, 2008).

The real contact area in an ant, *Oecophila smaragdina*, dynamically depended on the applied load: it comprised only 15–20% of its maximal value during unloaded upside down walking on smooth glass and reached 40% or more upon loading (Federle and Endlein, 2004). These values were 20–25 and >60% in standing ants. Rate of contact formation was recorded with the aid of high speed videorecording. The whole stance phase lasted about 600 ms; the contact area increased to its maximum within 300 ms. Detachment from 2/3 to 1/3 of the maximal contact area was short, only 20 ms. Claws parted faster than the arolium unrolled: 50% of the maximal retraction was completed in 50 ms. In honey bees, full retraction was completed within 30–120 ms (Baur and Gorb, 2000). The dynamics of arolium contact area and claw flexion were directly correlated (Federle et al., 2004).

In a hornet, walking on a ceiling, the stance phase lasted 120–2120 ms (median 520 ms). High speed video recording revealed that attachment and detachment were extremely swift and happened within 1 ms (Frantsevich and Gorb, 2002). Such prompt attachment was not so very astonishing, since the leg had to move the pad only several micrometers farther from the first contact area. However, the time was, most probably, underestimated, because only a silhouette of the expanded arolium was measured, not the real contact area.

Sticky pads, hairy or smooth, are both efficient on the smooth surface. The contact area of a hairy pad is obviously subdivided into hundreds and thousands of small contact points. Even without any fluid in contact, multiple branching hairs may hold onto the surface with the aid of dry adhesion. Tenacity of the seta in the dry adhesion system (van der Waals' interactions) of the lizard *Gekko gecko* was estimated to be 100 mN/mm^2(Autumn et al., 2000). However, the adhesion measured at the level of a single spatula in the gecko resulted in the tenacity of about $5 \cdot 10^5 mN/mm^2$ (Huber et al., 2005a). Branching spatulate setae of the claw scopula in the salticid spider, *Evarcha* exhibit a tenacity of 224 mN/mm^2 (Kesel et al., 2003).

Splitting the contact surface of the polyvinylsiloxane plate into numerous subcontacts (almost flat oval pins) of 0.25 by 0.125 mm size and 0.4 mm high provided better tenacity to the glass plate (Peressadko and Gorb, 2004a). In surfaces with subdivided contacts, catastrophic propagation of cracks is hampered, because a crack has to be initiated at the interface of each single contact (Kendall, 2001), and this would require more energy. The same, probably, holds true for capillary adhesion. The structured surface of some smooth pads may reinforce adhesion. Additionally, microscopic texture of pads may prevent hydroplaning and contribute to molecular adhesion between the solid material of the pad and the substrate (Ohler, 1995; Barnes et al., 2002; Gorb, 2008; Varenberg and Gorb, 2008).

The bottom surface of euplantulae in the bush cricket, *Tettigonia viridissima*, consists of polygonal elements of 4 μm size divided by deep slits (>1.5 μm). The surface of polygons is slightly and irregularly tuberous, presumably because of high deformability of the pad material (Fig. 3.13A, B) (Gorb et al., 2000). The bottom surface of arolii, as it is seen in SEM, is textured into primary folds spaced by 1–2 μm with irregular secondary folds spaced by about 0.2 μm in the crane fly, *Tipula*. In the scorpion fly, *Panorpa communis,* the surface is subdivided into primary longitudinal channels spaced 1–3 μm part and transverse secondary channels spaced by a distance of about 0.25 μm apart, sometimes tiled above one another (Beutel and Gorb, 2001). In the honey-bee, *Apis mellifera*, the arolium surface contains longitudinal folds of 3–5 μm width and anastomizing ridges between them, separating irregular cells of about 1 μm size between the folds (Conde-Beutel et al., 1989; Beutel and Gorb, 2001). Arolium surface texturing in the hornet, *Vespa crabro*, resembles that in *Panorpa*. It consists of longitudinal slits spaced by 5 μm and a fine secondary network of transverse slits spaced by about 0.1–0.2 μm (Fig. 3.13C, D). The arolium of the grasshoppers *Schistocerca gregaria* and *Melanoplus differentialis* has a heterogeneous bottom surface. Most distal and most proximal borders are equipped with sensory hairs. The main area of the arolium is divided into the distal crescent zone, covered with tiny knobs (about 2 μm in diameter) and the

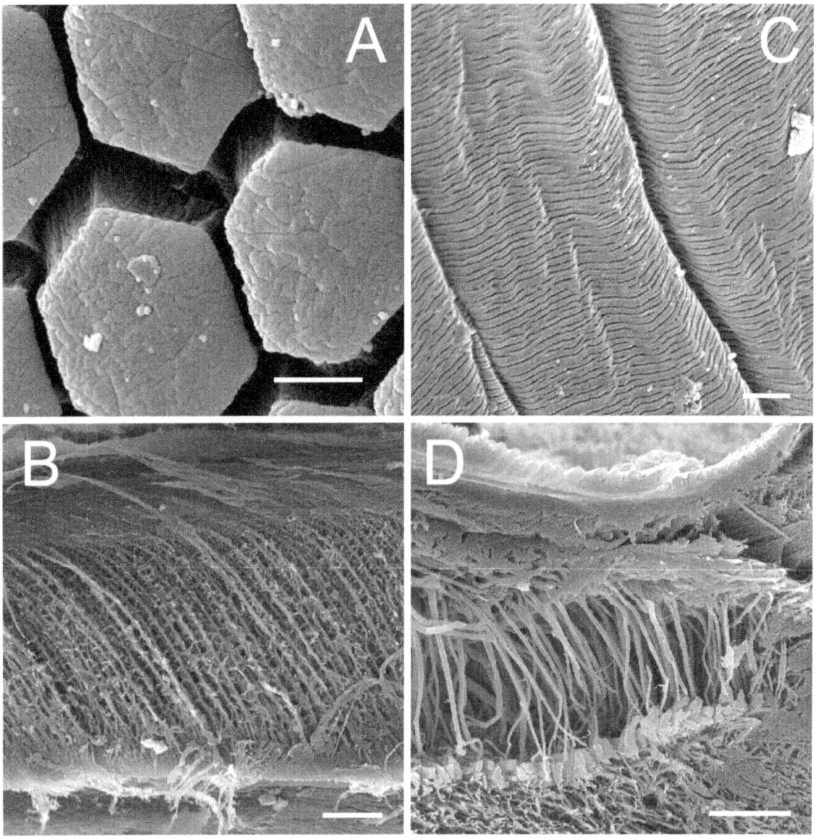

Fig. 3.13 Texture of the bottom surface and internal structure of the material in smooth sticky pads. **A**, **B**. Euplantula in the bush cricket, *Tettigonia viridissima*. **C**, **D**. Arolium in the hornet, *Vespa crabro*. **A**. Polygonal texture. **B**, **D**. Dendritic bottom cuticle. **C**. Orthogonal texture. Scale bars: A, C – 2 μm; B, D – 20 μm

proximal corrugated zone of longitudinally oriented cells, divided by anastomizing ridges (Slifer, 1950; Kendall, 1970). The papillose zone, similar to the tuberous zone named above, has also been described for cockroaches (Roth and Willis, 1952). The surface of the pulvillus in the bugs *Carpocoris pudicus* and *Coreus marginatus* was shown to be almost smooth (Ghazi-Bayat, 1979; Gorb and Gorb, 2004).

A model of biological adhesive contact, inspired by observations cited above, has been made out of a thin smooth glass plate attached to a smooth or textured surface made of polydimethylsiloxane (Ghatak et al., 2004). The thin plate was peeled off the basal plate. On peeling, a band of regular cavities emerged at some distance from the edge (evidently, the maximum concentration of strain was there). Further invaginations of the border appeared opposite the voids and progressed in the direction of the voids until merged with them, thus moving the border deeper between the plates. Texture containing orthogonal incisions led to enhancement of the interfacial fracture toughness. Probably, hexagonal incisions would provide better isotropy of the effect. It has been concluded that the size of texture elements must be less than the length constant of strain (Ghatak et al., 2004).

High strength of the adhesive contact causes an opposite problem: how to detach the pad at the beginning of the swing phase while walking? It has been previously observed how cockroaches, with their middle or hind legs spread on glass, suffered difficulties in detaching their euplantulae (Roth and Willis, 1952). We have video recorded a one minute long attempt of a hornet to detach its hind leg stuck to the smooth polystyrene. Difficulties presumably arose from an inappropriate direction of the leg lift. While proximal sliding, in parallel to the surface, improved the contact of the whole pad area, detachment is easier by peeling, which acts only at the edge of the contact area (Roth and Willis, 1952; Kendall, 1975; Gao et al., 2005). Peeling is even more efficient, if the force is applied at an acute angle to the substrate plane.

Flies, detaching their pulvilli from a smooth surface, used four methods: (1) pulling proximally, when the proximal part of the pulvillus was released hair by hair, while the distal part slipped on the substrate; (2) shifting the distally compressing the pulvillus and releasing its middle part; (3) twisting to release hairs; (4) peeling by lifting the proximal part of the pulvillus (Niederegger and Gorb, 2003). Pegs in the basal zone of the bee's arolium provide the aerial interface and facilitate peeling (Conde-Beutel et al., 1989).

3.7 Adaptation to the Surface

Another advantage of the textured contact surface of hairy pads is an ability to adapt to the surface unevenness of the substrate. However, the smooth pads can do the same. Their surface is underlain by composite dendrite-like structures connected together by thin filaments and spongy layers in grasshoppers (Gorb et al., 2001; Perez Goodwyn et al., 2006) (Fig. 3.13B), hornets (Fig. 3.13D), cicadas (Beutel and Gorb, 2001; Scherge and Gorb, 2001; Frantsevich et al., 2008), thrips (Heming, 1971), aphids (Lees and Hardie, 1988), and the honey bee (Beutel and Gorb, 2001).

Stems of dendrites, called rods, with intermittent filaments, were discovered in the locust, *Locusta migratoria*, long ago (Dahl, 1884). This observation was confirmed by later investigations (Perez Goodwyn et al., 2006). The arolium in the grasshopper, *Melanoplus differentialis* (Slifer, 1950), and euplantulae in the grasshopper, *Omocestus viridulus* (Schwarz and Gorb, 2003), have a thick multi-layered cuticle (200 μm) underneath. The innermost layer is thin and filamentous; the main layer consists of many parallel rods, about 1 μm thick, tied together by filaments. They are immersed in a fluid. Close to the surface of the arolium, the rods branch into fans of fine fibers, which support the outer thin epicuticular layer (Slifer, 1950). The same structure has been observed in related groups: stone-flies (Plecoptera) (Gorb, 2001), several species of cockroaches (Roth and Willis, 1952) and locusts, *Schistocerca gregaria* (Kendall, 1970). Each rod supports one knob in the papillose zone (Kendall, 1970).

Interestingly, a similar composite structure is inherent in euplantulae. The bottom side in the bush cricket, *Tettigonia viridissima*, is made of the thinnest epicuticular film (0.18 μm), underlain by dense rods, about 1 μm thick, branching into filaments 0.08 μm thick (Fig. 3.13B). Rods remain at a slant with respect to the bottom surface (Gorb and Scherge, 2000). Dendrites were considered to be either of exocuticular (Roth and Willis, 1952) or endocuticular origin (Beutel and Gorb, 2001).

Smooth pulvilli in the bug, *Coreus marginatus*, also consist of a foam-like cuticle (Gorb and Gorb, 2004). However, pulvilli in another bug, *Carpocoris pudicus*, have no dendritic structure. Their cuticle consists of horizontally oriented layers (Ghazi-Bayat, 1979).

All investigators agree that the dendritic or spongy structure ensures pliancy of the arolium and its adaptation to an uneven surface. The spongy, soft arolium in thrips is even able to embrace a single plant trichome (Heming, 1971). Freezing substitution experiments on the euplantulae of the grasshopper *Tettigonia viridissima* have clearly demonstrated that pad material is able to adapt to substrate unevenness at the micro- and nanoscale (Gorb et al., 2000). Upon applied pressure, the rods in the bushcricket's euplantula get compressed; their orientation angle relative to the surface diminishes (Gorb and Scherge, 2000). Their visco-elastic recoil has a rather long time constant (> 2 s) (Gorb et al., 2000), and it is still unclear whether a pad can replicate the relief during the swift walking. Beutel and Gorb (2001) suggested that rapid application of force causes elastic deformation of the dendrite-like material. Due to anisotropy of the oblique composite, static friction was 1.5–3 times larger if the pad moving distad was compared with that measured in the proximal movement (Gorb and Scherge, 2000).

3.8 Fluid in the Contact Area

Evidently, the nature of adhesion in smooth attachment pads is complex. Some observers reported that arolia left wet footprints on the smooth substrate: in locusts (Vötsch et al., 2002), cockroaches (Roth and Willis, 1952), aphids (Lees and Hardie, 1988; Dixon et al., 1990), ants (Federle et al., 2002), and in a hornet (personal unpublished observations). Desiccation of arolia by walking on silica gel impaired climbing on glass (Roth and Willis, 1952; Dixon et al., 1990). Frozen droplets of locust secretion contained in turn, nano droplets; probably, the fluid was a kind of viscous emulsion (Vötsch et al., 2002).

Tarsal or tibial glands and glandular epithelium have been suggested as the source of the secretory fluid. Transport pathways of the secretion are unclear. Slifer (1950) has admitted that epidermal cells at the bottom of the arolium in the grasshopper, *Melanoplus differentialis*, secreted fluid permeating through the cuticle, but found no channels responsible for its transport. Glands in adult thrips open into the cavity of the MRU apodeme. It was suggested that the secretion may flow to the arolium lumen and then be transported to the surface (Heming, 1972). No opening of the tarsal gland in the honey bee was found in the area of the unguitractor plate (Conde-Beutel et al., 1989). Pore channels were found in the region of the tarsal gland in the terminal tarsomere of the ant, *Amblyopone reclinata* (Billen et al., 2005), but this gland is not connected to the arolium. The aroliar gland in the bee, *Melipona seminigra*, is connected with the lumen of the arolium, but it bears neither openings outside nor pores responsible for the secretion transport through the cuticle (Jarau et al., 2005).

In the euplantulae of the bush cricket, *Tettigonia viridissima*, extremely thin pores, penetrating the pad cuticle, were reported, using special TEM techniques (Jiao et al., 2000; Schwarz and Gorb, 2003). It was hypothesized that they are responsible for transporting the secretion from the pad lumen to the surface. In the fly's pulvillus, large epidermal cells with lipid vesicles release the secretion into the lumen of the pulvillus, where it is further distributed through the spongy cuticle to the bottom of the pulvillus. Pore channels were found connecting the spongy layer with the bases of the setae and with the surface (Bauchhenss, 1979). Additionally, syrphid flies possess larger pores under the spatulae of their tenent hairs (Gorb, 1998b). Thus, the that the sticky fluid was transported over the corrugated dorsal side of the pulvillus in the house fly, *Musca domestica* (Hasenfuss, 1977, 1978) was disputed (Gorb, 1998b). The transporting of secretion along the dorsal integumental folds in bug's pulvilli (Ghazi-Bayat, 1979; Ghazi-Bayat and Hasenfuss, 1980a, b, 1981) should also be reconsidered.

The microcapillarity of slits of the spongy cuticle's textured surface and the presence of channels and pores is the ultrastructural background for the ability of squeezing secretion upon pressure at the moment of contact and during further sliding. It was previously hypothesized, at least for the fly system, that part of the secretion could be saved by the same capillary force on detachment (Bauchhenss and Renner, 1977), however, this hypothesis has been not experimentally proven.

Little is known about the chemical nature of secretion. The first hypothesis was that the fluid contains the same compounds as the wax cover of the body (Roth and Willis, 1952; Hasenfuss, 1978). However, Bauchhenss (1979) stated that the viscous secretion in flies' pulvilli was not identical to the lipids of the body integument. Beetle pad secretion is mostly composed of hydrocarbons and waxes (Ishii, 1987; Kosaki and Yamaoka, 1996; Attygalle et al., 2000). The detailed study of the chemical composition of the pad secretion in *Locusta migratoria* shows that the fluid contains long-chained fatty acids and alcohols, saturated and non-saturated fatty acids with 16–20 carbon atoms (in the free state or as glycerides), aminoacids and carbohydrates (Vötsch et al., 2002).

Physicochemical properties of the fluid were studied only in some insect species. The insects were able to walk with the aid of arolii over hydrophilic glass as well as over hydrophobic wax paper (Roth and Willis, 1952; Dixon et al., 1990). At a temperature increase to 41–43°, cockroaches failed to stick to glass presumably due to the phase transition of the fluid (Roth and Willis, 1952). In the pad of the ant, *Oecophila smaragdina*, sliding over the glass, friction increased at lower temperatures, which seemed to be consistent with the viscosity effect of the thin liquid film in the contact zone. However, strong, temperature-independent static friction in the same pad was inconsistent with a fully lubricated contact. Adhesive secretion alone

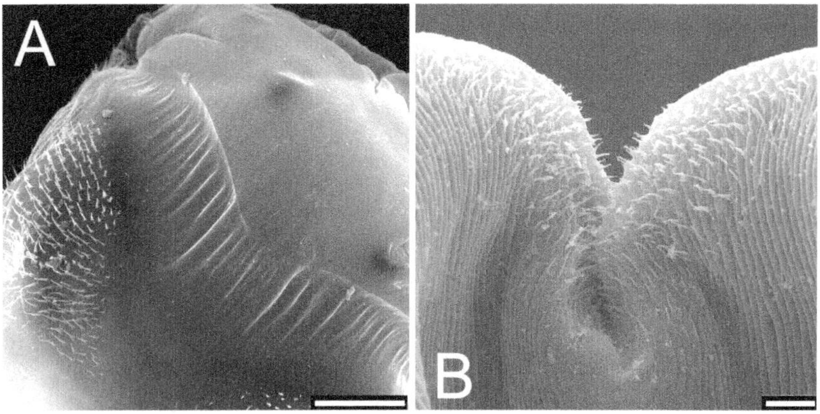

Fig. 3.14 Microtrichia protecting from sticking. **A**. At the basal area of the arolium in *Tenthredo scrophulariae* (Tenthredinidae). **B**. At the distal rim of the arolium in *Vespa crabro* (Vespidae). Scale bars: A – 50 μm; B – 20 μm

was insufficient to explain the observed forces. Direct interaction of the soft cuticle with the surface ("rubber friction") was suggested (Federle et al., 2004). A similar conclusion about a dry component of adhesion was previously expressed by Jiao et al. (2000).

It is worth mentioning larval thrips have no tibial gland and seem to attach without any fluid (Heming, 1972). If it is true, a scale effect on the role of fluid in the smooth attachment systems may be suggested. Animals with a very low mass, may presumably rely solely on the van der Waals interaction between soft smooth pads and the substrate, however, the contribution of the adsorbed water layer that is present on any surface under ambient conditions can not be completely excluded (see Huber et al., 2005b).

Adhesive contact must be only moderately strong, to hold the body's weight with some reserve, but this excess of generated adhesion must not prevent detachment. In addition, some mechanisms must exist in order to protect the folded arolium from sticking to itself. In the folded arolium of a hornet, side lobes touch each other anteriorly. They are protected from tight contact and from sticking by small microtrichia protruding from the surface (Fig. 3.14). Small hairs, 8–12 μm long and 2–4 μm in diameter have also been found at the distal side of the arolium of a honey bee. However, the exact purpose of these hairs was not experimentally proven (Conde-Beutel et al., 1989).

3.9 Sensory Equipment

The control over attachment to the substrate is ultimately important for insect locomotion. Absence of contact to the substrate during leg depression in a dung beetles, *Geotrupes* (Coleoptera, Scarabaeidae), switched the motor program from walking

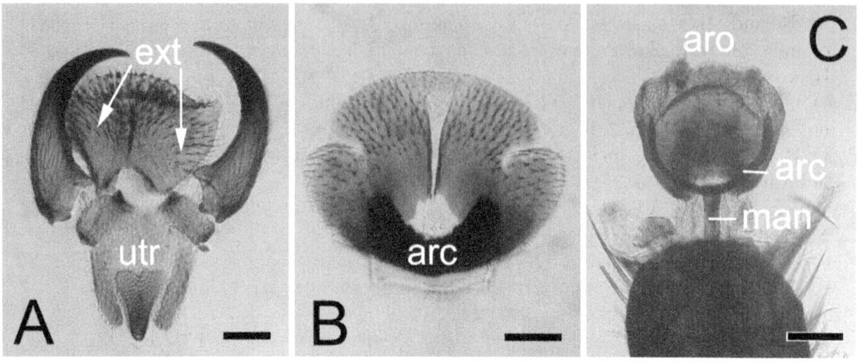

Fig. 3.15 Details of arolia in reprsentatives of different insect orders. **A**. Pretarsus in the crane fly, *Tipula hortulana* (Diptera). **B**. Separated arolium in the scorpion fly, *Panorpa communis* (Mecoptera). **C**. Expanded arolium in the caddis fly, *Limnophilus* sp. (Trichoptera). arc, arcus; aro, arolium; ext, extenders; man, manubrium; utr, unguitractor. Scale bars: A – 100 μm; B, C – 50 μm

to righting (Frantsevich and Mokrushov, 1980). Touching on any obstacle in a freely searching leg of the stick insect, *Cuniculina impigra*, exerted strong arching of the tarsus, evidently caused by reflective contraction of the MRU (Bässler et al., 1991). Posture and walking in grasshoppers was controlled, in particular, by reflexes from the tarsal trichoid sensilla onto the tibial extensor(Runion and Usherwood, 1968). Displacement of the claws caused a reflex in the ipsilateral tibial flexor and in the contralateral extensor of a cockroach (Gregory et al., 1997).

Reflective control over the arolii was deduced from behavioral observations. The size of contact area in the ant, *Oecophila smaragdina*, and in the fly, *Calliphora vicina*, depended on the load (Niederegger et al., 2002; Federle and Endlein, 2004). The extension of the arolium in thrips was completely controllable by leg contact with a smooth substrate (Heming, 1971, 1972).

The pretarsus enjoys diverse sensory equipment. A chordotonal organ between the terminal tarsomere and the unguitractor plate controls the extent and force of claw retraction. It is universally present in insects, e.g. in dragonflies (Mill and Pill, 1981), cockroaches (Gregory et al., 1997), locusts (Kendall, 1970), thrips (Heming, 1972), bugs (Wiese and Schmidt, 1974), chironomid midges (Seifert and Heinzeller, 1989), and moths (Faucheux, 1985).

Hymenopteran claws are equipped with chemo- and mechanoreceptors. Especially prominent are long mechanoreceptive bristles (macrosetae) protruding ventrad, two in number per each claw in ants, *Pachycondyla*, (Orivel et al., 2001), four in the honey bee (Conde-Beutel et al., 1989), one in the hornet (Frantsevich and Gorb, 2002). One or two of the bristles were found on the claw in chalcidoid parasitic wasps (Gladun and Gumovsky, 2006). Innervations and reflexes of the bristles have not been previously studied. Probably, they are responsible for controling of the contact between the pad and the ground.

The hymenopteran manubrium is equipped with two types of trichoid sensilla called STA and STB. STA sensilla are articulated with the underlying cuticle through sockets. The sensilla are long, thin, and cristate, bent down around the folded arolium, usually present of twos, but in some cases from 6 (in *Sirex*) to a complete absence (in females of *Typhia*). STB sensilla are short, with an inconspicuous socket, sometimes tapered to the base, in numbering three or more, seldom less. They are present in many representatives of Apocryta, seldom in Symphyta (Basibuyuk et al., 2000). A pair of long curved bristles (STA) was found in ants, *Pachycondyla* (Orivel et al., 2001), and *Oecophila smaragdina* (Federle et al., 2004), 5–6 in females and 9–12 in males of the honey bee. Receptor cells are situated in the proximal half of the manubrium (Conde-Beutel et al., 1989). A single STA sensillum was found in chalcids, but in scelionids two; 2–5 STB sensilla were also present (Gladun and Gumovsky, 2006). Judging by the shape, size, and position of STA bristles in the hornet (Frantsevich and Gorb, 2002), they contact the ground when the arolium is bent down.

Several small campaniform sensilla, 3–5 μm diameter, occupy the distal half of the manubrium in hymenopterans (Basibuyuk et al., 2000; Federle and Endlein, 2004; Gladun and Gumovsky, 2006; Gladun, 2008). Evidently, they monitor the strain in the elastic manubrium, when it presses on the arcus from above. The planta in hymenopterans is covered with many bristles, all or some of them presumably have a mechanosensory function (Conde-Beutel et al., 1989; Orivel et al., 2001; Gladun and Gumovsky, 2006).

The ventral surface of the grasshopper arolium is richly equipped with mechano- and chemosensory sensilla. At the distal rim, articulated hairs occupy the specific crescent-like area, with long trichoid sensilla bearing single neuron mechanoreceptors, smaller trichoid sensilla with 1–6 neurons each, and basiconic sensilla with about 6 neurons each. The proximal zone bears small basiconic sensilla (Slifer, 1950; Kendall, 1970). The contact areas of the bottom (the papillose zone and the zone of anastomizing ridges) completely lack sensilla, but for the former zone, some cells connected to nerves, but lacking their own hairs, were reported (Kendall, 1970). While euplantulae in cockroaches possess rows of coeloconic sensilla, their arolii have none (Roth and Willis, 1952).

The arolium in Diptera and Mecoptera bears no hairs on the bottom (Röder, 1980). The arolium in thrips also has no sensilla (Heming, 1971). The arolium in some suprafamilies of Auchenorrhyncha is equipped either with ventrolateral sclerites bearing two rows of bristles (in Cercopoidea) or with a few hairs among the smooth leather-like ventral face (in Fulgoroidea and Cicadelloidea) (Emeljanov, 1987).

The soft bottom surface of the hymenopteran arolium is sensory deprived. We think that the reason for this deprivation is that the sensilla named above need to emerge from the sclerotized cuticle which provides a reference for the mechanoreceptive gauges. The pliable bottom provides no such reference. Thus this "deaf" surface participates in purely mechanical responses to external forces, so called preflexes, while contacting the ground. Load is presumably monitored by measuring strain with sensilla attached to the pretarsal sclerites.

3.10 Conclusions

Arolia in insects are of two main types: those with constant shape and those able to fold and spread. Thysanoptera, Diptera-Tipulomorpha, Mecoptera, Hymenoptera, and Trichoptera enjoy the second type arolia. The last two orders have astonishingly similar arolia with the most elaborated aggregation of sclerites. Interestingly, among hymenopterans, arolia in primitive sawflies are no less elaborated than in higher lineages (Ichneumonidae, Vespidae, Sphecidae).

The internal sclerite, the arcus, is least variable compared to other aroliar sclerites throughout the entire order. We suggest that such a low variability in the arcus morphology emphasizes its critical necessity for the arolium kinematics. Other features, such as shapes of manubria, extenders, plantae, and type of bottom texture, are more diverse. The role of hydraulics still remains unclear in arolium kinematics. Remaining especially enigmatic is the action of the nanocomposite of elastic cuticular dendrites and a viscous fluid accommodating the dendrites. Tight contact of the arolium bottom with the 3D profile of the ground is ensured by a fluid of rather complex and, probably, even biphasic (emulsion) composition.

All motions of the arolium are multiply repeatable and kinematically and hydraulically reversible. The state of the arolium is dynamically controlled, depending on properties of the ground and the applied load. The reflex chain, from the ganglion to the very leg tip is the longest and the slowest in the body segment. Control is achieved in two ways: by neural reflexes and by mechanical feedback (preflexes). The largest area of the arolium, its bottom part, responsible for the main function of this organ, is, however, sensory deprived. Hence, attachment is controlled by indirect sensory signals from the neighbouring sclerites. The arolium unfolding is achieved by the action of a single muscle through a long tendon. Complex movements of pretarsal structures during unfolding, spreading and folding are provided by a multisclerite mechanism, in which simple neural control is compensated for by a complex structure and material properties.

There are several unsolved problems in functions of arolii:

(i) hydraulic models of the arolium; sources of hydraulic fluids, location of pumps; connections and crosstalk between fluid compartments; dynamics of flows.
(ii) kinematics of chains the links of which are not articulated directly but separated by "empty" spaces filled with hydraulic fluid; cooperation of hydraulics, claw-extender and manubrium-arcus-planta chains in spreading.
(iii) origin and transport routes of the secretion; its physico-chemical properties, especially regarding components of the emulsion; diversity of contact phenomena (capillary adhesion, viscose forces, van der Waals interactions, friction).
(iv) dependence of peeling on direction of applied force.
(v) MRU reflexes elicited by pretarsal sensilla; sensing of slipping and peeling.
(vi) construction, elasticity and function of the primitive arcus in representatives of Mecoptera.

Appendix 1. List of Abbreviations

apo	apodeme of musculus retractor unguis
aro	arolium
arc	arcus
BWU	body weight unit (ratio between the load and body weight)
con	condyles
cr	cristae
ext	extender
man	manubrium
MRU	musculus retractor unguis
ms	macroseta
pl	planta
un	claw, ungius
utr	unguitractor plate

Other abbreviations are indicated in figure legends.

Appendix 2. List of Insect Species

Insecta

Zygentoma
Lepismatidae: *Thermobia domestica* (Packard, 1873)
Blattodea
Blattelidae: *Periplaneta americana* (Linnaeus, 1758).
Phasmatodea
Bacillidae: *Carausius morosus* (Brunner 1907); *Cuniculina impigra* (Brunner, 1907).
Ensifera
Tettigoniidae: *Tettigonia viridissima* Linnaeus, 1758.
Caelifera
Acridiidae: *Schistocerca gregaria* (Forskål, 1775); *Locusta migratoria* Linnaeus, 1758; *Chorthippus apricarius* (Linnaeus, 1758); *Melanoplus differentialis* (Thomas, 1865).
Thysanoptera
Phlaeothripidae: *Haplothrips verbasci* (Osborn, 1897).
Heteroptera
Coreidae: *Coreus marginatus* (Linnaeus, 1758); Reduviidae: *Rhodnius prolixus* Stål,1859; Pentatomidae: *Carpocoris pudicus* (Poda, 1761).
Sternorrhyncha
Aphididae: *Aphis fabae* Scopoli 1763.
Coleoptera
Scarabaeidae: *Geotrupes* spp., *Lethrus apterus* (Laxman, 1770); Chrysomelidae: *Chrysolina polita* (Linnaeus, 1758).

Hymenoptera

Xyelidae: *Xyela jullii* (Brebisson, 1818); Tenthredinidae: *Tenthredo scrophulariae* Linnaeus, 1758; Diprionidae: *Diprion similis* (Hartig, 1837); Pamphiliidae: *Caenolyda reticulata* (Linnaeus, 1767); Cephidae: *Hartigia linearis* Schrank, 1781; Siricidae: *Sirex noctilio* Fabricius, 1793; Eulophidae: *Sympiesis* sp.; Trichogrammatidae: *Trichogramma semblidis* (Aurivillius, 1897); Ormyridae: *Ormyrus* sp.; Torymidae: *Torymus* sp.; Perilampidae: *Perilampus* sp.; Ichneumonidae: *Pimpla* sp.; Sphecidae: *Philanthus coronatus* Fabricius, 1790; *Bembix rostrata* (Linnaeus, 1758); *Crabro peltarius* (Schreber, 1784); Apidae: *Apis mellifera* Linnaeus, 1758; *Melipona seminigra* Friese 1903; Typhiidae: *Typhia* sp.; Formicidae: *Formica cunicularia* Latreille, 1798; *Amblyopone reclinata* (Mayr, 1879); *Oecophila smaragdina* Fabricius, 1775; *Amblyopone reclinata* Mayr, 1879; *Pachycondyla goeldii* Forel, 1912; Eumenidae: *Eumenes coarctatus* (Linnaeus, 1758); Vespidae: *Paravespula germanica* (Fabricius 1793); *Vespa crabro* Linnaeus, 1758.

Trichoptera

Limnophilidae: *Limnophilus* sp.

Lepidoptera

Nymphalidae: *Argynnis lathonia* (Linnaeus 1758).

Mecoptera

Panorpidae: *Panorpa communis* Linnaeus, 1758.

Diptera

Tipulidae: *Tipula hortulana* Linnaeus, 1758; *Tipula juncea* Meigen, 1818; Calliphoridae: *Calliphora vicina* Robineau-Desvoidy, 1830; Muscidae: *Musca domestica* Linnaeus, 1761.

References

Attygalle, A.B., Aneshansley, D.J., Meinwald, J., and Eisner, T. (2000) Defence by foot adhesion in a chrysomelid beetle (*Hemisphaerota cyanea*): characterization of the adhesive oil. *Zoology* 103: 1–6.

Autumn, K., and Peattie, A.M. (2002) Mechanisms of adhesion in geckos. *Integr. Comp. Biol.* 42: 1081–1090.

Autumn, K., Liang, Y.A., Hsieh, S.T., Zesch, W., Chan, W.P., Kenny, T.W., Fearing, R., and Full, R.J. (2000) Adhesive force of a single gecko foot hair. *Nature* 405: 681–685.

Autumn, K., Sitti, M., Liang, Y.A., Peattie, A.M., Hansen, W.R., Sponberg, S., Kenny, T.W., Fearing, R., Israelachvili, J.N., and Fall, R.J. (2002) Evidence for van der Waals adhesion in gecko setae. PNAS 99(19): 12252–12256.

Barnes, J., Smith, J., Oines, C., and Mundl, R. (2002) Bionics and wet grip. *Tire Technol. Int.* 2002(12): 56–61.

Basibuyuk, H.H., Quicke, D.L.J., Rasnitsyn, A.P., and Fitton, M.G. (2000) Morphology and sensilla of the orbicula, a sclerite between the tarsal claws, in the Hymenoptera. *Ann. Entomol. Soc. Amer.* 93: 625–636.

Bässler, U., Rohrbacher, J., Karg, G., and Breatel, G. (1991) Interruption of searching movements of partly restrained front legs of stick insects, a model situation for the start of the stance phase. *Biol. Cybern.* 65: 507–514.

Bauchhenss, E. (1979) Die Pulvillen von *Calliphora erythrocephala* Meig. (Diptera, Brachycera) als Adhäsionsorgane. *Zoomorphologie* 93: 99–123.

Bauchhenss, E., and Renner, M. (1977) Pulvillus of *Calliphora erythrocephala* Meig. (Diptera; Calliphoridae). *Int. J. Insect Morph. Embryol.* 6: 225–227.

Baur, F., and Gorb, S.N. (2000) How the bee releases its leg attachment devices. *Biona Rep.* 15: 295–297.

Beutel, R.G., and Gorb, S.N. (2001.) Ultrastructure of attachment specializations of hexapods (Arthropoda): evolutionary patterns inferred from a revised ordinal phylogeny. *J. Zool. Syst. Evol. Res.* 39: 177–207.

Billen, J., Thijs, B., Ito, F., and Gobin, B. (2005) The pretarsal footprint gland of the ant *Amblyopone reclinata* (Hymenoptera, Formicidae) and its role in nestmate recognition. *Arthropod Struct. Dev.* 34: 111–116.

Brown, J.E., and Loeb, G.E. (1999) A reductionist approach to creating and using neuromusculoskeletal models. In: *Biomechanics and Neural Control of Posture and Movement*, ed. by Winter, J.M., and Crago, P.E. New York: Springer verlag, pp. 148–163.

Conde-Beutel, R., Erickson, E.H., and Carlson, S.D. (1989) Scanning electron microscopy of the honeybee, *Apis mellifera* L. (Hymenoptera: Apidae) pretarsus. *Int. J. Insect Morph. Embryol.* 24: 59–69.

Crossland, M.W.J., Su, N.-Y., and Scheffrahn, R.H. (2005) Arolia in termites (Isoptera): functional significance and evolutionary loss. *Insect. Soc.* 52: 63–66.

Dahl, F. (1884) Über den Bau und die Functionen des Insektenbeins. *Zool. Anz.* 7: 38–41.

Dai, Z., Gorb, S.N., and Schwarz, U. (2002) Roughness-dependent friction force of the tarsal claw system in the beetle *Pachnoda marginata* (Coleoptera, Scarabaeidae). *J. Exp. Biol.* 205: 2479–2488.

Dashman, T. (1953) Terminology of the pretarsus. *Ann. Enlomol. Soc. Amer.* 46: 56–62.

Dixon, A.F.G., Croghan, P.C., and Gowing, R.P (1990) The mechanism by which aphids adhere to smooth surfaces. *J. Exp. Biol.* 152: 243–253.

Domenichini, G. (1994) I pretarsi delle zampe cursorie in insetti olometaboli: strutture e funzioni. *Atti Acad. Nat. Ital. Entomol. Rendiconti* 42: 19–127.

Eigenbrode, S.D., Castagnola, T., Roux, M.-B., and Steljes, L. (1996) Mobility of three generalist predators is greater on cabbage with glossy leaf wax than on cabbage with a wax bloom. *Entomol. Exp. Appl.* 81: 335–343.

Emeljanov, A.F. (1982) Structure and evolution of tarsi in Dictyopharidae (Homoptera) (in Russ.). *Entomologicheskoe Obozrenie* 61: 501–516.

Emeljanov, A.F. (1987) Phylogeny of cicads (Homoptera, Cicadina) by comparative morphological data (in Russ.). *Trudy Vsesoyuznogo Entomologicheskogo Obschestva* 69: 19–109.

Endlein, T., and Federle W. (2007) Walking on smooth or rough ground: passive control of pretarsal attachment in ants.*J. Comp. Physiol.* A 194: 49–60.

Faucheux M.J. (1985.) Structure of the tarso-pretarsal chordotonal organ in the imago of *Tineola bisselliella* Humm. (Lepidoptera: Tineidae). *Int. J. Insect Morph. and Embryol.* 14: 147–154.

Federle, W., and Endlein, T. (2004) Locomotion and adhesion: dynamic control of adhesive surface contact in ants. *Arthr. Struct. Dev.* 33: 67–75.

Federle, W., Baumgarner, W., and Hölldobler, B. (2004) Biomechanics of ant adhesive pads: frictional forces are rate- and temperature-dependent. *J.Exp. Biol.* 207: 67–74.

Federle, W., Brainerd, E.L., McMahon, T.A., and Hölldobler, B. (2001) Biomechanics of the movable pretarsal adhesive organ in ants and bees. *Proc. Nat. Acad. Sci.* 98: 6215–6220.

Federle, W., Riehle, M., Curtis, A.S.G., and Full, R.J. (2002) An integrative study of insect adhesion: mechanics and wet adhesion of pretarsal pads in ants. *Integr. Comp. Biol.* 42: 1100–1106.

Fennah, R. (1945) Character of taxonomic importance in the pretarsus of Auchenorrhyncha (Homoptera). *Proc. Entomol. Soc. Wasgington* 47. 120–128.

Foelix, R. (1982) *The biology of spiders*, Massachusetts and London: Harvard University Press.

Frantsevich, L., and Cruse, H. (1997) The stick insect, *Obrimus asperrimus* (Phasmida, Bacillidae) walking on different substrates. *J.Insect Physiol.* 43: 447–455.

Frantsevich, L.I., and Gorb, S.N. (2002) Arcus as a tensegrity structure in the arolium of wasps (Hymenoptera: Vespidae). *Zoology* 105: 225–237.

Frantsevich, L., and Gorb, S. (2004) Structure and mechanics of the tarsal chain in the hornet, *Vespa crabro* (Hymenoptera: Vespidae): implications on the attachment mechanism. *Arthr. Struct. Dev.* 33: 77–89.

Frantsevich, L.I., and Mokrushov, P.A. (1980) Turning and righting in *Geotrupes* (Coleoptera, Scarabaeidae). *J. Comp. Physiol.* 136: 279–289.

Frantsevich, L., Govardovski, V., Gribakin, F., Nikolajev G., Pichka, V., Polanovsky, A., Shenchenko, V., and Zolotov, V. (1977) Astroorientation in *Lethrus* (Coleoptera, Scarabaeidae). *J.Comp.Physiol.* 121: 253–271.

Frantsevich, L., Ji, A., Dai, Z., Wang, J., Frantsevich, L., and Gorb S.N. (2008) Adhesive properties of the arolium of a lantern-fly, *Lycorma delicatula* (Auchenorrhyncha, Fulgoridae). *J. Insect Physiol.* 54: 818–827.

Frazier, S.F., Larsen, G.S., Neff, D., Quimby, L., Carney, M., Di Caprio R.A., and Zill, S.N. (1999) Elasticity and movements of the cockroach tarsi in walking. *J. Comp. Physiol.* A 185: 157–172.

Full, R.J., and Koditschek, D.E. (1999) Templates and anchors: Neuromechanical hypotheses of legged locomotion on land. *J. Exp. Biol.* 202: 3325–3332.

Gannon, A.J., Bach, C.E., and Walker, G. (1994) Feeding patterns and attachment ability of *Altica subplicata* (Coleoptera: Chrysomelidae) on sand-dune willow. *Great Lakes Entomologist* 27: 89–101.

Gao, H., Wang, X., Yao, H., Gorb, S., and Arzt, E. (2005) Mechanics of hierarchical adhesion structures of geckos. *Mech. Mater.* 37: 275–285.

Ghatak, A., Mahadevan, L., Chung, J.Y., Chaudhury, M.K., and Shenoy, V. (2004) Peeling from a biomimetically patterned thin elastic film. *Proc. R. Soc. Lond. A.* 460: 2725–2735.

Ghazi-Bayat, A. (1979) Zur Oberflächenstruktur der tarsalen Haftlappen von *Coreus marginatus* (L.) (Coreidae, Heteroptera). *Zool. Anz.* 203: 345–347.

Ghazi-Bayat, A., and Hasenfuss, I. (1980a) Zur Herkunft der Adhäsionsflüssigkeit der tarsalen Haftlappen bei den Pentatomidae (Heteroptera). *Zool. Anz.* 204: 13–18.

Ghazi-Bayat, A., and Hasenfuss, I. (1980b) Die Oberflächenstrukturen des Prätarsus von *Elasmucha ferrugaia* (Fabricius) (Acanthosomatidae, Heteroptera). *Zool. Anz.* 205: 76–80.

Ghazi-Bayat, A., and Hasenfuss, I. (1981) Über den Transportweg der Haftflüssigkeit der Pulvilli bei *Coptosoma sculellatum* (Geoffr.) (Plataspididae, Heteroptera). *Nachrichtenblatt Bayer. Entomol.* 30: 58–58.

Gladun, D.V. (2008) Morphology of the pretarsus of the sawflies and horntails (Hymenoptera: 'Symphyta'). *Arthropod Struct. Dev.* 37: 13–28.

Gladun, D., and Gorb, S.N. (2007) Insect walking techniques on thin stems. *Arthropod-Plant Interact.* 1: 77–91.

Gladun, D., and Gumovsky, A. (2006) The pretarsus in Chalcidoidea (Hymenoptera Parasitica): functional morphology and possible phylogenetic implications. *Zool. Scr.* 35: 607–626.

Gorb, S.N. (1996) Design of insect unguitractor apparatus. *J. Morphol.* 230: 219–230.

Gorb, S. N., (1998b) The design of the fly adhesive pad: distal tenent setae are adapted to the delivery of an adhesive secretion. *Proc. Roy. Soc. Lond.* B 265: 747–752.

Gorb, S.N. (2001) *Attachment devices of insect cuticle*. Dordrecht, Boston, London: Kluwer Academic Publishers.

Gorb, S.N. (2005) Uncovering insect stickiness: structure and properties of hairy attachment devices. *Am. Entomol.* 51: 31–35.

Gorb, S.N. (2008) Smooth attachment devices in insects. In: *Advances in Insect Physiology, Volume 34: Insect Mechanics and Control*, edited by Casas, J., and Simpson, S. J., London: Elsevier Ltd., pp. 81–116.

Gorb, S., and Scherge, M. (2000) Biological microtribology: anisotropy in frictional forces of orthopteran attachment pads reflects the ultra-structure of a highly deformable material. *Proc. R. Soc. London* B 186: 821–831.

Gorb, E.V., and Gorb, S.N. (2002) Attachment ability of the beetle *Chrysolina fastuosa* on various plant surfaces. *Entomol. Exp. Appl.* 105: 13–28.

Gorb, S.N., and Gorb, E.V. (2004) Ontogenesis of the attachment ability in the bug *Coreus marginatus* (Heteroptra, Insecta). *J. Exp. Biol.* 207: 2917–2924.

Gorb, S.N., Beutel, R.G., Gorb, E.V., Jiao, Y., Kastner, V., Niederegger, S., Popov, V.L., Scherge, M., Schwarz, U., and Vötsch, W. (2002) Structural design and biomechanics of friction-based releasable attachment devices in insects. *Integr. Comp. Biol.* 42: 1127–1139.

Gorb, S.N., Jiao, Y., and Scherge, M. (2000) Ultrastructural architecture and mechanical properties of attachment pads in *Tettigonia viridissima* (Orthoptera Tellgoniidae). *J. Comp. Physiol.* A 186: 821–831.

Gregory, S., Larsen, S., Frazier, F., Zill, S.N. (1997) The tarso-pretarsal chordotonal organ as an element in cockroach walking. *J. Comp. Physiol. A* 180: 683–700.

Hasenfuss, I. (1977) Die Herkunft der Adhäsionsflüssigkeit bei Insekten. *Zoomorphology* 87: 51–64.

Hasenfuss, I. (1978) Über das Haften von Insekten an glatten Flächen – Herkunft der Adhäsionsflüssigkeit. *Zool. Jahrb. Anat.* 99: 115–116.

Hasenfuss, I. (1999) The adhesive devices in larvae of Lepidoptera (Insecta, Pterygota). *Zoomorphology* 119: 143–162.

Hassenstein, B., and Hustert, R. (1999) Hiding responses of locusts to approaching objects. *J. Exp. Biol.* 202: 1701–1710.

Heming, B. S. (1971) Functional morphology of the thysanopteran pretarsus. *Can. J. Zool.* 49: 91–108.

Heming, B. S. (1972) Functional morphology of the pretarsus in larval Thysanoptera. *Can. J. Zool.* 50: 751–766.

Hiller, U. (1968) Untersuchungen zum Feinbau und zur Funktion der Haftborsten von Reptilien. *Z. Morphol. Tiere* 62: 307–362.

Homann, H. (1957) Haften Spinnen an einer Wasserhaut? *Naturwissenschaften* 44: 318–319.

Huber, G., Gorb, S.N., Spolenak, R., and Arzt, E. (2005a) Resolving the nanoscale adhesion of individual gecko spatulae by atomic force microscopy. *Biol. Lett.* 1: 2–4.

Huber, G., Mantz, H., Spolenak, R., Mecke, K., Jacobs, K., Gorb, S.N., and Arzt, E. (2005b) Evidence for capillarity contributions to gecko adhesion from single spatula nanomechanical measurements. *PNAS* 102: 16293–16296.

Ishii, S. (1987) Adhesion of a leaf feeding ladybird *Epilachna vigintioctomaculata* (Coleoptera: Coccinellidae) on a vertically smooth surface. *Appl. Ent. Zool.* 22: 222–228.

Jarau, S., Hrncir, M., Zucchi, R., and Barth, F.G. (2005) Morphology and structure of the tarsal glands of the stingless bee *Melipona seminigra*. *Naturwiss.* 92: 147–150.

Jiao, Y., Gorb, S., and Scherge, M. (2000) Adhesion measured on the attachment pads of *Tettigonia viridissima* (Orthoptera, Insecta). *J. Exp. Biol.* 203: 1887–1895.

Jordan, K. (1888) Anatomie und Biologie der Physapoda.*Z. wiss. Zool.* 47: 541–620.

Kendall, K. (1975) Thin-film peeling – the elastic term. *J. Phys. D: Appl. Phys.* 8: 1449–1452

Kendall, K. (2001) *Molecular Adhesion and its Applications.* New York: Kluwer Academic Publishers.

Kendall, M.D. (1970) The anatomy of the tarsi of *Schistocerca gregaria* Forskål. *Z. Zellforsch.* 109: 112–137.

Kesel, A.B., Martin, A., and Seidl, T. (2003) Adhesion measurements on the attachment devices of the jumping spider *Evarcha arcuata*. *J. Exp. Biol.* 206: 2733–2738.

Kosaki, A., and Yamaoka, R. (1996) Chemical composition of footprints and cuticula lipids of three species of lady beetles. *Jpn. J. Appl. Entomol. Zool.* 40: 47–53.

Kubow, T.M., and Full, R.J. (1999) The role of the mechanical system in control: a hypothesis of self-stabilization in hexapodal runners. *Phil. Trans. R. Soc. Lond.* B 354: 849–862.

Langer, M.G., Ruppersberg, J.P., and Gorb, S. (2004) Adhesion forces measured at the level of a terminal plate of the fly's seta. *Proc. R. Soc. Lond. B* 271: 2209–2215.

Lees, A.M., and Hardie, J. (1988) The organs of adhesion in the aphid *Megoura viciae*. *J. Exp. Biol.* 136: 209–228.

Maderson, P.F.A. (1964) Keratinized epidermal derivatives as an aid to climbing in gekkonid lizards. *Nature* 203: 780–781.

Manton, S.M. (1972) The evolution of arthropodan locomotory mechanisms. Part 10. Locomotory habits, morphology and evolution of hexapod classes. *Zool. J. Linn. Soc.* 51: 203–400.

Mill, P.J., and Pill, C.E.J. (1981) The structure and physiology of the tarso-pretarsal chordotonal organ on the larva of *Anax imperator* Leach (Anisoptera: Aescchnidae). *Odonatologica* 10: 29–37.

Nachtigall, W. (1996) Locomotory behaviour in a population of the tiger beetle species Cicindella hybrida on a small, hot, sandy area (Coleoptera, Cicindellidae). *Entomol. Gener.* 20: 241–248.

Neff, D., Frazier, S.F., Quimby, L., Wang, R.-T., and Zill, S. (2000) Identification of resilin in the leg of cockroach, *Periplaneta americana*: confirmation by a simple method using pH dependence of UV fluorescence. *Arthr. Struct. and Devel.* 29: 75–83.

Niederegger, S., and Gorb, S. (2003) Tarsal movements in flies during leg attachment and detachment on a smooth substrate. *J. Insect Physiol.* 49: 611–620.

Niederegger, S., Gorb, S., and Jiao, Y. (2002) Contact behaviour of tenent setae in attachment pads of the blowfly *Calliphora vicina* (Diptera, Calliphoridae). *J. Comp. Physiol.* A 187: 961–970.

Nielsen, E.S., and Common, I.F.B. (1991) Lepidoptera. In: *The Insects of Australia*, New York: Cornell University Press, pp. 817–916.

Ohler, A. (1995) Digital pad morphology in torrent-living Ranid frogs. *Asiat. Herpetol. Res.* 6: 85–96.

Orivel, J., Malherbe, M.C., and Dejean, A. (2001) Relationships between pretarsus morphology and arboreal life in poberine ants of the genus *Pachycondyla* (Formicidae: Ponerinae). *Ann. Entomol. Soc. Amer.* 94: 449–456.

Peressadko, A., and Gorb, S.N. (2004a) When less is more: experimental evidence for tenacity enhancement by division of contact area. *J. Adhes.* 80: 1–15.

Peressadko, A., and Gorb, S.N. (2004b) Surface profile and friction force generated by insects. In: *Fortschritt-Berichte VDI*. Eds. I. Boblan and R. Bannasch. Düsseldorf: VDI Verlag, vol. 249, pp. 257–261.

Perez Goodwyn, P.J., Peressadko, A., Schwarz, H., Kastner, V., and Gorb, S. (2006) Material structure, stiffness, and adhesion: why attachment pads of the grasshopper (*Tettigonia viridissima*) adhere more strongly than those of the locust (*Locusta migratoria*) (Insecta: Orthoptera). *J. Comp. Physiol. A* 192: 1233–1243.

Radnikov, G., and Bässler, U. (1991) Function of a muscle whose apodeme travels through a joint moved by other muscles: why the retractor unguis muscle in stick insects is tripartite and has no antagonist. *J. Exp. Biol.* 157: 87–99.

Röder, G. (1986) Zur Morphologie des Praetarsus der Diptera und Mecoptera. *Zool. Jahrb. Abt. Anat. Ontog. Tiere* 114: 465–502.

Roth, L. M., and Willis, E. R. (1952) Tarsal structure and climbing ability of cockroaches. *J. Exp. Biol.* 119: 483–517.

Runion, H.J., and Usherwood, P.N.R. (1968) Tarsal receptors and leg reflexes in the locust and grasshopper. *J. Exp. Biol.* 49: 421–436.

Scherge, M., and Gorb, S.N. (2001) *Biological Micro- and Nanotribology*. Berlin: Springer.

Schulmeister, S. (2003) Morphology and evolution of the tarsal plantulae in Hymenoptera (Insecta), focussing on the basal lineages. *Zool Scr.* 32: 153–172.

Schwarz, H., and Gorb, S. (2003) Method of platinum-carbon coating of ultrathin sections for transmission and scanning electron microscopy: An application for study of biological composites. *Microsc. Res. Techn.* 62: 218–224.

Seifert, P., and Heinzeller, T. (1989) Mechanical, sensory and glandular structures in the tarsal unguitractor apparatus of *Chironomus riparius* (Diptera, Chironomidae). *Zoomorphology* 109: 71–78.

Slifer, E. H. (1950) Vulnerable areas on the surface of the tarsus and pretarsus of the grasshopper (Acrididae, Orthoptera) with special reference to the arolium. *Ann. Entomol. Soc. Amer.* 43: 173–188.

Snodgrass, R. E. (1956) *Anatomy of the Honey Bee*. New York: Comstock Publishing Associates.

Stork, N. E. (1980) Experimental analysis of adhesion of *Chrysolina polita* (Chrysomelidae, Coleoptera) on a variety of surfaces. *J. Exp. Biol.* 88: 91–107.

Stork, N. E. (1983) The adherence of beetle tarsal setae to glass. *J. Nat. Hist.* 17: 583–597.

Tanaka, Y., and Hisada, M. (1980) the hydraulic mechanism of the predatory strike in dragonfly larvae. *J. Exp. Biol.* 88: 1–19.

Varenberg, M., and Gorb, S.N. (2008) Hexagonal surface micropattern for dry and wet friction. *Adv. Mater.* 20: 1–4.

Vötsch, W., Nicholson, G., Müller, R., Stierhof, Y.-D., Gorb, S., and Schwarz, U. (2002) Chemical composition of the attachment pad secretion of the locust *Locusta migratoria*. *Insect Biochem. Mol. Biol.* 32: 1605–1613.

Weber, H. (1933) *Lehrbuch der Entomologie.* Stuttgart: Gustav Fischer Verlag.

Wiese, K., and Schmidt, K. (1974) Mechanorezeptoren im Insektentarsus. Die Konstruktion des tarsalen Scolopidialorgans bei *Notonecta* (Hemiptera, Heteroptera). *Z. Morphol. Tiere* 79: 47–63.

Wigglesworth, V.B. (1987) How does a fly cling to the under surface of a glass sheet? *J. Exp. Biol.* 129: 373–367.

Zielinska, T. (2001) Synthesis of control system – gait implementation problems. In: *CLAWAR 2001: 4th Internat. Conference on Climbing and Walking Robots.* London: Professional Engineering Publishing, pp. 489–496.

Chapter 4
Organs of Adhesion in Some Mountain-stream Teleosts of India: Structure-Function Relationship

Debasish Das and Tapas C. Nag

4.1 Introduction

The sub-Himalayan streams and rivers of India are known to be inhabited by a number of teleost fishes specialized for living in rapids (Day, 1958; Jayaram, 1981). Ecologically, these are perennial shallow-water bodies, characterized by having low water temperature, high turbulent current and sandy-rocky substratum (Das and Nag, 2004). In order to thrive against the adverse habitat situations, many species show a number of unique adaptive specialisations. One notable morphological feature shown by them is an adhesive organ (AO), by which the species adhere to submerged rocks and stones of streams. The AO is ventral in position and located either at the thoracic region behind the opercular openings or encircling the mouth opening. In mountain-stream catfishes (Sisoridae), additional adhesive devices are present on the ventral surface of the paired pectoral and pelvic fins. In these fishes, the skin of the outer rays of those fins is formed into a series of alternate ridges and grooves (Hora, 1922, 1930).

4.2 Historical

Long ago, the famous Indian ichthyologist Hora (1922) described the gross appearance of the AO (occurring in the head-thorax axis) in mountain-stream teleosts. He also observed the adhesive nature of the paired pectoral- and pelvic fins in certain species of sisorids. In both situations, some common structural features of the AO were noted, and based upon the gross anatomical features and minute structural details, he postulated a possible mechanism of adhesion by the AO in those stream fishes (Hora, 1922). His monumental work (Hora, 1930) nicely illustrates the evolution of adaptive strategies shown by the stream fishes of India, especially with regard to adhesion in fast-flowing torrents.

T. C. Nag (✉)
Department of Anatomy, All India Institute of Medical Sciences, New Delhi, India
e-mail: tapas_nag@hotmail.com

S.N. Gorb (ed.), *Functional Surfaces in Biology*, Vol. 2,
DOI 10.1007/978-1-4020-6695-5_5, © Springer Science+Business Media B.V. 2009

Later on, Saxena and Chandy (1966) examined in detail in the light microscope the organization of the AO in certain stream fishes. However, due to inherent limitations, studies on the ultrastructural organizations of the AO could not be carried out until the end of the 80's. Following this period, reports on the surface features of the AO of mountain-stream fishes of India, utilizing scanning electron microscopy (SEM) had begun to be available (Sinha et al., 1990; Singh and Agarwal, 1991, 1993; Ojha and Singh, 1992; Pinky et al., 2002; Johal and Rawal, 2003). Das and Nag (2004) studied the surface view of the adhesive fins of the sisorid catfish *Pseudocheneis sulcatus*. Detailed studies on the sub-microscopic organization of the AO utilizing transmission electron microscopy (TEM) are now available for two species (Das and Nag, 2005, 2006).

In this review, we present available data on the structure and ultrastructure of AO in mountain-stream teleosts of India. Also, we describe these aspects in one additional species (the snow-trout, *Schizothorax richardsnoii* Gray) by SEM, light microscopy (LM), TEM and immunohistochemistry. The species reviewed in this paper are two sisorids (*Pseudocheneis sulcatus* McClelland and *Glyptothorax pectinopterus* McClelland) and two cyprinids [*Schizothorax richardsnoii* Gray and *Garra gotyla gotyla* (Hamilton, 1822)]. A wealth of information is available for these species, which are considered representatives of the mountain-stream species of India with AO (Day, 1958; Jayaram, 1981).

4.3 Habit and Habitat of the Fishes

The sisorid catfishes *Pseudocheneis sulcatus* and *Glyptothorax pectinopterus* are bottom-dwellers, predominantly nocturnal in habit and carnivorous in nature. They prefer living in fast-flowing streams having rocky-gravel substratum and are reputed to swim against the strong water current by their well-expanded pectoral and pelvic fins (Fig. 4.1). The two cyprinids *Schizothorax richardsnoii* and *Garra gotyla gotyla* differ in their habitat preference. While the former lives in places where the velocity

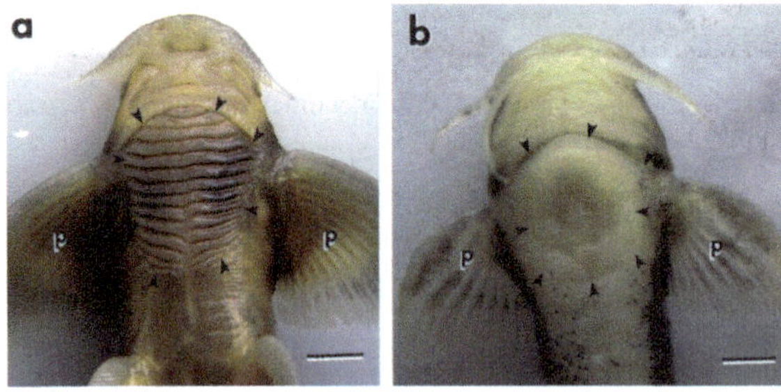

Fig. 4.1 AO (*arrowheads*) of the mountain-stream catfishes, *Pseudocheneis sulcatus* (**a**, standard length: 13.7 cm) and *Glyptothorax pectinopterus* (**b**, standard length: 14 cm) located between the base of the pectoral fins (p). Bars, 5 mm. [(Figure **a** is from Das and Nag (2005) and shown with permission of the Blackwell Science Publishers, Oxford)]

of the water is moderate, the latter, with its well-developed fins, is found to inhabit fast-flowing streams and adjacent torrential rivers (Hora, 1922). Both are diurnally oriented and predominantly herbivorous in food habit.

4.4 The AO of Mountain-Stream Catfishes

In both catfishes, the AO is disc-like in appearance and pale-yellowish in color in the live state. It is situated at the gular region of the body, immediately behind the opercular openings and between the bases of the pectoral fins. In *Pseudocheneis sulcatus*, the AO is nearly triangular in outline with the anterior portion being wider than the posterior portion (Fig. 4.1a). The integument is transformed prominently into alternate transverse ridges and grooves. The AO in *Glyptothorax pectinopterus* is nearly circular in outline and has a concave depression in the central area. The integument at the periphery of the AO is transformed into alternate ridges and grooves that appear to converge centrally at the circular depression (Fig. 4.1b).

SEM: Surface architecture of the AO of both species reveals the ridges to bear numerous elongated spines (Fig. 4.2), which are absent in the grooves. No mucous pores (outlets of mucous cells) are observed in the ridge or groove areas.

LM organization: Transverse sections of the AO show the two basic layers as seen in the skin, viz., the epidermis and the dermis. The epidermis is stratified into

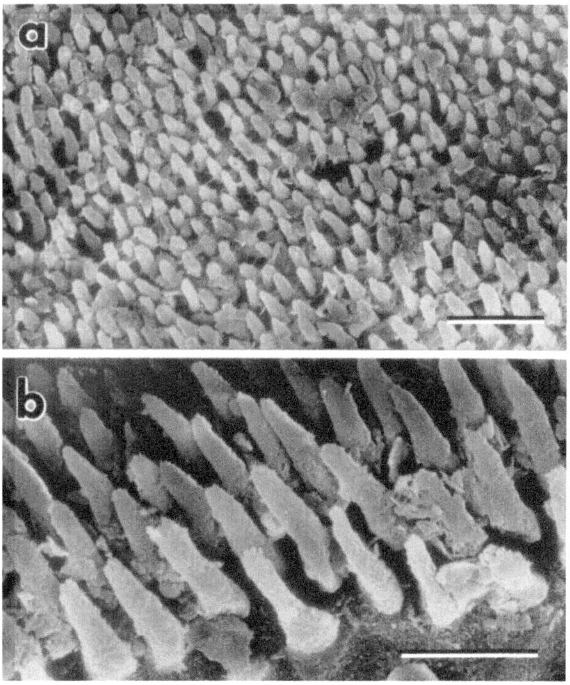

Fig. 4.2 (a, b). Scanning electron micrographs of the AO surface in stream-catfishes, showing spines of the ridges. Bars, 15 μm

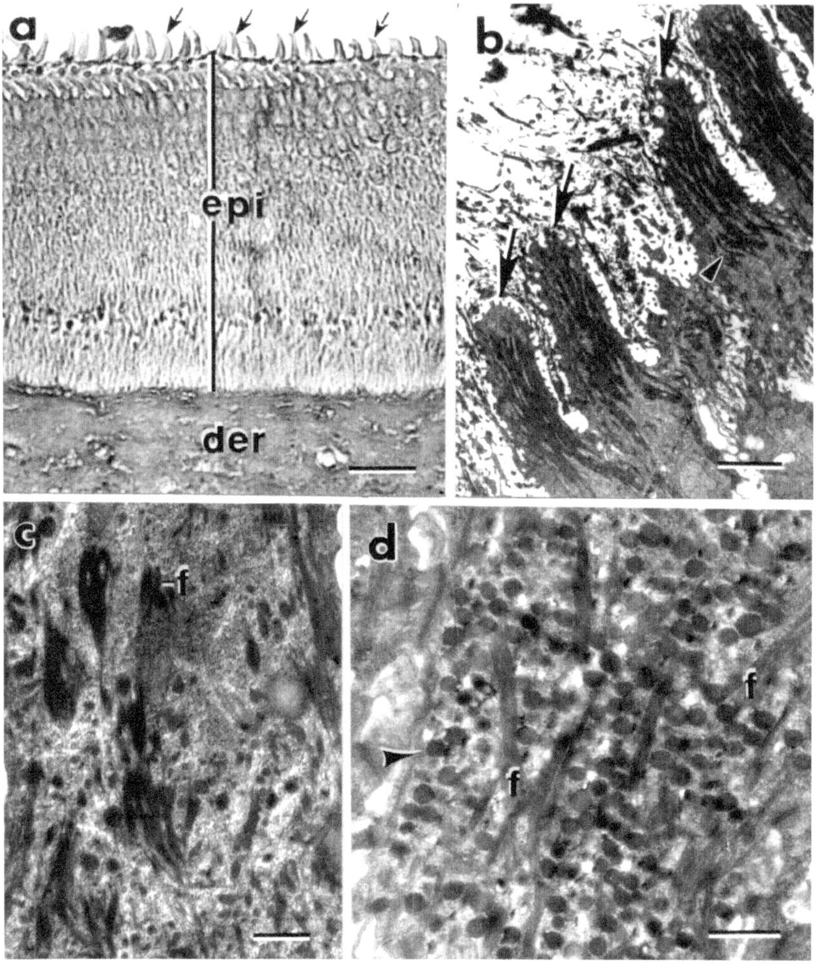

Fig. 4.3 (a) Light micrograph of AO of *Pseudocheneis sulcatus* (standard length: 16 cm; Haematoxylin-eosin stained), showing the thick epidermis (epi) and fibrous dermis (der). *Arrows* indicate spines. **(b–d)** Transmission electron micrographs of spines and filament cells. The spines (**b**, *arrows*) possess bundles of tonofilaments (*arrowhead*). The filament cells (**c, d**) are replete with bundles of clumped tonofilaments (f) and small mucous granules (*arrowhead*, in **d**). Bars, 30 μm **(a)**, 1 μm **(b–d)**

multiple rows of cells. It is relatively thicker (about 210 μm) at the ridges and made up of 18–20 rows of cells of similar appearance. The cells of the surface row project spines that curve near the extremities (Fig. 4.3a). The second row cells are elongated, and possess immature spine-like structures. The deeper part of the epidermis consists of closely packed, elongated cells. At the basal level, the cells are columnar in appearance. The epidermis of the ridges is devoid of mucous cells. The dermis is compact, and contains the usual connective tissue elements and nerve fibres. The epidermis in the adjacent pectoral region is thin (thickness: 50–60 μm) and 4–5 cell rows thick, containing mainly filament cells, occasional mucous cells and many large club cells. In the dermis, connective tissue fibres are loosely organized.

The AO epidermis at the grooves is also stratified in nature and consists of four to five rows of cells. Unlike the ridges, the outer row cells are devoid of spines. Mucous cells are absent also in this epidermal region. In *Glyptothorax pectinopterus*, the epidermal organisation at the circular depression of the AO is comparable to that of the groove area seen in *Pseudocheneis sulcatus*.

TEM: Sections of the AO epidermis reveal that the outer row of cells project spines containing bundles of dense, clumped tonofilaments (Fig. 4.3b). The spines are lined with a thick outer plasma membrane (Das and Nag, 2005). Their cytoplasm is compacted with similar bundles of tonofilaments and large mucous-like granules (Das and Nag, 2005), but clearly no organelles. The remainder of the epidermal cells possesses tonofilaments of a nature (Fig. 4.3c, d), similar to that in the spines. These cells also lack major organelles such as mitochondria, ribosomes and endoplasmic reticulum. They possess electron-dense mucous-like granules (diameter: 0.1–0.3 μm) that are often entangled with the tonofilaments (Fig. 4.3d). As described elsewhere in the literature (e.g., Parakkal and Matoltsy, 1964; Budtz and Larsen, 1975), they resemble the small mucous granules of teleost and amphibian epidermal cells. The cells in the deeper rows (sixth to tenth) possess relatively fewer filament bundles than in the upper row cells, but no mucous granules.

4.5 The AO of the Cyprinids

In both cyprinid species, the mouth is sub-ventral in position. The AO is located at the ventral head region, encircling the mouth opening. In *Schizothorax richardsnoii*, the epidermis of the AO is raised into prominent tubercles (Fig. 4.4a, b). The tubercles vary considerably in size; in the center of the tuberculated region, they measure about 180–200 μm in diameter, whereas in the periphery (3–4 mm away from the center), their size is reduced to 70–80 μm (standard length of fish: 12–15 cm, N= 20). In the center of the AO, there is a callus part (Fig. 4.4a) that is devoid of tubercles. The tubercles show signs of desquamation. There are no spines on the surface of the tubercles, nor in the callus part (see below).

The AO of *Garra gotyla gotyla* (Das and Nag, 2006) consists of a central callus part and an upper and a lower jaw sheath and resembles that of *Schizothorax richardsnoii*. SEM reveals the jaw sheaths to bear prominent tubercles, as shown for *Schizothorax richardsonii* (Fig. 4.4a). Each tubercle possesses about 20–25 curved spines (Fig. 4.4c). Tubercles are absent in the callus part of the AO.

4.5.1 Organisation of AO of *Schizothorax richardsnoii*

LM: In *Schizothorax richardsnoii*, the epidermis of the AO is raised into prominent tubercles (Fig. 4.5a, b) and consists of cells having a similar appearance. The cells are arranged in 15–20 rows. The dermal tissue often penetrates the large tubercles

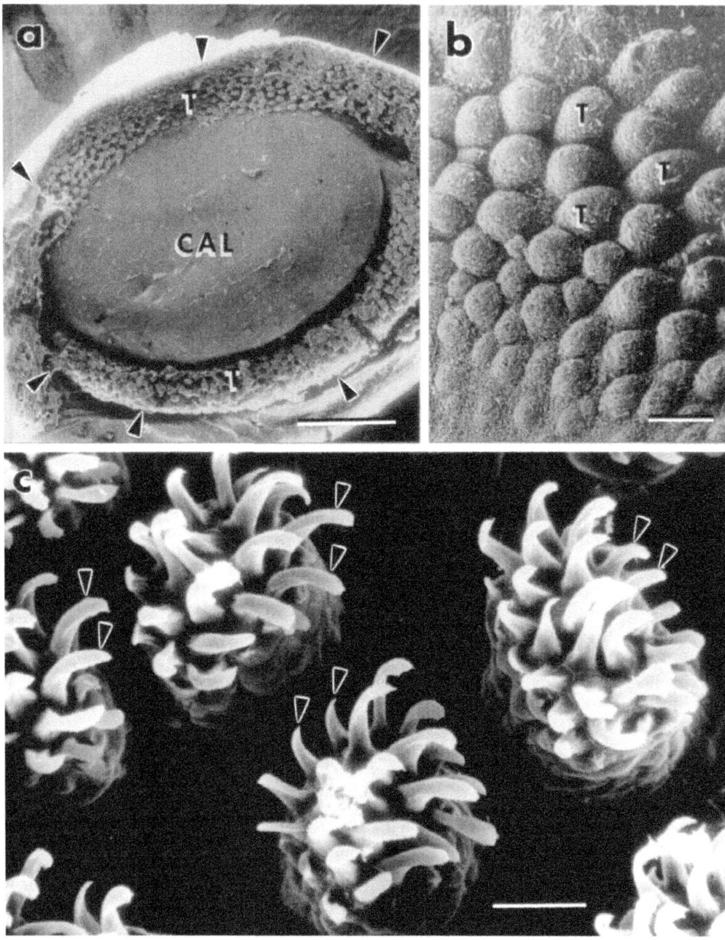

Fig. 4.4 (**a**) Scanning electron micrographs of the AO (*arrowheaded*) of *Schizothorax richardsonii* (standard length: 13 cm) showing its tuberculated parts (T) and central callus part (CAL). (**b**) Magnified view of tubercles (T). (**c**) From *Garra gotyla gotyla* showing spines (*arrowheads*) on the tubercles. Bars, 1 mm (**a**), 200 μm (**b**) and 20 μm (**c**)

(Fig. 4.5b), and mimics the dermal investment found inside the pulp cavity of mammalian teeth. There is evidence of keratinisation of the superficial epidermal rows. The thickness of the keratinised layer measured in seven fish specimens varied from 4 to 11 μm. The degree of keratinisation is, however, more marked in the callus part (Fig. 4.5c) than in the tuberculated epidermis. In the callus part, beneath the keratinised part, the epidermis has cells that are rich in filaments (Fig. 4.5d). There are neither mucous cells nor any other cell types present in this part of the AO. Dead cells, macrophages and mitoses are often encountered in the epidermis of the AO.

TEM observations: The outer 5–6 rows of cells of the AO epidermis show some degree of keratinisation. These cells, which appear as electron-dense sheets

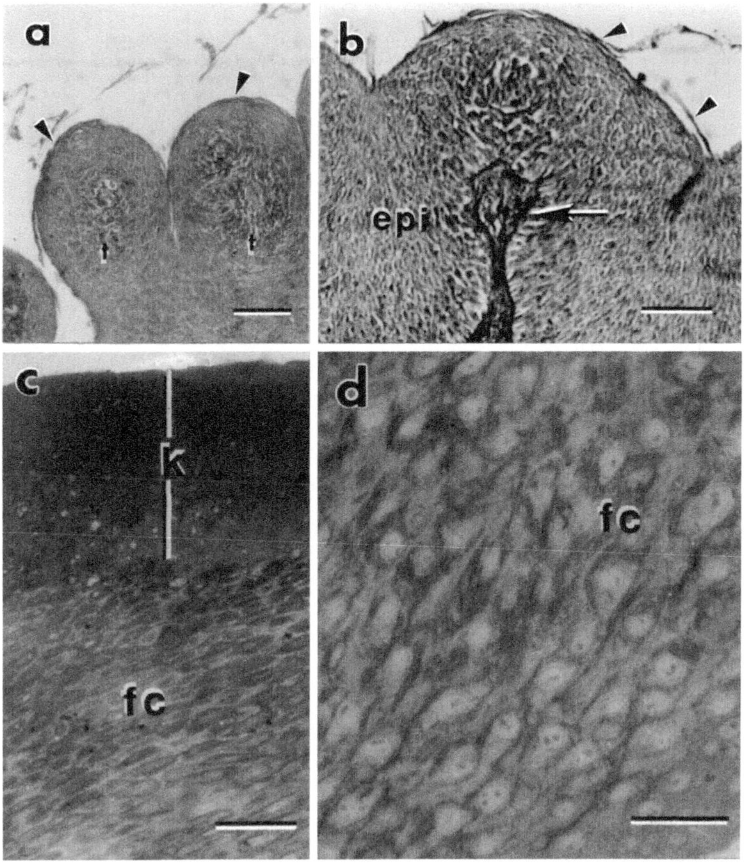

Fig. 4.5 (**a**) Light micrograph of AO of *Schizothorax richardsonii* (standard length: 15 cm) showing (**a**) epidermal tubercles (t). *Arrowheads* indicate keratinised layer. (**b**) Enlarged view of a tubercle, showing cells of similar appearance in the epidermis (epi). The tubercle is penetrated by dermal tissue (*arrow*). Arrowheads, keratinised layer. (**c, d**) Callus part of AO showing (**c**) outer keratinised layer (k) and (**d**) filament cells of epidermis (fc). Bars, 30 μm (**a, b**; haematoxylin and eosin-stained paraffin sections) and 10 μm (**c, d**; toluidine blue-stained resin sections)

(5–6 μm in thickness), contain tightly arranged bundles of tonofilaments. Desmosomal connections are present between the keratinised layers (Fig. 4.6a). No nuclei or organelles are seen therein, except some occasional lipid inclusions. Beneath this layer, the cells possess bundles of tonofilaments, which are more loosely arranged (Fig. 4.6b) than that in the uppermost rows. The cells lying immediately beneath the keratinised layers (in mid-epidermis) are somewhat oval in shape, possessing round euchromatic nuclei with 2–3 prominent nucleoli and tonofilament bundles that occupy the entire cytoplasm (Fig. 4.6c, d). The plasma membrane of the filament cells shows extensive finger-like processes. Bundles of tonofilaments often run

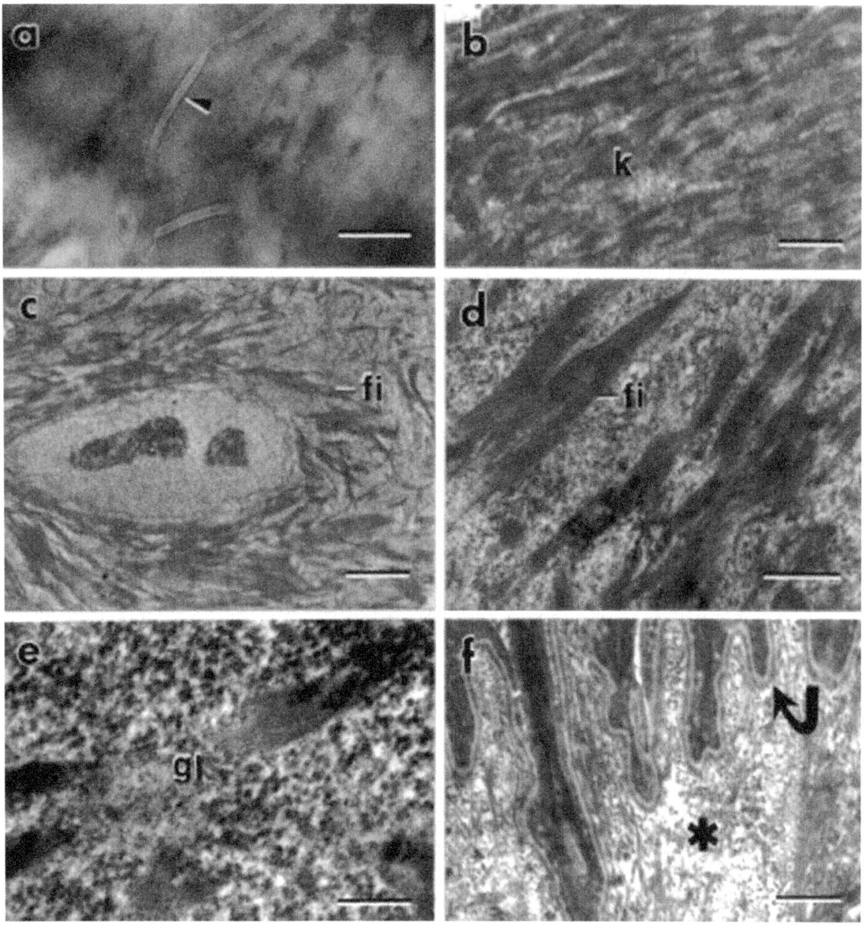

Fig. 4.6 Transmission electron micrographs of keratinised layers and filament cells of AO of *Schizothorax richardsonii*. **(a, b)** part of keratinised layers (k). Desmosomal connections (*arrowshead*, **a**) are present between the cytoplasmic processes. In **(b)** loose filament bundles are seen in the keratinised cytoplasm (k). **(c, d)** Filament cells with tonofilaments (fi). **(e)** Glycogen granules (gl) of filament cell. **(f)** Highly infolded basal lamina (*curved arrow*) of epidermis. Asterisk, dermis. Bars, 1 μm **(a–d)**, 0.5 μm **(e, f)**

through these processes. The latter appear to establish connections between the adjacent cells via desmosomes. A remarkable feature of the filament cells in this species is the absence of cytoplasmic mucous granules. Also, organelles therein are rarely encountered. Diffuse, glycogen granules are abundantly present in the cytoplasm of the filament cells. These granules often connect filament bundles (Fig. 4.6e). The distribution of glycogen granules is similar to that of the filaments: they are less abundant in the deeper cell rows lying close to the basal row. The basal lamina that separates the epidermis from the dermis shows deep infoldings (Fig. 4.6f). The dermis shows the usual connective tissue elements.

4.5.2 Organisation of AO of *Garra gotyla gotyla*

LM: As in *Schizothorax richardsonii*, in *Garra gotyla gotyla*, the epidermis of the jaw sheaths is raised into tubercles, creating grooves between them. It consists of 10–12 rows of cells of similar appearance (Fig. 4.7a). Spines are projected from the cells of the outermost row. A thin outer keratinised layer is present in the fully-differentiated tubercles of the epidermis (Das and Nag, 2006). The tuberculated epidermis is devoid of mucous cells and the groove regions between the tubercles are devoid of spines and mucous cells. The dermis possesses the usual connective tissue elements.

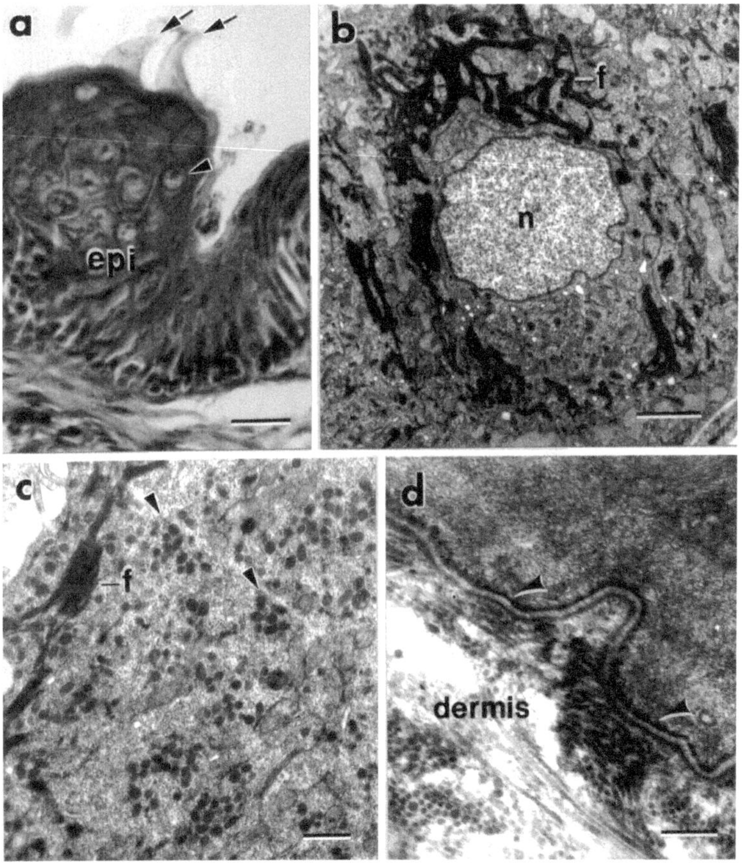

Fig. 4.7 **(a)** Light micrograph of AO of *Garra gotyla gotyla* (standard length: 9 cm). The epidermis (epi) possesses filament cells (*arrowhead*). Spines are present in outer row cells (*arrows*). **(b–d)** Transmission electron micrographs of filament cells and epidermis-dermis interface. **(b)** A filament cell with euchromatic nucleus (n) and cytoplasmic tonofilaments (f). **(c)** Mucous granules (*arrowheads*) of the filament cell. f, A bundle tonofilament. **(d)** Hemidesmosomes (*arrowheads*) of the basal lamina. Bars, 50 μm **(a)**, 1 μm **(b)**, 0.5 μm **(c)** and 0.25 μm **(d)**

The callus part of the AO shows the epidermal cells to be stratified into 10–15 rows of cells of similar appearance. Unlike *Schizothorax richardsonii*, there is no sign of keratinisation of the superficial layers. The epidermis consists of filament cells and large mucous cells.

TEM: The epidermal cells of the outermost row are apparently keratinised and bear spines that are lined with thick plasma membrane envelope and packed with clumped tonofilaments (Das and Nag, 2006), as shown for *Pseudocheneis sulcatus*. The cells of the remainder of the epidermis possess round euchromatic nuclei and thick discrete bundles of tonofilaments in their cytoplasm (Fig. 4.7b). Their cytoplasm possesses mitochondria, ribosomes and rough endoplasmic reticulum. Electron-dense granules, each measuring about 0.2–0.3 μm in diameter, are present in their cytoplasm (Fig. 4.7c). The cells in the deeper part of the epidermis show a few bundles of tonofilaments, and no mucous granules. The basal lamina of the epidermis shows the presence of hemidesmosomes (Fig. 4.7d). The dermis possesses the usual connective tissue elements, such as collagen fibrils, fibrocytes, blood vessels and nerve axons.

The epidermis of the callus part is lined with pavement cells, which are rich in fine disperse filaments. Internally, they are joined to the outer row filament cells by desmosomes. These cells possess abundant intermediate filaments in their cytoplasm and mucous granules (diameter: 0.1–0.3 μm). Large mucous cells are present amongst the epidermal cells in the callus part (Das and Nag, 2006). They are absent in the tuberculated regions of the AO, as in *Schizothorax richardsonii*.

In all four species examined, the epidermis of the pectoral region contains filament cells and large mucous cells. In addition, numerous lipid droplets are seen in the epidermal filament cells, especially in the catfishes. The surface of the filament cells bears prominent micro ridges (Das and Nag, 2006, *Garra gotyla gotyla*), which occur more frequently than those in the AO counterparts. Their cytoplasm is filled with disperse, fine filaments and organelles, but no mucous granules.

4.6 Immunohistochemical Observation

In the AO epidermis of the two catfishes examined, an anti-pan cytokeratin antibody mainly labels the outer row cells (Fig. 4.8a). The spines are weakly labeled, which is perhaps due to the poor penetration of the reagents through the thick plasma membrane. Actin immunoreactivity is moderately present in the cells of the inner rows, but is clearly absent from those of the outer row (Das and Nag, 2005). In *Garra gotyla gotyla*, the antibody labels the spines and cytoplasm of the outer row cells of the AO epidermis (Das and Nag, 2006). No significant immunoreactivity is found in the inner part of the epidermis. The filament cells of the callus part show diffuse labeling (not shown). In *Schizothorax richardsonii*, the antibody labels the outer keratinised layer of the AO epidermis. No reaction is seen in the remainder of the epidermis (Fig. 4.8b). No immunoreactivity is detected in the integument of the pectoral region in any of the four species examined (not shown).

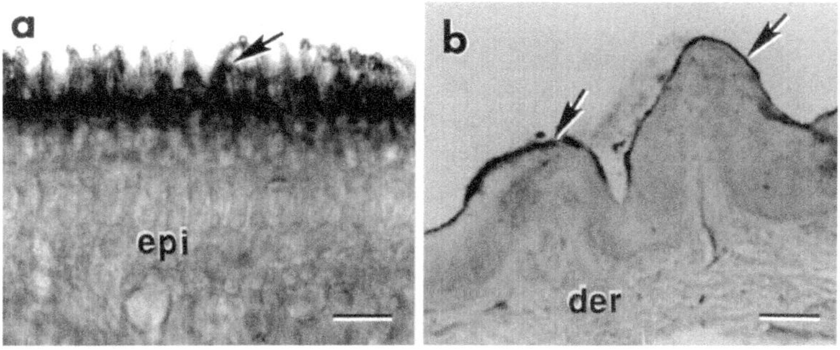

Fig. 4.8 Immunohistochemical demonstration of pan-cytokeratin (Sigma, USA; dilution 1: 100) in AO epidermis of *Pseudocheneis sulcatus* (**a**) and *Schizothorax. Richardsonii* (**b**). Cytokeratin is present in the outer tier cells with spines (*arrow*, **a**) and in superficial outer layer of epidermis (*arrows*, **b**). epi, epidermis; der, dermis. Avindin-biotin immunoperoxidase method. Bars, 30 μm

The use of additional cytokeratin antibodies may provide some information about a broad spectrum of keratins including the AE2 antibody, which is a specific marker of keratinised layers. AE1 antibody stains basal cells and no pre-keratinised and keratinised layers and AE3 antibody stains all epidermal layers.

4.7 Adhesive Devices on Paired Fins in Sisorids

The integument covering the outer rays' undersurfaces of paired pectoral and pelvic fins of the stream-catfishes *Pseudocheneis sulcatus*,*Glyptothorax pectinopterus* and *Bhavania annadalei* and *Glyptosternum labiatum* possess adhesive devices (Hora, 1922). In these species, these fins are unusually large, expanded, and fan-shaped in appearance. Instead of being situated on the undersurface of the body, they are pushed outward and placed horizontally on the sides of the body. The outer rays of these fins are lamellated, bearing transverse ridges and grooves. SEM reveals that in *Pseudocheneis sulcatus*, the outer epidermal cells at the ridges are modified into elongated spines. Mucous pores with entangled mucous materials are frequently encountered in the epidermis of the ridges. Spines as well as mucous pores are absent in the cells lining the groove. The inner rays of these fins are inconspicuous and do not show any such modifications for attachment (Das and Nag, 2004).

Generally speaking, in teleosts, the pectoral and pelvic fins are used for various functions, in addition to swimming and maintenance. In the flying- fish, *Exocoetus volitans*, the modified pectoral fins are used for gliding above the water's surface, and in the suckerfish, *Remora remora*, the first dorsal fin is utilized for adhesion (Migdalski and Fichter, 1983). In the sisorid catfishes, the pectoral- and pelvic fins are effectively used for adhesion, at rest, besides their common use during swimming (Hora, 1930). It is possible that the outer rays work on the principle of suction for adhesion.

4.8 Nature of Epidermal Modifications in AO

In some mountain-stream fishes of India, the epidermis of the AO shows prominent morphological and sub-cellular changes in order to perform adhesive functions. In all four species described here, it is noticeably thicker than seen elsewhere, e.g., in the adjacent pectoral region. This is due to a higher content of cells in the AO epidermis, being arranged compactly into multiple rows (10–25). LM revealed that only one type of cells (filament cells) is present in the AO epidermis. On the other hand, the pectoral epidermis contains a variety of cells (filament cells, mucous cells and club cells), all being arranged in 4–6 rows. Filament cells are the principal cell type in teleost epidermis (Henrickson and Matoltsy, 1968a; Hawkes, 1974; Leonard and Summers, 1976; Whitear, 1986; Abraham et al., 2001). Due to adaptation, an increase in the number of filament cells has taked place in the AO epidermis. Ultrastructural examinations have revealed that they are modified remarkably compared to the features observed in their skin counterparts. They possess an extensive cytoskeleton made up of bundles of clumped tonofilaments, whereas in their skin counterparts, the filaments (10 nm thick keratin fibrils) are dispersed and mainly occupy the peripheral part of the cytoplasm (Henrickson and Matoltsy, 1968a; Harris and Hunt, 1975). Their aggregation into discrete bundles, as seen in the AO filament cells is rare and species-specific.

The pectoral epidermis of the four species examined in this study possesses large mucous cells, which is a characteristic feature for teleost epidermis in general (Henrickson and Matoltsy, 1968b; Whitear, 1977, 1986; Abraham et al., 2001). A remarkable feature of the AO is the absence of mucous cells in its epidermis. In their absence, the filament cells of the mid- to upper epidermis are uniquely specialized for the synthesis of keratin and mucous granules. This is likely due to adaptation of the AO epidermis to accommodate an increased number of filament cells involved in adhesion. The main cytological features of the AO are compared in four species in Table 4.1.

Electron microscope studies have shown that mucous granules are characteristically found in the epidermis of amphibians and reptiles (Parakkal and Matoltsy, 1964;

Table 4.1 Comparison of various features of AO in four species

		Features of filament cells				
Name of species	Organisation	Spines	Keratinisation of outer row cells	Tonofilament bundles	Mucous granules	Callus part
Pseudocheneis sulcatus	Epidermis with ridges	☼☼☼	***	***	***	–
Glyptothorax pectinopterus	Epidermis with ridges	☼☼☼	***	***	***	–
Schizothorax richardsonii	Epidermis with tubercles	–	**	***	–	+[a]
Garra gotyla gotyla	Epidermis with tubercles	☼	**	**	**	+[b]

+ present; – absent; ☼ few; ☼☼☼ numerous; ** moderate amount; *** high amount; [a] epidermis without mucous cells; [b] epidermis with mucous cells

Matoltsy and Huszar, 1972; Whitear, 1977; Fox, 1986; Alibardi, 2001). On the other hand, in birds and mammals, epidermal cells (keratinocytes) with mucous granules are generally not formed (Matoltsy, 1986; Menon and Menon, 2000).

4.9 Keratinisation of the Outer Rows of Cells in AO

Keratinisation is common in terrestrial vertebrates, such as reptiles, birds and mammals (Spearman, 1977). Although limited in teleosts, this process is however, reported to occur in isolated groups (Aspredinidae; Friel, 1989) and species, viz. the feather-back, *Notopterus notopterus* (Mittal and Banerjee, 1974), the sisorid catfish *Bagarius bagarius* (Mittal and Whitear, 1979), the cyprinid *Garra lamta* (Pinky et al., 2002) and the Australian lungfish, *Neoceratodus forsteri* (Alibardi, 2001). Electron microscope and histochemical examinations revealed that certain regions of the epidermis in those teleosts show keratin formation. Additionally, Mittal and Whitear (1979) found no succinic dehydrogenase activity in the keratinised layer in *Bagarius bagarius*, which is consistent with the absence of mitochondria in the outer keratinised layers of AO epidermis seen by us. In aspredinid catfishes, the skin is completely keratinised and covered with tubercles (Friel, 1989). These findings show that the whole or specific regions of the teleost epidermis can undergo to a process of keratinisation. In the AO of four species examined, the outer epidermal cells, which are rich in tonofilaments, have the potential for undergoing keratinisation. By In immunohistochemistry, these filaments are found to be cytokeratin positive. In three of the species, spines are differentiated from the outer row cells of the AO epidermis. The mature spines possess clumped tonofilaments and are lined with a thick plasma membrane envelope. The matrix substances that glue the filaments are at present unknown; it is presumed that they may be derived from the large mucous-like granules present in the outer row cells (Das and Nag, 2005). These granules are believed to be involved in keratinisation of the fish and amphibian epidermis (Budtz and Larsen, 1975; Fox, 1986; Whitear, 1986). Resing and Dale (1991) reported that in mammals, proteins like filaggrins are involved in aggregation of keratin into dense tonofilaments.

By immunohistochemistry, previous workers have demonstrated cytokeratin localisation in teleost epidermis (Bunton, 1993; Ainis et al., 1995; Alibardi, 2001; Martorana et al., 2001; Alibardi and Joss, 2003). These results indicate a complex pattern of cytokeratin expression in teleost epidermis. Under sub-teleost taxons, cytokeratins (subtypes 7, 8, 18 and 19) are present in the skin and horny teeth of the adult sea lamprey, *Petromyzon marinus* (Zaccone et al., 1995). In zebrafish (*Danio rerio*), cytokeratin 8 is expressed in the epidermis during regeneration of the caudal fin (Martorana et al., 2001). The exact cytokeratin types present in the AO epidermis of the four species examined remain to be seen. Markl et al. (1989) and Conrad et al. (1998) reported that the teleost epidermis in general lacks basic keratin.

An inverse relationship appears to exist between the degree of keratinisation and the abundance of actin. In the AO of *Pseudocheneis sulcatus*, we found no actin immunoreactivity in the epidermal outer row cells, which are keratinised (Das and Nag, 2005). This indicates that keratinisation of the outer row cells occurs in the absence of actin filaments, supporting the view of Budtz and Larsen (1975). Similarly, in regenerated teleost fin rays, keratin upregulation is associated with actin downregulation during cell differentiation (Santos-Ruiz et al., 2005).

The epidermis of birds and mammals shows formation of keratohyalin granules, which are generally considered a major event of keratinisation (Matoltsy, 1969, 1986; Lavker and Matoltsy, 1970). It has been reported that these granules are not formed in teleost and amphibian epidermis (Parakkal and Matoltsy, 1964; Parakkal and Alexander, 1972; Lavker, 1973, 1974). Although cytokeratin filaments are present in cells of the AO, the absence of keratohyalin granules therein would indicate that they are not comparable to the keratinocytes of the avian and mammalian epidermis.

In teleosts, the bulk of the mucous substances is produced by goblet cells, whereas the upper epidermal cell layer also contributes to such secretion, forming the glycocalyx. Besides adhesive structures described in this report, specialisations of the glycocalyx of the superficial epithelial cells in certain teleosts are reported. The flame cell caps of sea horse, the products of large spindle cells of the free pectoral fins of gurnards and stonefish, and the thick cuticle of sucking disc of the clingfish *Lepadogaster candolei* are regarded different types of adhesive structures due to specialisations of the glycocalyx of the superficial epidermal cells (see review in Zaccone et al., 2001). Similar to the observation in *Lepadogaster candolei*, in another clingfish *Gobiesox maendricus*, an extracellular cuticle covers the epidermal papillae (Whitear, 1986).

4.10 Mechanism of Adhesion

Based upon the structural organization of the AO, Hora (1922) commented that this device in the sisorid catfishes functions by way of suction for adhesion. Upon stimulus, the dermal muscles attached to the ridges and grooves of the AO are contracted. Probably this process results in the creation of some vacuum, allowing the spines of the outer epidermal cells to press against the substratum. The spines then assist in anchorage to the organic growth on the submerged rocks. Mucous secretion from the AO causes a weak adhesion and prepares the substratum for the action of the spines. In *Garra gotyla gotyla,* the epidermal tubercles are provided with spines, and mucous secretion from the callus part may afford protection to the spines from abrasion during adhesion.

Unlike the sisorid catfishes, the cyprinids possess a callus part in their AO. A differential mechanism of adhesion is likely to be operative in the cyprinids due to the presence of this specialised part in their AO. It is possible that the callus part plays a significant role in the process of adhesion in the cyprinids. The callus part is drawn and a cavity is produced that is surrounded by the tuberculated borders. Adhesion results due to friction between tubercles and the surface of the substratum (in *Schizothorax richardsonii*). As mucous cells are absent in the callus part of *Schizothorax richardsonii*, it appears that the process of adhesion in this species is mucous-independent, which is contrary to the mucous-dependent functioning of the callus part in *Garra gotyla gotyla*. The keratinisation of the epidermal spines (in sisorids and the cyprinid *Garra gotyla gotyla*) and the outer row cells (in all four species) are perhaps involved in protecting the AO surface against mechanical damage during adhesion.

Teleost skin shows remarkable adaptation and also plasticity in response to diverse habit and habitat. In the AO, LM and SEM observations suggest that the outer row epidermal cells (in *Schizothorax richardsonii*) as well as the spines (in the other three species) are desquamated and replaced upon damage. In the sisorid AO, there is evidence that the second row epidermal cells containing immature spines replace the old/worn-out spines of the outer row cells. Being a usual phenomenon in amphibian and reptilian skin (Parakkal and Alexander, 1972; Fox, 1986), this sort of shedding of the epidermal cells is also known for some teleosts (Friel, 1989), especially when exposed to stressors (reviewed in Wendelaar Bonga, 1997).

Various types of epidermal outgrowths are reported to occur in the teleosts. The formation of breeding tubercles in many fishes (Collette, 1977), the hooks of the male nurseryfish, *Kurtus gulliveri* (Berra and Humphrey, 2002), and the highly vascularised cotylephores of the brooding female ghost pipefish (*Solenostomus* sp.) and the catfish, *Platystacus* sp. (Wetzel et al., 1997) represent such epidermal outgrowths, which are used for courtship and/or for attachment of eggs. In hagfish, stressed individuals release two types of cells, gland thread cells and gland mucous cells that rupture on contact with water. This results in the formation of a mass of viscous mucus (Koch et al., 1991). In both cell types, keratin like components have been demonstrated by immunohistochemistry. It has been suggested that the gland thread cells interact with the mucosubstances released from the gland mucous cells and thereby form a viscous slimy substance in water (Spitzer and Koch, 1998). All these examples are of importance in understanding the comparative morphology and physiology of the teleost integument and its plasticity in different species. In this study, the epidermal modifications seen with the AO of some teleosts are the result of unique evolutionary adaptations for anchorage in fast-flowing mountain-streams and rivers.

The subject of this review is a virgin field for study. Both metabolic and functional studies including the use of a panel of cytokeratin antibodies and the determination of amino acid analysis of AO cytokeratins await further investigation. These results may provide newest information on cytokeratin evolution and the cell type-related cytokeratin functions.

Acknowledgments The authors sincerely thank Mr Sunil Pradhan, Fisheries Development officer, Fisheries Office, Jorethang (Government of Sikkim, India), for his assistance in fish capture. The electron microscope work was carried out at the Sophisticated Analytical Instrumentation Facility (DST), Department of Anatomy, All India Institute of Medical Sciences, New Delhi, India. Financial assistance in part was received from The University Grants Commission, eastern regional chapter, Kolkata (PSW 039/01-02, DD).

References

Abraham, M., Iger, Y., and Zhang, L. (2001) Fine structure of the skin cells of a stenohaline freshwater fish *Cyprinus carpio* exposed to diluted seawater. *Tissue and Cell* 33: 46–54.

Ainis, L., Tagliafierro, G., Mauceri, A., Licata, A., Ricca, M.B., and Fasulo, S. (1995) Cytokeratin type distribution in the skin and gill epithelia of the Indian freshwater catfish, *Heteropneustes fossilis* as detected by immunohistochemistry. *Folia Histochemica et Cytobiologica* 33: 77–81.

Alibardi, L. (2001) Keratinization in the epidermis of amphibians and the lungfish: comparison with amniote keratinization. *Tissue and Cell* 33: 439–449.

Alibardi, L., and Joss, J.M. (2003) Keratinization of the epidermis of the Australian lungfish *Neoceratodus forsteri* (dipnoi). *Journal of Morphology* 256: 13–22.

Berra, T.M., and Humphrey, J.D. (2002) Gross anatomy and histology of the hook and skin of forehead brooding male nurseryfish, *Kurtus gulliveri*, from Northern Australia. *Environmental Biology of Fishes* 65: 263–270.

Budtz, P.E., and Larsen, L.O. (1975) Structure of the toad epidermis during the moulting cycle. II. Electron microscopic observations on *Bufo bufo* (L.). *Cell and Tissue Research* 159: 459–483.

Bunton, T.E. (1993) The immunocytochemistry of cytokeratin in fish tissues. *Veterinary Pathology* 30: 418–425.

Collette, B.B. (1977) Epidermal breeding tubercles and bony contact organs in fishes. *Symposium of the Zoological Society of London* 39: 225–268.

Conrad, M., Lemb, K., Schubert, T., and Markl, J. (1998) Biochemical identification and tissue-specific expression patterns of keratins in the zebrafish *Danio rerio*. *Cell and Tissue Research* 239: 195–205.

Das, D., and Nag, T.C. (2004) Adhesion by paired pectoral and pelvic fins in a mountain-stream catfish, *Pseudocheneis sulcatus* (Sisoridae): A scanning electron microscope study. *Environmental Biology of Fishes* 71: 1–5.

Das, D., and Nag, T.C. (2005) Structure of adhesion organ of the mountain-stream catfish, *Pseudocheneis sulcatus* (Teleostei: Sisoridae). *Acta Zoologica (Stockholm)* 86: 231–237.

Das, D., and Nag, T.C. (2006) Fine structure of the organ of attachment of the teleost, *Garra gotyla gotyla (Ham)*. *Zoology* 109: 300–309.

Day, F. Sir (1958) *The fishes of India*. Vol. 1, London: William Dawson.

Fox, H. (1986) The skin of amphibians-epidermis. In: *Biology of the Integument*, Vol 2, Vertebrates, ed. by Bereiter-Hahn, J., Matoltsy, A.G. and Sylvia-Richards, K.S., Berlin: Springer-Verlag, pp. 78–110.

Friel, J. (1989) Epidermal keratinization and moulting in the banjo catfishes (Siluriformes: Aspredinidae). *Abstract, Annual Meeting of the American Society of Icthyologists and Herpetologists*, San Francisco, p. 89.

Hamilton, F.B. (1822) *An Account of Fishes Found in the River Ganges and its Branches*. London: Edinburgh.

Harris, J.E., and Hunt, S. (1975) The fine structure of the epidermis of two species of salmonid fish, the Atlantic salmon (*Salmo salar* L.) and the brown trout (*Salmo trutta* L.). I. General organization and filament-containing cells. *Cell and Tissue Rsearch* 157: 553–565.

Hawkes, J.W. (1974) The structure of fish skin. I. General organization. *Cell and Tissue Research* 149: 147–158.

Henrickson, R.C., and Matoltsy, A.G. (1968a) The fine structure of teleost epidermis. I. Introduction to filament cells. *Journal of the Ultrastructural Research* 21: 201–212.

Henrickson, R.C., and Matoltsy, A.G. (1968b) The fine structure of teleost epidermis. II. Mucous cells. *Journal of the Ultrastructural Research* 21: 213–221.

Hora, S.L. (1922) Structural modifications of fishes of mountain torrents. *Records of the Indian Museum* XXIV: 46–58.

Hora, S.L. (1930) Ecology, bionomics and evolution of torrential fauna with special reference to the organs of attachments. *Philosophical Transaction Of the Royal Society London B* 218: 171–282.

Jayaram, K.C. (1981) *The freshwater fishes of India. A handbook.* Calcutta: Zoological Survey of India. 475pp.

Johal, M.S., and Rawal, Y.K. (2003) Mechanism of adhesion in a hillstream fish, *Glyptothorax garhwali* Tilak, as revealed by scanning electron microscopy of adhesive appárartus. *Current Science* 85: 1273–1275.

Koch, E.A., Spitzer, R.H., and Pithavalla, R.B. (1991) Structural forms and possible roles of aligned cytoskeletal biopolymers in hagfish (slime eel) mucus. *Structural Biology* 106: 205–210.

Lavker, R.M. (1973) A highly ordered structure in the frog epidermis. *Journal Of the Ultrastructural Research* 45: 223–230.

Lavker, R.M. (1974) Horny cell formation in the epidermis of *Rana pipiens*. *Journal of Morphology* 142: 365–378.

Lavker, R.M., and Matoltsy, A.G. (1970) Formation of horny cells: the fate of cell organelles and differentiation products in ruminal epithelium. *Journal of Cell Biology* 44: 501–512.

Leonard, J.B., and Summers, R.G. (1976) The ultrastructure of the integument of the American eel, *Anguilla anguilla. Cell and Tissue Research* 171: 1–30.

Markl, J., Winter, S. and Franke, W. (1989) The catalog and the expression complexity of cytokeratins in a lower vertebrate: biochemical identification of cytokeratins in a teleost fish, the rainbow trout. *European Journal of Cell Biology* 50: 1–16.

Martorana, M.L., Tawk, M., Lapointe, T., Barre, N., Imboden, M., Joulie, C., Geraudie, J., and Vriz, S. (2001) Zebrafish keratin 8 is expressed at high levels in the epidermis of regenerating caudal fin. *Internal Journal of Developmental Biology* 45: 449–452.

Matoltsy, A.G. (1969) Keratinization of the avian epidermis. An electron microscope study of the newborn chick skin. *Journal of the Ultrastructural Research* 29: 438–458.

Matoltsy, A.G. (1986) The skin of mammals. Structure and function of the mammalian epidermis. In: *Biology of the Integument*, vol 2, Vertebrates, ed. by Bereiter-Hahn, J., Matoltsy, A.G. and Sylvia-Richards, K.S., Berlin: Springer-Verlag, pp. 255–271.

Matoltsy, A.G., and Huszar, T. (1972) Keratinization of the reptilian epidermis: An ultrastructural study of the turtle skin. *Journal of the Ultrastructural Research* 38: 87–101.

Menon, G.K., and Menon, J. (2000) Avian epidermal lipids: Functional considerations and relationship to feathering. *American Zoologist* 40: 540–552.

Migdalski, E.C., and Fichter, G.S. (1983) *The Fresh Water and Saltwater Fishes of the World.* New York: Greenwich House.

Mittal, A.K., and Banerjee, T.K. (1974) Structure of keratinization of the skin of a freshwater teleost *Notopterus notopterus* (Notopteridae, Pisces). *Journal of Zoology (London)* 174: 341–355.

Mittal, A.K., and Whitear, M. (1979) Keratinization of the fish skin with special reference to the catfish *Bagarius bagarius. Cell and Tissue Research* 202: 213–230.

Ojha, J., and Singh, S.K. (1992) Functional morphology of the anchorage system and food scrapers of the hill stream fish, *Garra lamta* (Ham.) (Cyprinidae, Cypriniformes). *Journal of Fish Biology* 41: 159–161.

Parakkal, P.F. and Matoltsy, A.G. (1964) A study of the fine structure of the epidermis of *Rana pipiens. Journal of Cell Biology* 20: 85–94.

Parakkal, P.F., and Alexander, N.J. (1972) *Keratinization. A survey of vertebrate epithelia.* New York: Academic Press.

Pinky, Mitta, S., Ojha, J., and Mittal, A. (2002) Scanning electron microscopic study of the structures associated with lips of an Indian hill-stream fish *Garra lamta* (Cyprinidae, Cypriniformes). *European Journal of Morphology* 40:161–169.

Resing, K.A., and Dale, B.A. (1991) Proteins of keratohyalin. In: *Physiology, Biochemistry and Molecular Biology of the Skin*, ed. by Goldsmith, L.A. 2nd edition, New York: Oxford University Press, pp. 148–167.

Santos-Ruiz, L., Santamaria, J.A. and Becerra, J. (2005) Cytoskeletal dynamics of the teleostean fin ray during fin epimorphic regeneration. *Differentiation* 73: 175–187.

Saxena, S.C., and Chandy, S. (1966) Adhesive apparatus in certain hillstream fishes. *Journal of Zoology* 148: 315–340.
Singh, N., and Agarwal, N.K. (1991) SEM surface structure of the adhesive organ of the hillstream fish *Glyptothorax pectinopterus* (Teleostei: Sisoridae) from the Garhwal Hills. *Functional and Developmental Morphology* 1: 11–13.
Singh, N., and Agarwal, N.K. (1993) Organs of adhesion in four hillstream fishes. A comparative morphological study. In: *Advances in Limnology*, ed. by Singh, H.R. New Delhi: Narendra Publishing House, pp. 311–316.
Sinha, A.K., Singh, I., and Singh, B.R. (1990) The morphology of the adhesive organs of the sisorid fish *Glyptothorax pectinopterus. Japanese Journal of Icthyology* 36: 427–431.
Spearman, R.I.C. (1977) Keratins and keratinization. *Symposiumof the Zoological Society of London* 39: 335–352.
Spitzer, R.H., and Koch, E.A. (1998) Hagfish skin and slime glands. In: *Biology of hagfishes*, ed by Jorgenson, J.M., Lomholt, J.P., Weber, R.E., and Malte, H. London: Chapman and Hall, pp. 109–132.
Wendelaar Bonga, S.E. (1997) The stress response in fish. *Physiological Review* 77: 591–625.
Wetzel, J., Wourms, J.P., and Friel, J. (1997) Comparative morphology of cotylephores in *Platystacus* and *Solenostomus*: modifications of the integument for egg attachment in skin-brooding fishes. *Environmental Biology of Fishes* 50: 13–25.
Whitear, M. (1977) A functional comparison between the epidermis of fish and of amphibians. *Symposium of the Zoological Society of London* 39: 291–313.
Whitear, M. (1986) The skin of fishes including cyclostomes. In: *Biology of the Integument*, vol 2, Vertebrates, ed. by Bereiter-Hahn, J., Matoltsy, A.G. and Sylvia-Richards, K.S., Berlin: Springer-Verlag, pp. 8–38.
Zaccone, G., Howie, A.J., Mauceri, A., Fasulo, S., Cascio, P., and Youson, J.H. (1995) Distribution patterns of cytokeratins in epidermis and horny teeth of the adult sea lamprey, *Petromyzon marinus. Folia Histochemica et Cytochemica* 33: 69–75.
Zaccone, G., Kapoor, B.G., Fasulo, S., and Ainis, L. (2001) Structural, histochemical and functional aspects of the epidermis of fishes. *Advances in Marine Biology* 40: 255–348.

Chapter 5
Surface Characteristics of Locomotor Substrata and Their Relationship to Gekkonid Adhesion: A Case Study of *Rhoptropus* cf *biporosus*

Megan Johnson, Anthony Russell* and Sonia Delannoy

5.1 The Gekkonid Adhesive System

Many taxa of gekkonid lizards bear adhesive toe pads encompassing a hierarchy of structure from the setae that contact the substratum (Autumn, 2006), through a tight epidermal-dermal junction (Russell, 1986) in the plantar regions of the exposed portions of the scansors (highly modified scales) (Russell, 1981) to an elaborate lateral digital tendon system (Russell, 1986) that links to muscular and skeletal systems through a complex of aponeurotic connections (Russell, 1981, 1986) (Fig. 5.1). There are common aspects to the overall assembly of the adhesive system (Russell, 1976) and various phylogenetic lineages have elaborated such a system independently (Russell, 1976, 1979). In some cases the system incorporates specialized elaborations of soft-anatomy that enhance compliance with the locomotor substratum (Russell, 1981; Vanhooydonck et al., 2005). Such adhesive systems have been borne by gecko lineages for at least the last 54 million years (Bauer et al., 2005), indicating that the integrative mechanical configuration that enhances the performance of the adhesive structures (Gay, 2002) is a fundamental component of setal deployment in gekkonids.

The entirety of this adhesive system is involved in controlling the contact and configuration of the subdigital setae (Ruibal and Ernst, 1965) as a field of fibrillar microstructures (Spolenak et al., 2005), to effect both attachment and release (Russell, 2002). The setae themselves may be elaborately and profusely branched, and their form varies by location on the subdigital pad (Delannoy, 2006) (Fig. 5.2). These epidermal outgrowths have received the majority of attention in studies of the adhesive system, as it is these structures that operate at the organism-environment interface to effect the very close contact necessary for adhesion (Yurdumakan

*A. Russell (✉)
Department of Biological Sciences, University of Calgary, Calgary, Canada
e-mail: arussell@ucalgary.ca

S.N. Gorb (ed.), *Functional Surfaces in Biology*, Vol. 2,
DOI 10.1007/978-1-4020-6695-5_6,

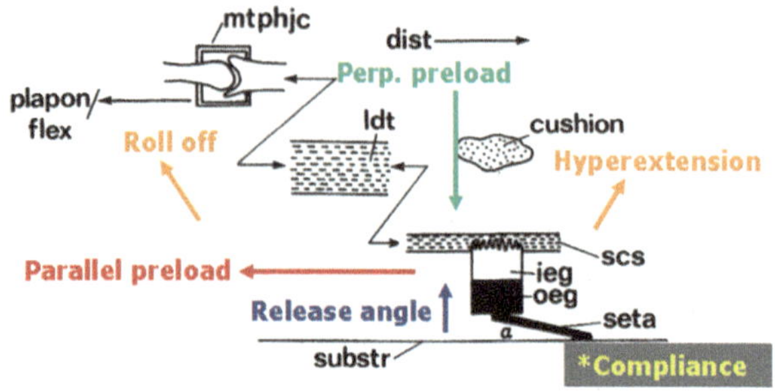

Fig. 5.1 Diagrammatic representation of the relationships between the seta/substrate contact pattern, the mechanical linkages from seta to skeleton and the application of forces bringing about and releasing the adhesive grip. The setal stalk changes its angle with respect to the substratum and ventral surface of the foot as substrate contact is made and released. At contact, the stalk angle (α) is depressed below that critical for maximal contact to be made. Perpendicular preloading (Perp. Preload) is applied by unfurling of the digit from its hyperextended position distally, or by rolling onto the more proximal scansors proximally, and is assisted by a distal cushion (such as a vascular sinus network). The parallel preload is applied and maintained by a combination of active loading through the tensile skeleton of the digit, mediated by flexor muscles acting through the plantar aponeurosis, and by gravitational loading (on vertical surfaces). Detachment is facilitated by hyperextension distally and roll off proximally, both of which raise the setal shaft above the critical angle threshold. The parallel and perpendicular preloads are diminished as a result of pressure changes and unloading of the tensile skeleton. Abbreviations: α, angle of setal stalk with substratum; dist., distal; ieg, inner epidermal generation; ldt, lateral digital tendon; mtphjc, metatarsophalangeal joint capsule; oeg, outer epidermal generation; plapon/flex, connection to the plantar aponeurosis and crural flexor muscles; scs, stratum compactum of the dermis of the scansor; substr, substratum (modified from Russell, 2002)

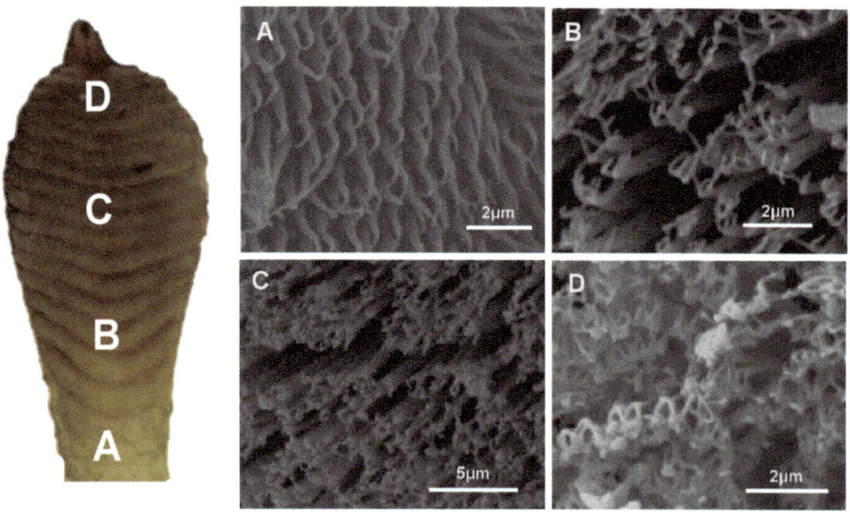

Fig. 5.2 Variation in setal size and complexity from the proximal (A) to the distal end of the digit (D) of the Tokay gecko (*Gekko gecko*). **A** curved spines carried on the basal scales of the digit **B** branched prongs carried on proximal lamellae **C** setae carried on proximally located scansors **D** setae carried on the distalmost scansors

et al., 2005; Autumn, 2006). The setae of gekkonids are themselves elaborations of the spinulate outer epidermal generation that was initially involved in facilitating skin shedding (Maderson, 1970). In the subdigital regions these outgrowths became modified for adhesion via hypertrophy and were probably initially involved in other frictionally-based, tractive interactions with the substratum (Delannoy, 2006).

5.1.1 Van der Waals Interactions

The functioning of the adhesive system at the organism-environment interface has been demonstrated to employ van der Waals forces between the setal tips and the substratum (Hiller, 1968, 1969; Autumn, 2002a, b), although capillary forces (Yamamoto et al., 1995) likely enhance the strength of the adhesive bond in certain situations (Huber et al., 2005a). Primarily, van der Waals interactions are highly dependent upon the amount of contact area that can be established, and the diminution of the separation distance between the contacting surfaces (Rimai and Quesnel, 2001). Secondarily there is a dependence upon the polarizability of the surfaces, whereby the strength of the bond increases with increasing polarizability, although this is not related to surface polarity (Rimai and Quesnel, 2001). Such characteristics render the seta-based adhesive system of geckos capable of adhesion to almost any surface.

5.1.2 Surface Characteristics and Adhesive Capabilities

Authors investigating the mechanisms involved in effecting adhesion in geckos have employed a variety of artificial surfaces (Hiller, 1968, 1969; Irschick et al., 1996; Autumn, 2002a, b; Vanhooydonck et al., 2005). Initially these were smooth surfaces such as glass and polyethylene film (Hiller, 1968, 1969; Ircshick et al., 1996; Autumn et al., 2000), that enabled the strength of the bond established by intact animals (Hiller, 1968, 1969; Irschick et al., 1996; Bergmann and Irschick, 2005) or individual setae (Autumn et al., 2000; Peressadko and Gorb, 2004; Sun et al., 2005) to be measured through force readings. Subsequently, the roughness of surfaces has been considered in the context of how it might affect adhesive capability. Such roughness has been considered across a spectrum of asperity sizes, from very fine-grained (Campolo et al., 2003; Persson, 2003; Persson, 2003; Meine et al., 2004; Huber et al., in press) to coarse (Zani, 2000; Vanhooydonck et al., 2005). At coarse levels of roughness (which should realistically be related to surface topology – see below), Vanhooydonck et al. (2005) demonstrated that clinging performance and acceleration are both diminished by gap-filled substrata, but their investigation did not relate setal structure or setal field patterning to their assessment of roughness (which consisted of structurally different materials with differing three-dimensional configurations that were only two-dimensionally characterized).

In general, the gekkonid adhesive system has been posited to have evolved to assist in the scaling of smooth surfaces (Autumn et al., 2000; Arzt et al., 2003), although no empirical evidence is available to suggest that this is the case. Indeed, observations on and characterization of locomotor surfaces employed in nature by gekkonid lizards are remarkably sparse (Gasc and Renous, 1980; Gasc et al., 1982) and refer only to a leaf-litter dwelling, very small-bodied sphaerodactyl gekkonid (*Coleodactylus amazonicus*) not known for its climbing capabilities. If the gekkonid adhesive system is routinely involved with the employment of rough surfaces with undulant topology (see below), which seems likely, then our conceptualization of the dimensions and structure of setal fields may have to be reassessed to permit a refined understanding of the scaling of available setal tip area to body size and mass. It is known that adhesion is influenced by dimensional scales from the atomic to the micro level (Chow, 2003), and the impact of this upon the hierarchical structure of the gekkonid adhesive system can be explored via the investigation of surfaces that are exploited in nature.

The addition of studying adhesion on naturally occurring substrata to the study of adhesive capabilities on artificial substrata has the potential to return the investigation of gekkonid adhesion to the evolutionary theatre, and to extend our investigations beyond the realm of function and into the arena of biological role (Bock and Von Wahlert, 1965).

Maderson (1970) raised some questions about setae which we herein begin to address. Noting that many geckos climb, but lack adhesive setae, Maderson (1970) asked (i) what precise locomotory advantage do setae provide that cannot be adequately met by claws or digital flexure, and (ii) is there any evidence that some or all seta-bearing taxa live on surfaces that have any particular physical characteristics that make setae particularly advantageous. These are complex and multifaceted questions that require an appropriately structured plan of attack.

5.1.3 Adhesive Performance and Its Relationship to Maderson's Questions

The apparent "overbuilding" of adhesive capacity, in relation to the body mass to be supported, of anywhere from several thousand-fold (Autumn, 2006), to several hundred-fold (Irschick et al., 1996), to many tens of times (Huber et al., 2005b), depending on how recordings are made, seem to be at odds with a system that may have evolved to exploit smooth surfaces. At the upper estimated limit of force output, a 50 g Tokay gecko would need only 0.04% of its setae to support its own mass (Autumn, 2006). Such capacity may be able to be more appropriately understood in the context of real world situations in which attachment opportunities on any given footfall may be patchy and sporadic, rather than maximal. Such recognition accords more appropriately with the idea that natural selection produces systems that function no better than they must (Bartholomew, 2005) and prompts us to explore how the adhesive system may actually meet the environment in situations to which it is adapted.

Adhesive effectiveness of the setae is increased by contact splitting, which both lowers the effective elastic modulus of the tip region, rendering this zone more deformable and increasing real contact (Arzt et al., 2003; Persson, 2003; Persson, 2003; Peressadko and Gorb, 2004; Spolenak et al., 2004), and providing more contact points (Scherge and Gorb, 2001; Peressadko and Gorb, 2004). These outcomes are likely enhanced by the quality of the ß-keratin of which the setae are composed, which is softer and more elastic than that associated with the mechanically-protective outer scale surfaces (Alibardi, 2003). Lowering the elastic modulus of the setal tips and providing more contact points have both been advocated to enhance contact on rough surfaces by conforming to surface irregularities (Jagota and Bennison, 2002; Campolo et al., 2003; Hui et al., 2004). Although this may be associated with defects of individual contacts related to nano-scale surface irregularities (Arzt et al., 2003), it will be less effective at the micro scale as topographical undulance increases. Beyond the size of tip and terminal branch dimensions, fibrillar microarrays are likely to be able to make only patchy contact (Briggs and Briscoe, 1976; Yu, 2004), with only localized areas of the setal field being able to make the intimate level of contact necessary for initiating van der Waals based interactions. For modeled fibrillar microstructures Jagota and Bennison (2002) found that soft materials could make intimate contact over an asperity range of 9 μm, whereas stiffer materials could only comply in the range of 9 nm.

For setae to generate their full load-bearing capacity not only a perpendicular, but also a parallel preload is required (Autumn et al., 2000), resulting in the setae taking up the appropriate contact angle. In this functional mode the setae act as tensile attachment devices (Peterson et al., 1982; Alibardi, 2003; Bergmann and Irschick, 2005). Assumption of the appropriate contact angle (Autumn et al., 2000) limits the vertical range over which the entire seta, and hence its tips, can be effective, with the entire setal field operating as a bed of springs with directional stiffness (Autumn, 2006). The setae act as a programmable adhesive (Autumn, 2006) able to be switched on and off by the hierarchy of control levels.

These attributes, not only of individual setae, but also of setal fields and their hierarchical control systems that activate and release the adhesive bond, suggest that contemplating them in the context of the structure of real locomotor surfaces is likely to be instructive. For that purpose we have selected the genus *Rhoptropus*, and in this first instance, one species in particular.

5.2 The Choice of an Exemplar Taxon

Selecting an initial example to investigate the relationship between setal field configuration and locomotor surface topology requires the fulfillment of a number of criteria. In a field situation locomotor behaviour must be observable (difficult for most geckos because of their nocturnality); locomotor substrata must be able to be easily and effectively sampled for subsequent structural observation (difficult for plant material that may undergo significant changes from sampling site to laboratory); and attachment to surfaces should be effected only by the seta-based adhesive system (the confounding effect of claws or other attachment devices should not be evident).

5.2.1 The Genus Rhoptropus

The *Rhoptropus* (Namib Day Geckos) radiation is endemic to northwestern Namibia and southwestern Angola, and forms part of the larger *Pachydactylus* radiation of southern Africa (Branch, 1988; Bauer, 1999; Bauer and Lamb, 2005). Periods of climate change in southern Africa have led to population isolation, substrate specificity, diversification and speciation among this radiation (Bauer, 1999; Bauer and Lamb, 2005). *Rhoptropus* is the sister taxon of all other members of the *Pachydactylus* radiation (Bauer and Good, 1996; Lamb and Bauer, 2001; Bauer and Lamb, 2005) and comprises six to eight (at least two new species await description) rupicolous, medium-sized (60–140 mm snout-vent length) species whose pattern of relationship is well-understood (Lamb and Bauer, 2000, 2001, 2002; Bauer and Lamb, 2002, 2005).

Rhoptropus is restricted to arid and hyper-arid regions of Namibia and Angola, and its included species occupy rocky outcrops and boulders, and are entirely dependent upon these rocky surfaces, which often exist in isolation in the middle of large expanses of sandy desert (Branch, 1988; Bauer and Good, 1996). As such, the included species are subject to microhabitat isolation and substrate specialization. The array of species exploit different locomotor substrata across a relatively small geographic range, enabling effective sampling and the investigation of intra- and interspecific variation in adhesive system morphology.

Rocky surfaces can be easily sampled and retain their physical characteristics from the site of sampling to the site of investigation in the laboratory. *Rhoptropus* fulfills the desiderata outlined above by also being diurnal and effectively lacking claws (Bauer and Good, 1996).

Rhoptropus is morphologically and behaviourally divergent from all other members of the *Pachydactylus* radiation (Bauer et al., 1996; Russell et al., 1997; Johnson et al., 2005), with relatively elongated distal limb segments, a reduced number of presacral vertebrae, and a propensity for rapid running and jumping (Bauer et al., 1996; Johnson et al., 2005). It also exhibits elongated digits that radiate less from each other than do those of other geckos (Bauer et al., 1996; Russell et al., 1997), and that have the adhesive subdigital pads displaced distally (Bauer et al., 1996). *Rhoptropus* is thus a diurnal, rupicolous, clawless, rapidly moving and jumping taxon that offers an excellent extremal situation for the investigation of the relationship between the adhesive system and the locomotor substratum.

5.2.2 Rhoptropus *cf* biporosus

In this initial investigation we confine our exploration to one species, *Rhoptropus* cf *biporosus* (an undescribed species which is the sister species of *R. biporosus*, (A.M. Bauer, personal communication) the Kaokoveld Namib day gecko), and one field locality, Gai'As, which represents one type of locomotor substratum, sandstone (Fig. 5.3). *R. biporosus* has an SVL of 60–75 mm and is restricted to rock outcrops in northwestern Namibia (Branch, 1988). The site at Gai'As is located

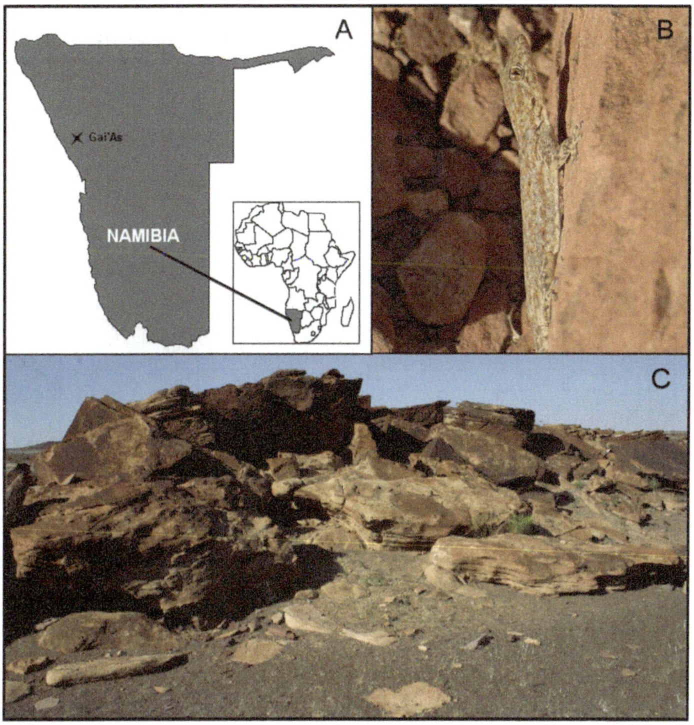

Fig. 5.3 The site of collection and habitat of *Rhoptropus* cf *biporosus*. **A** The location of the site of Gai'As where specimens were collected. **B** A specimen resting on a vertical sandstone rock face (photo courtesy of Tony Gamble) **C** The terrain at Gai'As, which consists of isolated piles of sandstone boulders and outcrops

in northwestern Nambia, just north of the Ugab River at the edge of the central Namib Desert (Fig. 5.3A). Geologically, the area is part of the Gai'As formation of the Karoo Sequence sandstones, formed during the Carboniferous through Jurassic ages about 300–130 million years ago (Grünert, 2000). The area consists of isolated rock outcrops and piles of boulders that are home to various species including *R.* cf *biporosus* (Fig. 5.3B), and are surrounded by a flat desert landscape (Fig. 5.3C). Typically *R.* cf *biporosus* moves rapidly on these rock surfaces and scales surfaces in all orientations from horizontal through vertical to inverted.

5.3 Materials and Methods

5.3.1 Rock Surfaces

Sandstone samples (flakes of rock exfoliated from the surface using a small cold chisel and geological hammer) were collected at Gai'As (20.46.45 S, 14.04.30 E) in June 2005 from surfaces across which individuals had run. The sandstone grain sizes ranged from fine to coarse.

The surface topologies of these sandstones were evaluated using scanning electron microscopy (SEM) (Phillips Environmental SEM XL30) and the computer software MeX 4.2 (Scherer, 2002). The StereoCreator module of MeX was used to generate 3-dimensional digital elevation models (DEMs) of the rock surfaces using SEM images taken at different tilt angles (a 22° tilt between left and right images was found by MeX to be optimal for most of the surfaces). Three DEMs from different areas of the surface were created for each of six sandstone samples. DEMs were also created for the underside of two samples of freshly exfoliated sandstone surfaces to evaluate potential differences between weathered and newly exposed substrates. These DEMs were used to characterize and analyze the surfaces of the rocks by evaluating their general surface topology, as well as the potential area available for contact by the setae of *R.* cf *biporosus*. For each DEM the Profile Module of MeX was used to generate a cross-sectional profile of the surface, in order to visually compare the topology of the surfaces.

Depth-coloured images of each DEM were also generated, which demonstrate visually the areas of the rock surface available in 50 μm increments through the entire depth of the exposed surface. In conjunction with the depth images, a 1 mm^2area of each DEM was examined using the Area Module of MeX. This area corresponds to the approximate area of a single subdigital pad of *R.* cf *biporosus* (see below, Fig. 5.4), and as such allows for the evaluation of potential contact with the surface at the level of entire setal fields, rather than individual setae. The Area Module of MeX was used to calculate the maximum height of each surface (from the lowest point of the deepest valley, to the top of the highest peak), as well as to calculate the percentage of the 1 mm^2 area that was available for contact in the uppermost 100 μm, 50 μm, 40 μm, 30 μm, 20 μm and 9 μm of each rock surface. The increments of 100 μm and 50 μm were chosen to bracket the length

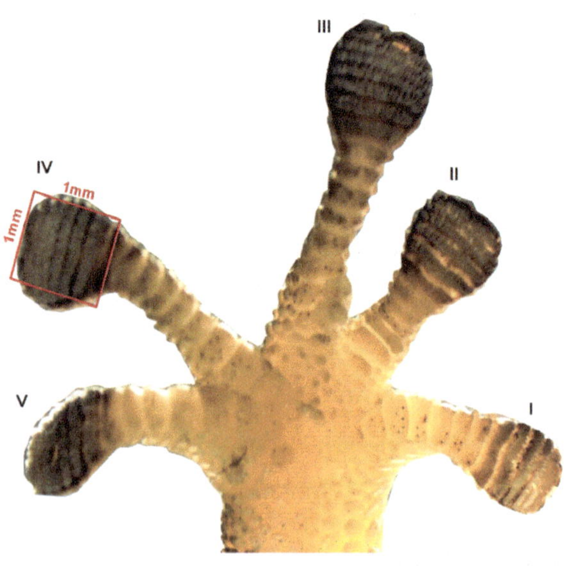

Fig. 5.4 The right manus of *Rhoptropus* cf *biporosus* showing the subdigital pad of digit IV and its association with the 1 mm^2 area that was examined on each of the sample surfaces. Digits are numbered I–V

of the setae of *Rhoptropus* cf *biporosus* (see below), as best case scenarios for potential attachment, and the value of 9 μm was chosen to approximate the asperity range deemed to be available for contact by soft microfibrillar structures (Jagota and Bennison, 2002), and as a worst case scenario for potential attachment. The remaining increments (40 μm, 30 μm, 20 μm) were used to provide an indication of the pattern of change in contact area available between 50 μm and 9 μm. The percent of available contact area was also converted to true area values to represent the area available for potential contact by the setae at each of these increments.

5.3.2 Additional Experimental Surfaces

In addition to the six sandstone samples, a variety of other surfaces were examined for their topographical characteristics. One cluster of these comprised surfaces that have been employed in previous studies of gekkonid adhesion: glass, plywood, 1 mm metal wire mesh, canvas, Plexiglas, hardwood (oak), cinderblock, acetate overhead transparency film and polishing film (asperity sizes of 0.5 μm and 12 μm) (Irschick et al., 1996; Zani, 2000; Vanhooydonck et al., 2005; Huber et al., 2007). A second set of surfaces (sandpaper 60 grit and 220 grit) and emery paper (600 grit) were examined as potential intermediates between smooth and nanoscale-rough surfaces on the one hand and unpredictable and micro to mega scale undulant (natural and artificial) surfaces on the other. This second set of surfaces was employed to determine how regularity of form (sandpaper and emery paper are designed to be non-patchy) and increasing levels of surface undulance may impact the potential to make adhesive contact. DEMs were created for these surfaces and were used to characterize and analyze the surfaces following the methods described above for the Gai'As sandstones. Measurements for maximal height and the percent of contact area available in the uppermost 100 μm, 50 μm, and 9 μm of these surfaces, as well as the sandstone surfaces, were compared using Single Factor Analysis of Variance (ANOVA) in order to evaluate differences between the samples. If a significant difference was found, Tukey's test was used to evaluate which surfaces differed.

Because of a lack of absolute flatness of the cross-sectional profiles of the DEMs of some of the smoothest surfaces (glass, Plexiglas, polishing film – see below), a smaller test area of 200 μm × 200 μm (0.04 mm^2), concordant with the scale of the surface irregularity, was used to re-evaluate the potential area available in the uppermost 9 μm of these surfaces. This was repeated for three rough or undulant surfaces (sandstone, 60 grit sandpaper, and canvas) for comparison of areas available for attachment at this scale.

5.3.3 Subdigital Pads and Setae of **Rhoptropus** ***cf*** **biporosus**

Twelve adult specimens of *R.* cf *biporosus* from Gai'As were collected under the auspices of a permit issued to Dr. A.M. Bauer, and now reside in the collections of the Museum of Comparative Zoology, Harvard University. They were weighed and their snout-vent lengths measured in order to establish an average size and mass for the sample. The subdigital pads of one adult female specimen (184180, MCZ) were then examined, with digit IV being chosen as a representative digit. The subdigital pad area of digit IV of both the manus and the pes was measured using the TeX module of MeX, and the average pad area calculated (Fig. 5.4). Digit IV was then removed from each manus and pes, sectioned longitudinally along the approximate midline of the digit, dehydrated and critically point dried (Seevac CO_2Critical Point Dryer). The samples were mounted with colloidal silver paste on Zeiss aluminum SEM mounts and coated with gold in a Hummer II Sputter Coater. The digital sections were examined using the GSE (gaseous) mode of the ESEM to determine the general pattern and arrangement of the setal fields. Measurements of digit depth, setal length and diameter, and setal density were taken from the digital SEM images using the TeX module of MeX. These density measurements were used in conjunction with area measurements from the rocks and other surfaces to evaluate the potential number of setae that could contact the surfaces.

5.4 Results

5.4.1 Surface Topology and Available Contact Area

The DEMs, depth-coloured images, and profiles of the various surfaces reveal a variety of patterns that typify smooth, rough (asperity sizes on the nano to micro scales) and undulant (asperities on the macro scale) topologies. The sandstone surfaces show little variation from each other in terms of their profiles (Fig. 5.5) and DEMs. This is true even for the freshly exfoliated surfaces, which do not differ substantially from the weathered surfaces. Cross sectional profiles of the sandstone surfaces (Fig. 5.5) reveal them to be both qualitatively and quantitatively different from the smooth and rough surfaces (Figs. 5.6 and 5.7) (acetate, glass, Plexiglas, polishing film) typically employed in force-based estimation of adhesive potential and capability. More undulant surfaces used in locomotor performance experiments (Fig. 5.6) (canvas, plywood, wire mesh, oak, cinder block) have much higher peaks and lower valleys than the smooth and rough surfaces, and lie between the latter and the sandstone in both asperity height and predictability of undulance. 60 grit sandpaper (Fig. 5.6) comes closest to the asperity height of sandstone (Fig. 5.5), but has much more predictable regularity of pattern.

Depth-coloured images of such surfaces show the same pattern of variance of depth across the materials examined, and reveal more graphically the relative predictability and evenness of surface irregularities (Figs. 5.8 and 5.9).

ANOVAs comparing the maximum heights attained in the 1 mm^2 areas of each surface (Table 5.1, Fig. 5.10A), reveal that, with the exception of Sandstone 4, which was found to have a significantly lower maximal height than Sandstones 1, 5, and the freshly exfoliated rock surfaces (Tukey's Test; $p<0.05$), the sandstone surfaces do not differ from each other (Tukey's Test; $p > 0.05$). Sandstone surfaces, however, displayed significantly greater heights than the smooth surfaces, the 600 grit emery

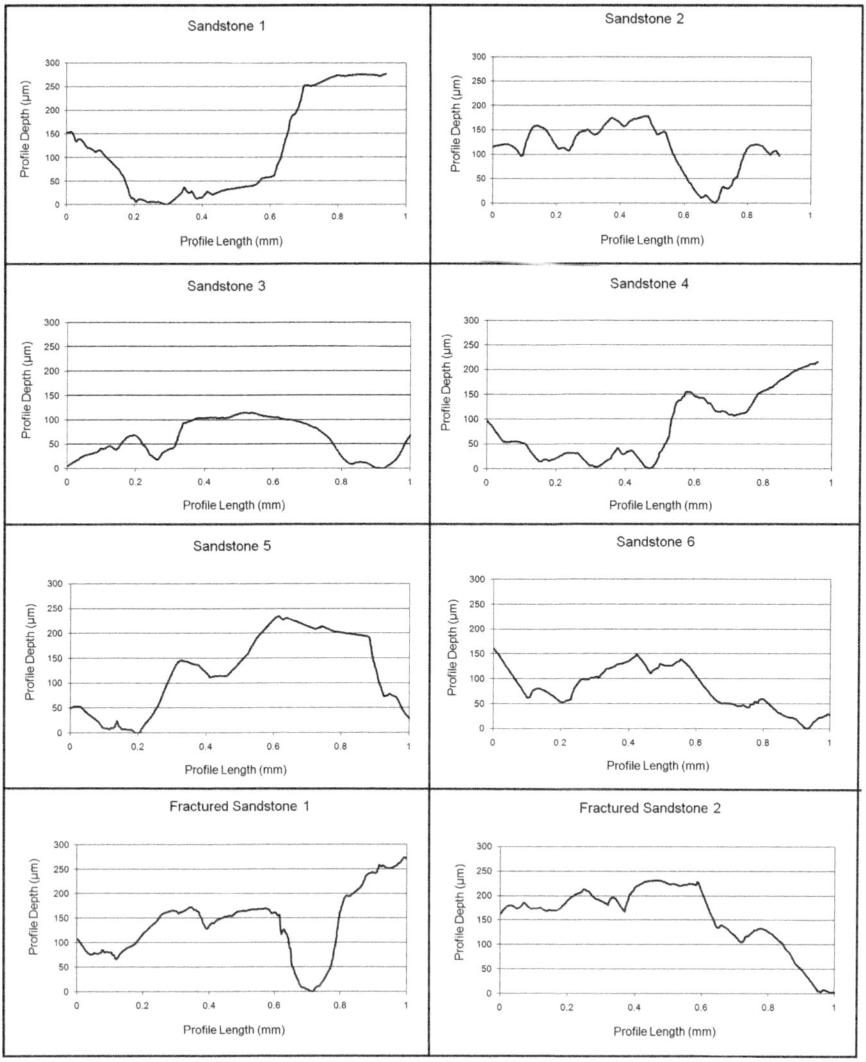

Fig. 5.5 Cross sectional profiles of sandstone surfaces (the natural surfaces employed by *R.* cf *biporosus* for locomotion). All profiles are presented at an identical scale

paper, and the polishing films (Fig. 5.10A) and all except Sandstone 4 also differed from wood and 220 grit sandpaper samples (Tukey's Test; $p<0.05$). Other undulant surfaces (wire mesh, canvas, oak, cinder block, 60 grit sandpaper) had greater maximal heights than the smooth surfaces (Fig. 5.10A) (Tukey's Test, $p<0.05$), and did not differ significantly from sandstone surfaces (Tukey's Test, $p>0.05$).

When examined for the amount of area available for contact in the uppermost 100 μm of the surface (Table 5.1; see also Fig. 5.11 for examples, and Fig. 5.10B), the smooth surfaces and the most regular of the undulant surfaces (600 grit emery paper, 220 grit sandpaper, polishing films) had 100%, or almost so, of the 1 mm^2

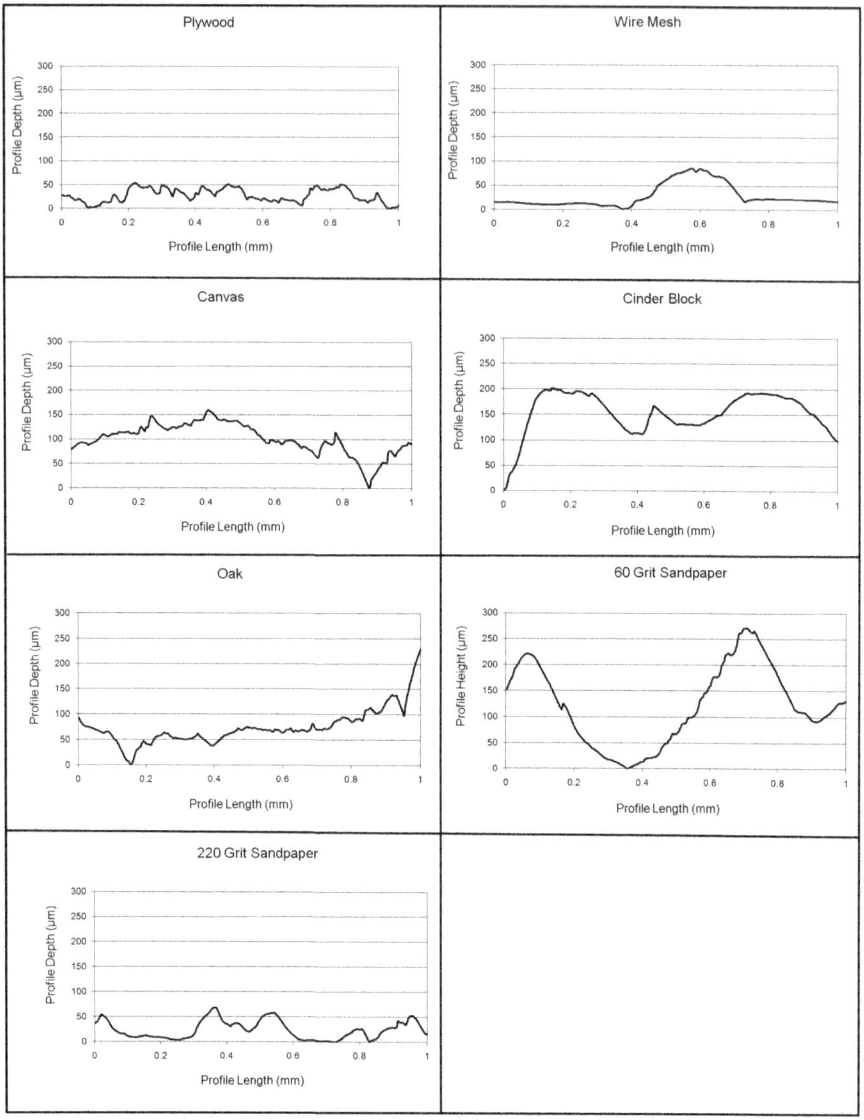

Fig. 5.6 Cross sectional profiles of undulant sample surfaces, some of which have been used in previous studies of gekkonid adhesive performance (plywood, wire mesh, canvas, cinder block, oak) and others which have not (60 and 220 grit sandpaper). All profiles are presented at an identical scale, and at the same scale as the samples depicted in Fig. 5.5

area available. Indeed, the topology of such surfaces had a total depth of less than 100 μm (Fig. 5.10B). This differs significantly from all other surfaces, which had between 5 and 50% of the 1 mm^2 test area (5000–50000 μm^2) available in that range (Fig. 5.10B) (Tukey's Test; $p<0.05$). Among the sandstone samples, sandstone 4 (the least undulant) had significantly more surface area available in the uppermost

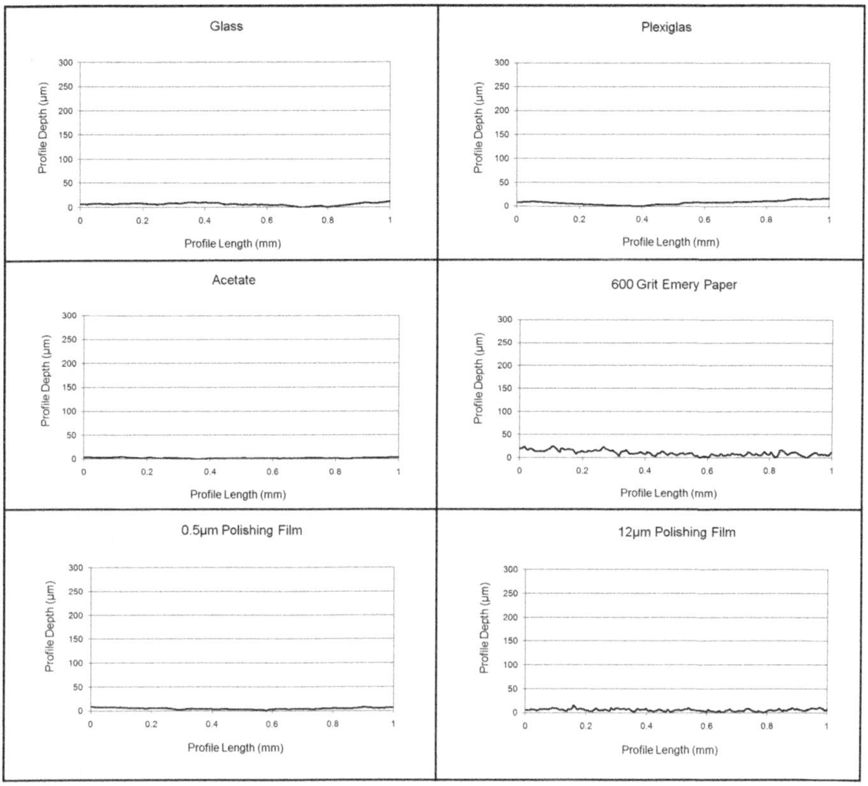

Fig. 5.7 Cross sectional profiles of smooth or rough (at the nano or micro scale) sample surfaces, some of which have been used in previous studies of gekkonid adhesive performance (glass, Plexiglas, acetate, polishing film) and others which have not (600 grit emery paper). All profiles are presented at an identical scale, and at the same scale as samples depicted in Fig. 5.5

100 µm than sandstone 5, as well as the canvas and 60 grit sandpaper samples (Tukey's Test, $p<0.05$) (Fig. 5.10B).

A similar trend was revealed for the uppermost 50 µm of the surfaces, where the smoother, less irregular surfaces (acetate, glass, Plexiglas, 600 grit emery paper, polishing papers) had significantly more area available for contact (88–100% of the 1 mm^2 test area) than all of the other surfaces, which had only between 2.5–10% of the 1 mm^2test area (2500–10000 μm^2) available in this range (Tukey's Test; $p<0.05$) (Fig. 5.10B).

For the uppermost 9 µm of the surfaces there was a great deal less variation between samples. Acetate, which was found to be the smoothest surface, with 86% of the 1 mm^2test area available for contact in the top 9 µm, differed significantly from all other surfaces (Tukey's Test; $p<0.05$), and was the only surface found to differ significantly at this level. However, although not significant, glass, Plexiglas, and the 0.5 µm and 12 µm polishing films had much more surface area available in the top 9 µm of the 1 mm^2test area (73000 μm^2, 63000 μm^2, 40000 μm^2 and

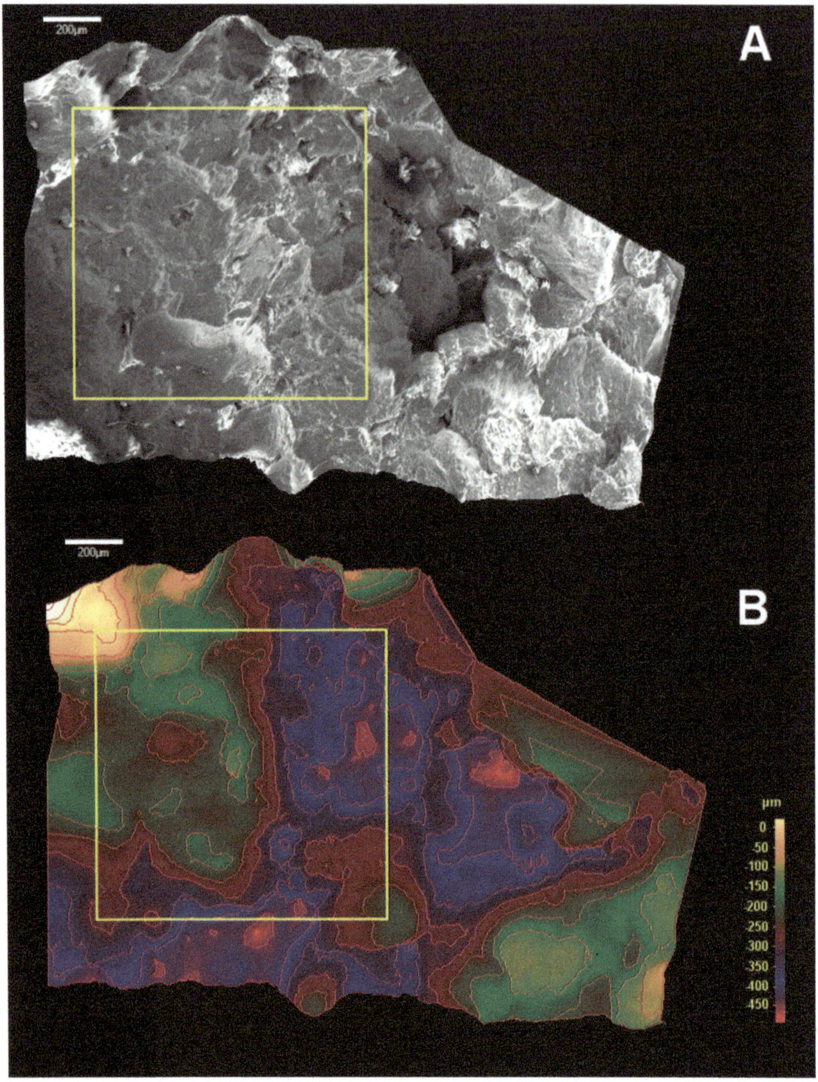

Fig. 5.8 The surface topology of one of the sandstone surfaces (sandstone 5, Fig. 5.5). **A** Digital Elevation Model of the surface showing its general topographical features. **B** Depth-coloured image of the surface showing the surface area available for contact in 50 μm increments from the highest to the lowest point of the sandstone. Both A and B show a 1 mm^2area of the surface analyzed as representing the approximate area of a subdigital pad of *Rhoptropus* cf *biporosus*

43000 μm^2 respectively) than the other surfaces examined, which all had less than 5000 μm^2 (0.5% of the 1 mm^2test area) available for contact (Fig. 5.10B).

Examination of the cross-sectional profiles of some of the smooth surfaces (glass, Plexiglas, polishing film) revealed that they show some degree of curvature or warping along the 1 mm transect, especially near the edges of the DEM (Fig. 5.7). This

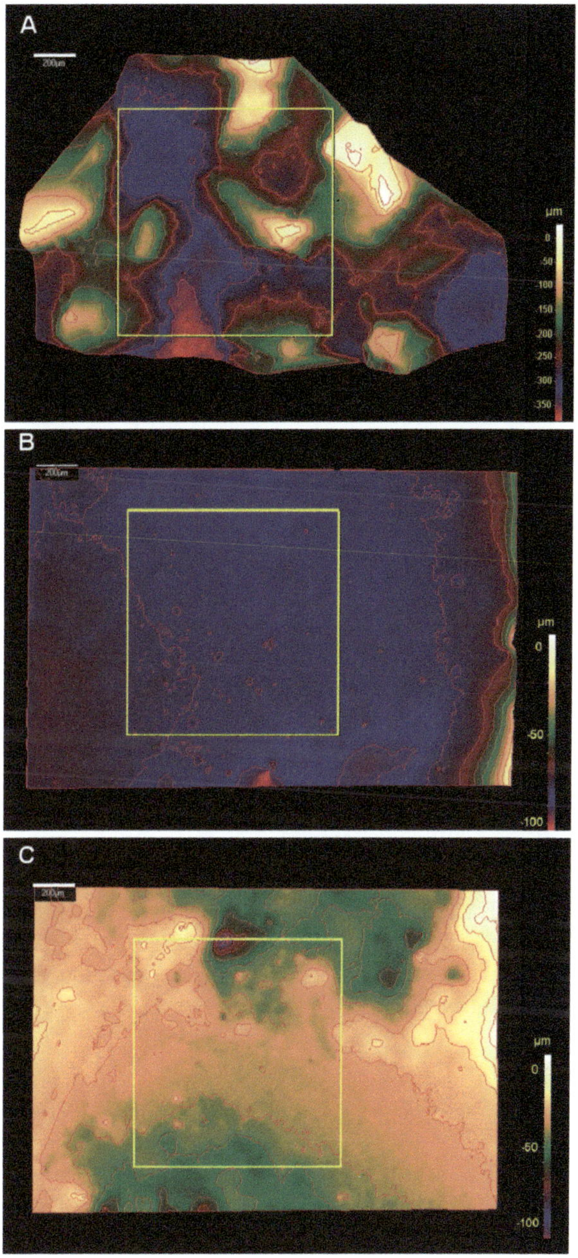

Fig. 5.9 Depth coloured images showing the topology and the area available for contact in 50 μm increments from the highest to the lowest point for three example surfaces **A** 60 Grit Sandpaper **B** Acetate Overhead Transparency **C** 0.5 μm Polishing Film. A, B, and C are all overlain with a 1 mm^2 area representing the approximate area of a subdigital pad of *Rhoptropus* cf *biporosus*

is possibly a result of the natural warp of these surfaces, or may be an artifact of the 3-dimensional reconstruction of the surfaces by MeX because of insufficient surface irregularities for a completely accurate reconstruction across such a broad span. Due to the extreme smoothness of these surfaces, and the small scale at which they were

Table 5.1 Results of ANOVAs comparing measurements taken from various surfaces, including the maximum height of the surfaces, and the area available for contact within the uppermost 100 μm, 50 μm and 9 μm of each surface. All ANOVAs revealed a significant difference between surfaces

Variable compared (μm)	df(groups)	df(error)	F	p
Maximum height	20	42	22.86	0.000
Top 100	20	42	32.07	0.000
Top 50	20	42	128.72	0.000
Top 9	20	42	134.61	0.000

being evaluated (the uppermost 9 μm), these elevational changes seemingly resulted in an underestimation of the area available for contact across the 1 mm^2 area at this level (Figs. 5.10 and 5.11). It seems highly likely that a digit could conform easily to a surface with a profile such as those observed for the smooth surfaces (Fig. 5.7). To rectify this, smaller sample areas of 0.04 mm^2 were evaluated for each of these surfaces. This revealed that the areas available for contact by setae in the uppermost 9 μm of these surfaces were substantially higher than initially calculated. For the 0.04 mm^2 areas of the smooth surfaces evaluated (glass, Plexiglas, polishing film) an average of approximately 75% of the surface was available for contact for glass and Plexiglas, 93% for the 0.5 μm polishing film and 38% for 12 μm polishing film (Fig. 5.11E). In contrast, a re-evaluation of several rough and undulant surfaces (sandstone, 60 grit sandpaper, canvas) using a 0.04 mm^2 test area yielded results that did not differ from the values obtained for the larger, 1 mm^2 area.

A very substantial drop in available contact area between the uppermost 50 μm and the uppermost 9 μm of the surface was observed for the majority of the samples examined. To explore the nature of this diminution we examined the area available for contact for each of the surfaces at depths of 20 μm, 30 μm, and 40 μm. This revealed that the reduction occurs fairly gradually for the sandstones and other undulant surfaces (plywood, wire mesh, canvas, cinder block, oak, sandpaper), with area values decreasing steadily at each level. For the smoother surfaces (glass, Plexiglas, acetate, emery paper, polishing papers) the reduction occurs much more suddenly, but is corrected for when smaller areas are assessed (see above for reasoning).

5.4.2 Subdigital Pads and Setae of Rhoptropus *cf* biporosus

The specimens of *R.* cf *biporosus* had an average mass of 2.11 ± 0.17 g, and an average snout-vent length of 42.55 ± 0.98 mm. The average area of the subdigital pad of one digit (all digits having approximately the same setal field area despite major differences in digit length) was 1.16 ± 0.08 mm^2 (Fig. 5.4), which corresponds roughly to the 1 mm^2 test area evaluated for each of the surfaces. The depth of the digits averaged 281 ± 17.5 μm. Examination of the subdigital pads in the GSE mode of the SEM revealed that the general pattern of setal arrangement of *R.* cf *biporosus* is similar to that observed for *Gekko gecko* (Delannoy, 2006) (Fig. 5.2). The subdigital pads of digit IV of the manus and pes of *R.* cf *biporosus* bear 6–8 scansors carrying

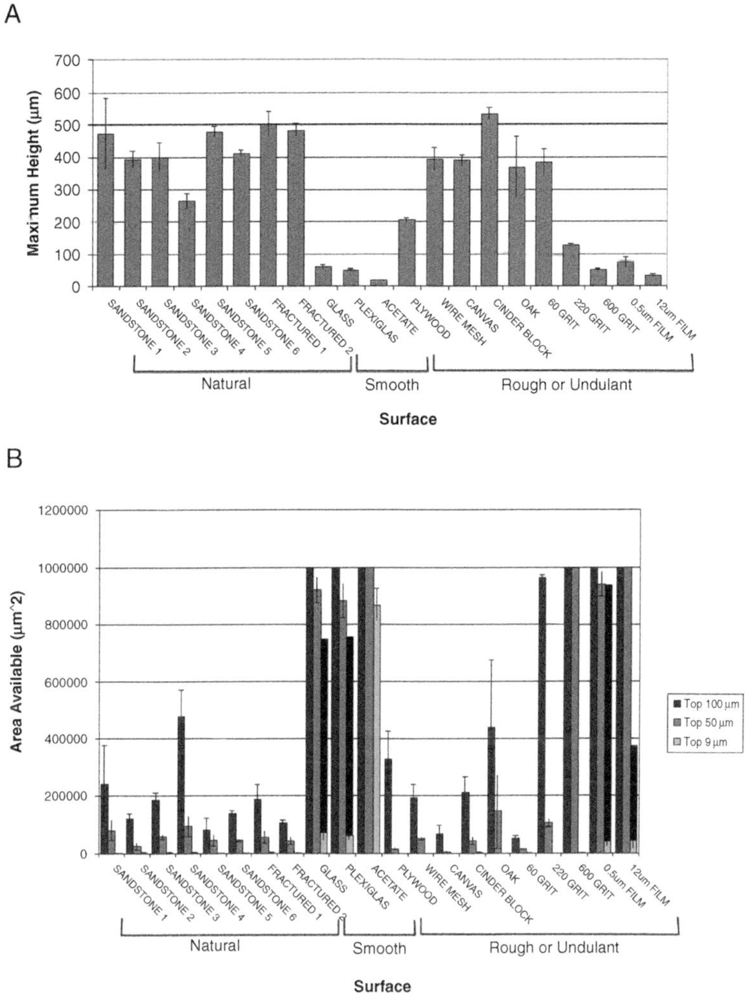

Fig. 5.10 Histograms showing **A** The maximum height of the Digital Elevation Model of a variety of surfaces within a 1 mm^2 evaluation area, and **B** the area available for contact by setae within the same 1 mm^2 area for the uppermost 100 µm, 50 µm and 9 µm of each surface. The *black bars* shown for glass, Plexiglas and the polishing papers for the 9 µm category represent the area available when extrapolated from 0.04 mm^2 evaluation areas (see text for explanation). The surfaces are arrayed in three categories: natural surfaces, smooth surfaces and rough or undulant surfaces

rows of setal elaborations (Fig. 5.12). These setae are highly branched structures, very similar in attributes to those of *Gekko gecko* (Delannoy, 2006), with an average length of 70.4 ± 2.2 µm on the manus and 60.4 ± 2.0 µm on the pes. The longest setae are located on the distalmost scansor, and the average length of the setae on each scansor decreases proximally (Fig. 5.12E). The setae have an average diameter of 3.15 ± 0.12 µm on the manus and 2.77 ± 0.10 µm on the pes, giving an

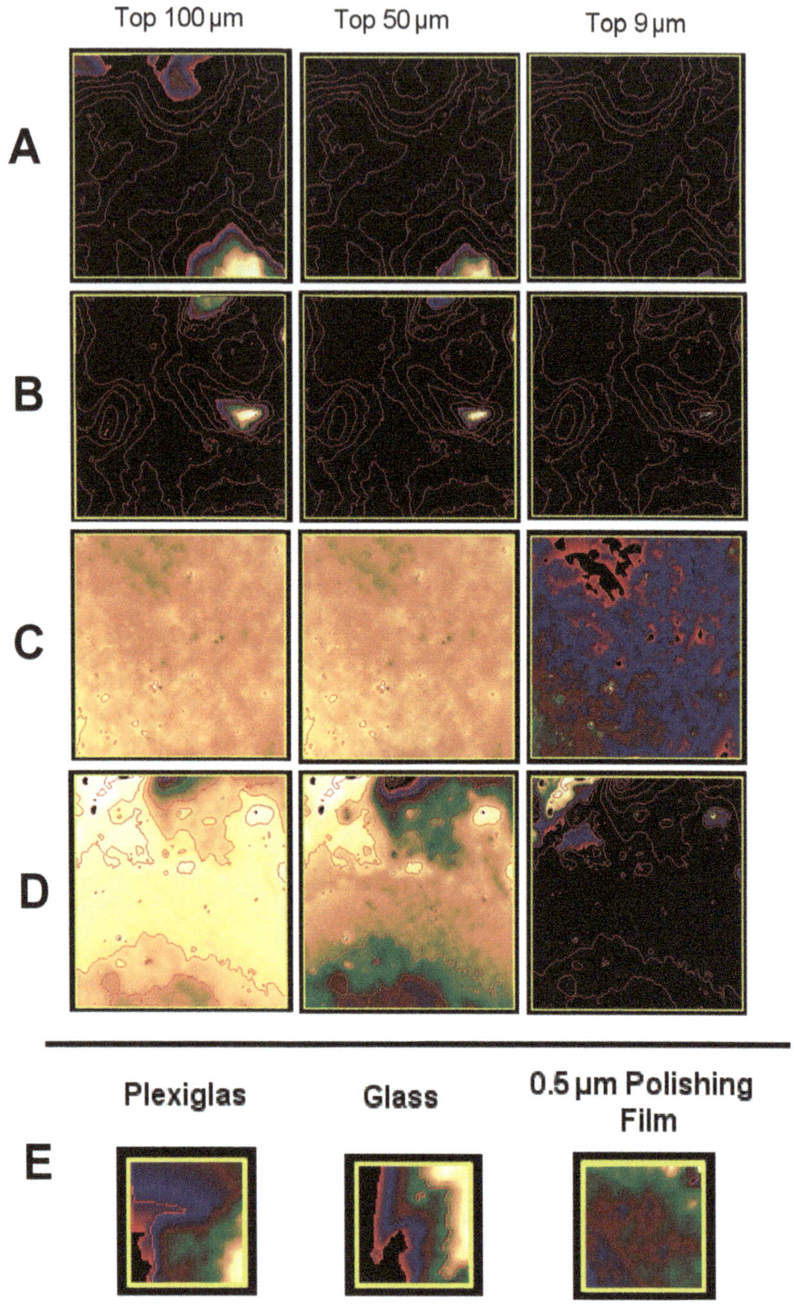

Fig. 5.11 (*Continued*)

aspect ratio of approximately 22 for both manus and pes. There was some variation in setal density between scansors of digit IV, ranging from a minimum of 16500 setae/mm^{2}to a maximum of 26000 setae/mm^2; but this did not follow any consistent pattern. Overall, the average setal density on the subdigital pads of digit IV of the manus and pes was 19822 ± 490 setae/mm^2, or approximately 23000 setae/digit.

5.4.3 Available Contact Area and Adhesive Force Estimates

The above estimated values of both setal density per mm^2 and surface area available for contact at various levels of the different test surfaces, in conjunction with published values for adhesive force (Autumn et al., 2000; Huber et al., 2005b), make it possible to estimate the number of setae that could potentially contact the surface on any given footfall, and the maximum adhesive force that could be generated in these situations.

Based on the assumption (Autumn et al., 2000) that one seta can produce an adhesive force of 200 μN, one fully attached digit of *R.* cf *biporosus* could produce approximately 4.6 N of force. This is more than 200-fold the force required (0.02 N total; 0.001 N per digit) to support an individual with a mass of approximately 2.0 g, and translates to a capability of 92 N for all 20 digits. However, on rough and undulant surfaces the area available for contact is much lower, and so, therefore, is the potential for the generation of adhesive force (Table 5.2).

Assuming that the uppermost 100 μm of the surface is available for contact and that the setae make perfect contact with the available area, one digit of *R.* cf *biporosus* could generate roughly 4 N (80 N for all 20 digits) of adhesive force on the smoother surfaces, such as glass, Plexiglas, acetate, 220 grit sandpaper, 600 grit emery paper, and the polishing films. On the remaining surfaces maximal adhesive force estimates ranged from a minimum of 0.20 ± 0.04 N (4 N total for all 20 digits) for 60 grit sandpaper, to a maximum of 1.89 ± 0.38 N (37 N total) for one of the sandstone samples (#4) (Fig. 5.13, Table 5.2).

For most of the smoother surfaces (glass, Plexiglas, acetate, 600 grit emery paper, and the polishing films) the available contact area, and thus the maximum adhesive force estimates, were similar for the uppermost 50 μm of the surface (Fig. 5.13). The estimates were, however, somewhat lower for the other surfaces, ranging from 0.026 ± 0.02 N (0.52 N total) for the canvas surface to 0.42 ± 0.05 N (8.4 N total) for the 220 grit sandpaper (Fig. 5.13, Table 5.2).

←

Fig. 5.11 Depth coloured images for a 1 mm^2 area of four example surfaces **A** Sandstone 6 (Fig. 5.5) **B** 60 grit sandpaper **C** Acetate and **D** 0.5 μm polishing film showing the area available for contact within the uppermost 100 μm, 50 μm and 9 μm of each surface. **E** Depth coloured images for a 0.04 mm^2 area of three smooth surfaces: Plexiglas, glass and 0.5 μm polishing film, showing the area available for contact within the uppermost 9 μm of each surface. These images show that the evaluation of 1 mm^2 areas for these surfaces may have resulted in an underestimation of the surface area available for contact at this level. Areas that are available for contact within each range are coloured and those that are not available are *black*

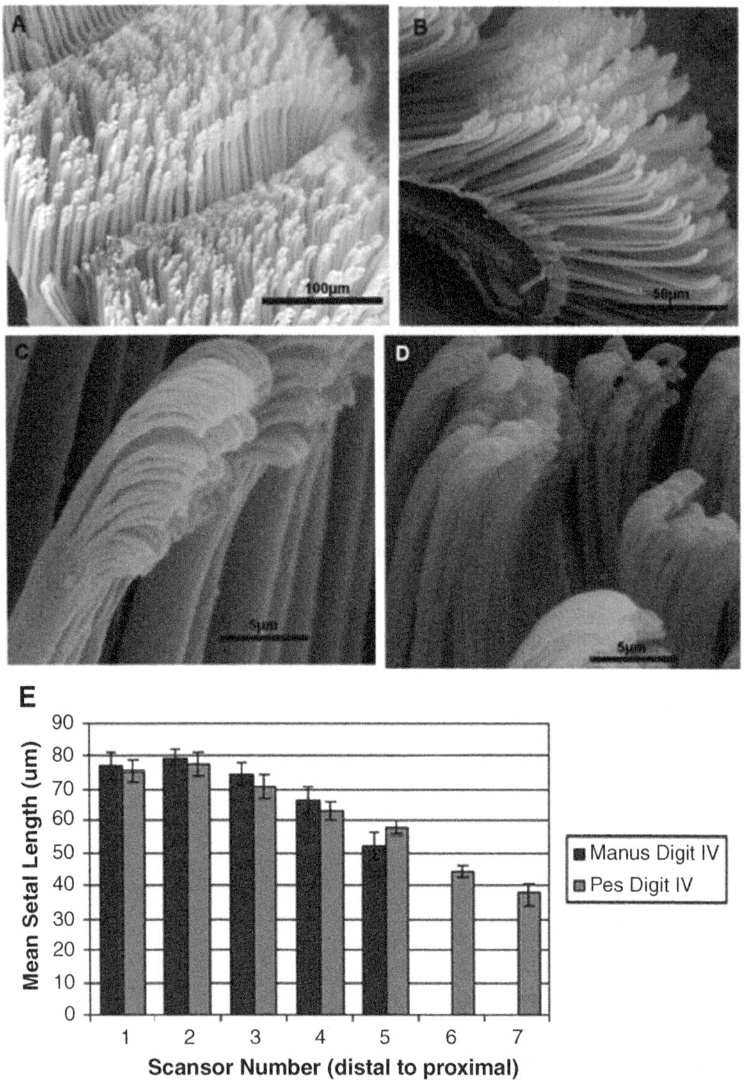

Fig. 5.12 SEM images of the setal fields and setae from digit IV of the pes of *Rhoptropus* cf *biporosus*. **A** Setal fields. **B** Edge-on view of the setal fields of one scansor (proximal – *left*, distal – *right*). **C** Tips of setae located at the proximal edge of a scansor. **D** Tips of setae located at the distal edge of a scansor. **E** Histogram showing the change in length of the setae along the length of the digit. For both the manus and pes average setal length decreases proximally

For the uppermost 9 µm of the surface the maximum force estimates were very low for most of the surfaces, ranging from 0.0005 N (0.01 N total) for the canvas surface to 0.015 ± 0.01 N (0.3 N total) for some of the sandstone surfaces (3 and 4) (Table 5.2, Fig. 5.13). The force estimates were slightly higher for very smooth surfaces like glass, Plexiglas, and the polishing papers (0.29 ± 0.11 N [5.8 N total], 0.24 ± 0.07 N [4.8 N total], 0.16± 0.07 N [3.2 N total], 0.17± 0.11 N [3.4 N total])

Table 5.2 Estimated values of the maximum adhesive force generatable by *Rhoptropus* cf *biporosus* for all 20 digits on a variety of surfaces. Estimates are calculated assuming contact is possible within the uppermost 100 μm, 50 μm and 9 μm of a 1 mm^2 area of each surface, and are calculated using literature values of adhesive force for an individual seta (Autumn et al., 2000) and individual spatulae (Huber et al., 2005b). *Asterisks* (*) denote values falling below the 0.02 N necessary to support the 2.0 g mass of an individual. *Crosses* (†) denote force values that have been re-estimated using a 0.04 mm^2 area, and are substantially higher than the estimates based on a 1 mm^2 area (see text for explanation)

	Adhesive force (N) (Autumn et al., 2000)			Adhesive force (N) (Huber et al., 2005)		
Surface	100 μm	50 μm	9 μm	100 μm	50 μm	9 μm
Sandstone 1	19.09	6.48	0.21	0.954	0.324	0.010*
Sandstone 2	9.54	2.02	0.24	0.477	0.101	0.012*
Sandstone 3	14.71	4.72	0.29	0.735	0.236	0.015*
Sandstone 4	37.79	7.47	0.29	1.889	0.373	0.014*
Sandstone 5	6.71	3.66	0.12	0.336	0.183	0.006*
Sandstone 6	10.93	3.65	0.19	0.547	0.182	0.009*
Fractured Sandstone 1	14.83	4.60	0.31	0.741	0.230	0.016*
Fractured Sandstone 2	8.42	3.54	0.19	0.421	0.177	0.009*
Glass	79.29	72.95	59.55†	3.964	3.648	2.977†
Plexiglas	79.29	70.08	61.52†	3.964	3.504	3.079†
Acetate overhead sheet	79.29	79.29	68.88	3.964	3.964	3.444
Plywood	26.12	1.38	0.09	1.306	0.069	0.005*
Wire mesh	15.25	4.19	0.06	0.762	0.209	0.003*
Canvas	5.40	0.52	0.01*	0.270	0.026	0.001*
Oak	34.94	11.56	0.03	0.844	0.175	0.012*
Cinder block	16.89	3.51	0.24	1.747	0.578	0.002*
60 Grit sandpaper	4.07	1.26	0.10	0.204	0.063	0.005*
220 Grit sandpaper	76.25	8.44	0.06	3.812	0.422	0.003*
600 Grit emery paper	79.29	79.13	0.18	3.964	3.956	0.009*
0.5 μm Polishing film	79.27	74.56	74.04†	3.963	3.728	3.702†
12 μm Polishing film	79.29	79.29	30.40†	3.964	3.964	1.520†

and especially for acetate (3.44 ± 0.23 N [68.8 N total]). When adjusted for the smaller 0.04 mm^2 area the whole animal force estimates for these materials rise to 59.6 N for glass, 61.5 N for Plexiglas, 74.0 N for 0.5 μm polishing film, and 30.4 N for 12 μm polishing film, with all but the latter having magnitudes similar to that observed for acetate (See above and Fig. 5.13, Table 5.2).

Using an alternative estimate that one spatula can produce approximately 10 nN of force (Huber et al., 2005b), and assuming that each seta possesses between 100 and 10000 spatulae, yields values for adhesive force approximately 1–2 orders of magnitude lower than those listed above (Table 5.2).

5.5 Discussion

The above results provide information about how surface topology influences the amount of area available for attachment by the intact gekkonid adhesive system (i.e. setal fields), as well as the impact this has on the adhesive force these animals

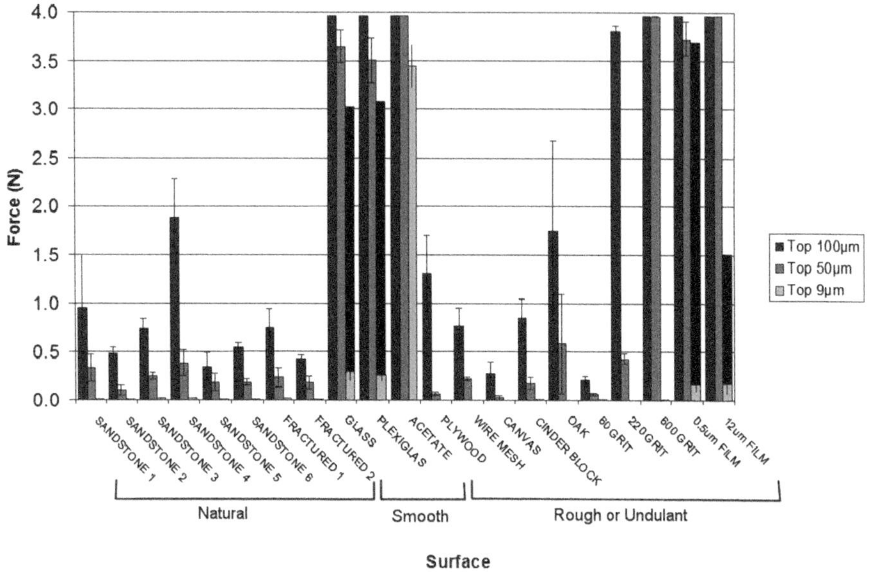

Fig. 5.13 Histogram showing the estimated maximal adhesive force that could be generated by a single digit of *Rhoptropus* cf *biporosus* on a variety of surfaces, assuming that contact of the setae is possible in the uppermost 100 μm , 50 μm and 9 μm of the surface. The black bars shown for glass, Plexiglas and the polishing papers for the 9 μm category represent the maximum force based on 0.04 mm^2 evaluation areas (see text for explanation). The surfaces are arrayed in three categories: natural surfaces, smooth surfaces, and rough or undulant surfaces

are capable of generating on a variety of natural and artificial surfaces. Setal fields constitute integrated systems and cannot be considered simply as clusters of setae. The entire field has limited global capabilities of compliance, but may be able to make local adjustments. The smoother the surface the greater the available contact area. Roughness has an impact on adhesive capability, but undulance has a much more trenchant effect and dramatically reduces the potential surface available to setal fields (Fig. 5.10) for van der Waals based contact. A subdigital pad would essentially bridge across high points rather than insinuating between them. These findings allow us to explore how subdigital pad design permits highly irregular surfaces to be exploited by the gekkonid adhesive system.

5.5.1 Adhesive System Design and Contact with Topographically Undulant Substrata

The adhesive system of gekkonids has often been modeled as a mat of fibrillar microstructures that is brought into contact with a surface (Jagota and Bennison, 2002; Glassmaker et al., 2004; Hui et al., 2004). The reliance of the system on van der Waals forces for adhesion (Autumn, 2002a, b) necessitates extremely close contact between the setae and the locomotor surface. The performance of such a system depends primarily on the mechanical control of the contacting structures as well as their geometry and material properties (Hui et al., 2004; Spolenak et al., 2005). The primary advantage of a fibrillar structure is that it allows for increased compliance between the fibres and the surface, leading to increased contact and adhesion on rough surfaces with a variety of wavelengths and asperity heights (Jagota and Bennison, 2002; Glassmaker et al., 2004; Hui et al., 2004; Spolenak et al., 2004). This interpretation can be extrapolated for relatively uniform subsrata, but will only be locally applicable for highly irregular and/or highly undulant surfaces as contacted by complete setal fields.

The general structure and geometry of the setae observed in *R.* cf *biporosus* corresponds well to these models in that they are arranged in evenly spaced rows, have a high aspect ratio of approximately 20, and show a high level of contact splitting at the terminal ends. However, these models generally simplify the complex mechanical control of setal fields and the way in which they contact the surface. The subdigital pads of gekkonids form part of a complex hierarchical and finely controlled system (Fig. 5.1). In models of setae as fibrillar mats, arrays of fibres are normally considered to contact the surface directly from above, being pushed into the surface by a perpendicular force (Jagota and Bennison, 2002; Glassmaker et al., 2004). However, in gekkonids setae contact the surface at an angle of approximately 45° (Autumn, 2006) and require both a perpendicular and a parallel preload force in order to establish an adhesive bond, with setae being held at a critical angle of less than 30° (Autumn et al., 2000). As such, setae act as tensile structures (Peterson et al., 1982) rather than as compressive, vertically oriented columns. This greatly influences how they may contact and adhere to topographically undulant surfaces.

Furthermore, models of the adhesive system and its ability to achieve contact with uneven surfaces have focused most extensively on the influence of roughness at the level of individual setal attachment (Campolo et al., 2003; Persson, 2003; Persson and Gorb, 2003; Meine et al., 2004; Huber et al., 2007), and have not considered the system at the whole-animal level. Examining the available contact area at three levels, the uppermost 100 μm, the uppermost 50 μm, and the uppermost 9 μm, takes into account the influence of both setal and digital compliance, and allows for the best and worst case scenarios for attachment to be considered.

Studies of fibrillar microstructures have indicated that soft, compliant structures, like setae (Alibardi, 2003; Autumn, 2006), are able to conform to, and make intimate contact with, asperities up to 9 μm (Jagota and Bennison, 2002). Thus, assessing the system at the setal level only, 9 μm may be considered an operational depth over which setal tips can attach, and the area available within the uppermost 9 μm can be regarded as a potential "worst case scenario" for adhesion. The adhesive system of gekkonids, is, however, a hierarchical system, and local compliance at the level of the digit may allow for increased contact of setae across a greater range of asperity heights. The digits of *R.* cf *biporosus* have a depth of approximately 280 μm; based on this, depths of 50 and 100 μm represent potential limits within which localized areas of the digits can deform to enhance setal contact with the surface. Setal length itself is not likely to be a principal factor in determining the asperity height over which setae can make contact, because, as discussed above, setae are finely controlled, tensile structures, which must contact the surface at a low angle to achieve contact (Autumn et al., 2000). Although a 70 μm seta may be able to contact a point 70 μm from the top of the surface, it is unlikely to be able to achieve the correct angle to enable the imposition of the parallel preload necessary to establish an adhesive bond.

5.5.2 Adhesive Force Estimates: The Influence of Surface Topology

Observations of very high adhesive capabilities in gekkonids has led to various statements that the adhesive system of geckos is "overbuilt" (Bauer and Good, 1986; Autumn et al., 2000, 2002a). A number of factors have been suggested to be mitigating factors that might help explain the presence of such a potentially powerful system. For instance, at any given time it may be necessary for one foot, or possibly even a single digit, to support a gecko's weight. During rapid vertical locomotion in *Gekko gecko* there are several short phases of the locomotor cycle where only one foot is in contact with the surface, and there are never more than two feet in contact with the substratum at a given time (Russell, 1975). Furthermore, based on the fine control required to achieve setal attachment, it is likely that the digits aligned along the direction of locomotion on vertical surfaces will bear the greatest load, and may be more optimally attached to the surface than digits arrayed in other directions (Russell et al., 1997; Russell, 2002). Another consideration is that geckos are not stationary animals. Their adhesive system must be capable not only of supporting their stationary mass, but also of maintaining attachment during periods of thrust and acceleration, which would increase the load on the setae (Vanhooydonck et al., 2005, but see below). Adhesive capacity may also be related to the ability to recover should the animal slip or fall from a surface (Autumn, 2006). Surface roughness and topology, and their effect on adhesive potential (Vanhooydonck et al., 2005) have also been suggested as elements related to the safety factor built in to whole animal adhesive capability, although this has heretofore not been tested.

The results presented above relating to available contact area and potential adhesive force suggest that the apparent "overbuilding" of the adhesive system and its ability to produce forces far exceeding those needed to support the animal's mass may be an interpretation derived from the smooth surfaces on which adhesive capacity has been tested (Autumn, 2006; Irschick et al., 1996), and not of surfaces actually exploited in nature. Consideration of the topology of natural rock substrates and the potential for adhesive contact on these surfaces in comparison with artificial, smooth surfaces indicates that the system is far less "overbuilt" than initial investigations suggest, and also points to the possibility that the adhesive system of geckos evolved in association with locomotion on undulant, unpredictable surfaces rather than smooth ones. For instance, the typical "smooth" surfaces examined (glass, Plexiglas and acetate) were found to have far more area available for adhesive contact at all of the stratified levels considered (Figs. 5.10 and 5.11). At both the 100 μm and 50 μm levels it was estimated that one digit of *R.* cf *biporosus* could generate approximately 4 N of adhesive force; this is 200-fold greater than the 0.02 N required to support the animal's mass (Fig. 5.13). Extrapolated to a whole animal, this works out to 80 N for four fully attached feet, or approximately 4000 times the amount of force required to support the animal's 2.0 g mass. This 4000-fold factor is similar in magnitude to estimates of maximum adhesive capacity in the much larger species *Gekko gecko* (Autumn et al., 2000). It is, however, much higher than the maximum clinging performance measured for *Gekko gecko* on a smooth surface, where two fully attached feet were shown to generate 20 N of force, which is several hundred-fold (rather than thousand-fold) the force required to support its mass (50 g) (Irschick et al., 1996). Even assuming attachment across only the uppermost 9 μm of the surface, adhesive force estimates for the smooth surfaces still far exceed those required to support the mass of *R.* cf *biporosus.* On acetate, an estimated 3.4 N of adhesive force per digit was possible (or 68.8 N total), and on glass and Plexiglas a maximum force of approximately 0.3 N per digit, or 6.0 N total, was estimated (Fig. 5.13), but when corrected for a smaller, flatter area (0.04 μm^2) (see above) these values rise to approximately 60 N for the whole animal.

In contrast, on natural, undulant, rocky surfaces, the results indicate that much less area is available for potential contact by the setae for either the 1 mm^2or 0.04 mm^2areas assessed (Figs. 5.10 and 5.11). Assuming that the setae are able to contact the surface anywhere in the uppermost 100 μm of the surface, the maximum adhesive force possible on sandstone ranged from approximately 0.3 N–2 N per digit, or 6.0–40 N total, which is roughly 100–1000 times greater than the force required (Fig. 5.13). These forces are substantially lower than those estimated for the top 9 μm of smoother surfaces (based on 0.04 mm^2 test areas), suggesting that the "best-case scenario" on the rough, natural surfaces may be worse than the "worst-case scenario" considered for the smooth surfaces. The results for the "worst-case" (uppermost 9 μm) of the sandstone surfaces suggest maximum adhesive force estimates that are in the realm of adequate to support the 2.0 g mass of *R.* cf *biporosus.* In this scenario, the maximum adhesive force for all the sandstone substrates was estimated to be approximately 0.01 N per digit, or 0.2 N total., which is only ten times greater than the required force to support the gecko's mass (Fig. 5.13). These estimates are much lower than those observed for smooth surfaces, and reveal the potential impact that surface topology and available contact area have on adhesive capacity in geckos. It is likely that compliance of the adhesive system allows for contact beyond only the uppermost 9 μm, but potential available contact area will still be severely limited. In such circumstances the adhesive system of these animals must manifest massive adhesive potential simply to ensure that adequate contact and attachment is achieved on any given footfall. An apparently extremely large number of setae is likely necessary in order to provide the capacity for adhering to challenging surfaces with an irregular structure (Vanhooydonck et al., 2005), while still maintaining some degree of a "safety-factor".

Studies of other tissues, such as bone, have revealed the importance of a "safety factor" in biological systems (Currey, 1984). However, the magnitude of this factor depends upon a balance between the costs associated with developing and maintaining larger, stronger bones, and the benefit of preventing their failure (Currey, 1984). A similar argument can be applied to setae, where having an abundance of setae may be highly beneficial to the animal, but incurs costs associated with their production, maintenance, and the operation of setal fields. This is especially true as the gekkonid adhesive system consists not only of keratinous setae, which may be relatively "cheap" to produce, but also of a complex hierarchical control structure. As such, it is unlikely that geckos would possess a system with a real safety factor on the order of 1000 or more (the safety factors measured for bone range from approximately 1–5, Currey, 1984). When surface topology is taken into account, a safety factor of approximately 10 is likely more realistic.

Examination of adhesive force estimates using the spatular pull-off forces measured by Huber et al. (2005b) reveal an even more challenging situation for these geckos. Huber et al. (2005b) report that their findings are "in agreement with earlier measurements by Autumn et al. (2000)"; however, extrapolations of maximal adhesive force based on their measurement of 10 nN per spatula are much lower than those obtained by the measurements of Autumn et al. (2000) (Table 5.2). In fact, for the worst case scenario (uppermost 9 μm of the surface) the maximum force estimates (based on either 1 mm^{2}or 0.04 mm^2 test areas) were lower than the adhesive force required by the animal to support it's own mass on all but the smoothest surfaces (glass, Plexiglas, acetate, polishing film) (Table 5.2). It may be that this "worst case" is actually worse than anything the animal would encounter in nature, or it may be that the measurements of Huber et al. (2005b) underestimated the potential force that can be produced by a spatula., although the spatulae were exposed to both perpendicular and parallel preload forces required to help maximize adhesion.

Comparisons of the natural sandstone surfaces to other rough surfaces that have been included in previous adhesion studies reveal that many of these surfaces are similar in terms of their overall topography and available contact area at various levels. In general, the range of measurements of available contact area and maximum adhesive force across the sandstone surfaces (both weathered and freshly exposed) are comparable to those observed for other undulant surfaces (wood, wire mesh, oak, cinder block and canvas) (Figs. 5.10, 5.11, 5.13). Some of the surfaces, are, however, more predictable than others. The wire mesh surface offers a linear continuum of support at the same level, although the surface is curved. This may allow for lateral compliance in a grip-like fashion (Robinson, 1975). The wood and canvas surfaces are less predictable, but offer a lesser level of undulance than do the sandstone surfaces (Figs. 5.5, 5.6 and 5.7). Overall, the sandstone surfaces were the most undulant, and yield profile heights similar to those of 60 grit sandpaper (Figs. 5.5 and 5.6).

In a more global sense the sandstone surfaces differ from sandpaper and other substrates considered in one very important way; they are much less regular in terms of the arrangement of surface features and potential contact points, and therefore also much less predictable. The depth-coloured image of 60 grit sandstone (Fig. 5.9A) shows that the potential areas of contact on this surface are more evenly distributed than for the sandstone surface (Fig. 5.8B) where the available areas tend to be randomly distributed, and concentrated into a single local region. There is also more variation between replicate measurements and estimates taken from individual sandstone samples than between those taken from the other surfaces, such as wood or sandpaper. This provides further evidence of the unpredictability of the natural surfaces that these animals encounter in nature. On any given footfall the amount of area available for contact could vary substantially, although this depends on the depth to which the system can comply with the surface and achieve contact. When a more limited depth of contact is assumed (the uppermost 9 μm as opposed to 50 μm or 100 μm) the amount of available contact is fairly similar for all of the sandstone surfaces (ranging from 0.15 to 0.37% of the surface available) even though they differ more substantially when a greater depth is considered (Fig. 5.10). This suggests that at some level, perhaps the level at which setal contact and attachment is occurring, even unpredictable, rough surfaces, may be fractally predictable, thereby allowing the development of an adhesive system with setal field-scale dimensions that provide adequate adhesive contact on any given footfall on any surface the animal may encounter in its environment.

The evaluation and characterization of other test surfaces also allows comparison of the results to those of previous studies examining surface characteristics and adhesion in geckos. Vanhooydonck et al. (2005) investigated the influence of available contact area of wood, canvas and wire mesh, on the acceleration and maximum speed of *Hemidactylus* geckos. However, their evaluation of the surfaces relied solely upon two dimensional measurements, and can be re-evaluated based on a three-dimensional consideration of their attributes. Vanhooydonck et al. (2005) estimated that 98% of a wooden surface was available for attachment by setae, 41% for a canvas surface, and 37% for a wire mesh surface. In contrast, three dimensional analysis of comparable surfaces revealed that much less area is likely available for contact by the setae. The results indicate that, if the setae of *Hemidactylus* are similar to those of other geckos examined, anywhere from 0.12 to 33% of the surface may be available for attachment on wood, 0.07–19% on wire mesh and 0.01–7% on canvas (Fig. 5.10). Even the highest estimates, lying within the top 100 μm, are much lower than those made using two-dimensional assessments, especially in the case of the wooden surface, which is not nearly as smooth as suggested by two-dimensional observation (Vanhooydonck et al., 2005).

5.5.3 Conclusions: A Return to Maderson's Questions

Maderson asked (i) what precise locomotory advantage do setae provide that cannot be adequately met by claws or digital flexure?, and (ii) is there any evidence that some or all seta-bearing taxa live on surfaces that have any particular physical characteristics that make setae particularly advantageous? This initial investigation of *Rhoptropus* cf *biporosus* cannot address the entirety of these issues, but by investigating the animal and its preferred locomotor substratum, some initial conclusions can be drawn, and some additional hypotheses erected for further testing.

In addressing the first question, *Rhoptropus* cf *biporosus* is informative in that it is rupicolous, lives on undulant, unpredictable surfaces, and lacks claws. The irregularity and randomness of asperities of the scale and distribution evident in the sandstone samples (Figs. 5.5, 5.8), and the unyielding nature of the locomotor substratum, all indicate that fields of setae may provide a locomotor advantage that individual setae alone might not suggest. Placement of digits onto such a locally unpredictable and unyielding surface would likely mean that adequate purchase provided by claws alone could not be attained without, at best, considerable slippage until an appropriate site of anchorage had been located. Fields of setae, given the random distribution of available support areas, allow islands of suitable substratum to be located and utilized. Furthermore, this will be the case for all body orientations, from vertical to overhanging, to inverted, a situation certainly not so for claws (Cartmill, 1985). From footfall to footfall the actual distribution of contact is likely to vary substantially within and between digits, but within the confines of the setal field our analyses of substratum pattern indicate that sufficient area is available for contact to support the mass of the individual during locomotion. This opportunistic and variable location of support, rather than a focal point of support, such as a claw, also helps to explain the apparent "overbuilding" of adhesive capability. In the real world situation of this species the available contact at any footfall will be far less than the maximal output that the digits could generate in ideal situations, but will likely always be sufficient to support the individual in question.

Furthermore, a seta-based adhesive system essentially allows static contact, so that patches of support are not compromised by distortion due to thrust production during the power stroke of locomotion. Claws engage in surface irregularities and remain in contact for as long as possible during the stance phase, often being subject to considerable rotational and shear loading as pedal plantarflexion raises the wrist and ankle to place loading on the digital tips (Arnold, 1998). In contrast to this, setal fields are subject to release via digital hyperextension, with the setal field relinquishing contact with the substratum prior to the commencement of pedal plantarflexion (Russell, 2002). Thus, during the period of their contact with the substratum, the setae are maintained in a static configuration under tensile loading (Fig. 5.1) and at an angle that makes adhesion possible. The limb passes over the anchored foot, providing forward thrust, but the setal field is unloaded and released through hyperextension (Fig. 5.1) prior to wrist and ankle extension beginning, with an alternate set of setal fields on another limb taking up the supporting role by the unfurling of digits to re-establish adhesive contact (Russell, 1975). Such utilization of static patches of support means that localized islands of intimate contact enable the programmable adhesive of the setal fields to exploit available patches and maximize the contact and bonding that is made. This localizaed, brief but static contact may be enhanced by intricate vascular compliance mechanisms (Russell, 1981) that assist in the establishment of patchy contact.

The observations made herein concerning the relationship of features of the locomotor substratum to the form of the setae and setal fields also provide further insights about the degree of contact splitting and setal dimensions seen in geckos.

The greater deformability and consequent increase in real contact area engendered by the extreme levels of contact splitting seen in geckos (Arzt et al., 2003; Persson, 2003; Persson and Gorb, 2003; Peressadko and Gorb, 2004; Spolenak et al., 2004) will enhance the adhesive capabilities of a system in circumstances in which only a small portion of the available adhesive contacts can be deployed. For geckos, setal fields are large in comparison with the physical attributes of the terrain they traverse (Fig. 5.8), and conformation to the minutiae of the interstices is not possible. Thus, for localized areas where contact can be established, maximization of contact and adhesive force generation is of major importance, and extensive contact splitting will assist this.

The second question pertains to the nature of surfaces exploited by setae. Here again it is likely that setal fields, their control and their deployment have a bearing on this. Although this single example of *Rhoptropus* cf *biporosus* cannot address a question about surfaces in general, it presents what might be an extreme case that can inform us about setal usage in a more general sense. Setal fields may be particularly useful on undulant and unpredictable surfaces where contact is likely to be patchy and limited. Subdigital pads of geckos exhibit a wide array of pattern and "design" variation (Russell, 1976) and across species there is a great deal of variation in subdigital pad area as it relates to overall body size (Bauer and Good, 1986). It is at the level of the setal fields, rather than the individual setae, that some understanding of this variation might be sought, as there may be an as yet unexplored relationship between pad area dimensions (available surface area) and substrate characteristics. This question remains for future explorations, and will be addressed through a more extensive investigation of substrate characteristics and setal field dimensions across the genus *Rhoptropus* and its sister taxon.

Rhoptropus cf *biporosus* manages to effect adhesion on highly irregular and undulant surfaces while moving rapidly and with seemingly equal facility in all orientations. The ability to exploit such irregular and challenging (from the perspective of adhesion) surfaces attests to the versatility and adaptability of the gekkonid adhesive system. It also helps to explain why setal fields have the dimensions and plethora of potential adhesive contacts that they do, and is germane to further understanding of why the system should be seemingly so over endowed with adhesive capability. Indeed, if uneven, relatively unpredictable surfaces were involved in tractive, frictional interactions initially, this may help to explain the variety of setal form, their placement, and the dimensions of the setal fields as they became adapted to a molecular bonding-based mode of attachment. Exploitation of smooth surfaces using such a system is likely a secondary eventuality and not a driving force in the evolution of the configuration and dimensions of the system.

Acknowledgments An NSERC Discovery grant to APR, NSERC studentship grants to MKJ and SMD, and an Alberta Ingenuity Fund grant to MKJ supported this research. We also thank Heather Jamniczky, Aaron M. Bauer and the field party in Namibia for their help collecting the material used in this research. We are grateful to the University of Calgary Microscopy and Imaging Facility where all SEM work was done. Vuong Bang and George Heil of Mipox International Corporation kindly made available samples of polishing film for examination.

References

Alibardi, L. (2003) Ultrastructural autoradiographic and immunocytochemical analysis of setae formation and keritinization in the digital pads of the gecko *Hemidactylus turcicus* (Gekkonidae, Reptilia). *Tissue Cell*, 35: 288–296.

Arnold, E. M. (1998) Structural niche, limb morphology and locomotion in lacertid lizards (Squamata: Lacertidae): A preliminary survey. *Bull. Nat. Hist. Mus. Lond. (Zool)*, 64: 63–89.

Arzt, E., Gorb, S. and Spolenak, R. (2003) From micro to nano contacts in biological attachment devices. *Proc. Nat. Acad. Sci.*, 100: 10603–10606.

Autumn, K. (2006) How gecko toes stick. *Am. Sci.*, 94: 124–132.

Autumn, K. and Peattie, A. M. (2002a) Mechanisms of adhesion in geckos. *Integr. Comp. Biol.*, 42: 1018–1090.

Autumn, K., Liang, Y. A., Hsleh, S. T., Zesch, W., Chan, W. P., Kenny, T. W., Fearing, R. and Full, R. J. (2000) Adhesive force of a single gecko foot-hair. *Nature*, 405: 681–685.

Autumn, K., Sitti, M., Liang, Y. A., Peattie, A. M., Hansen, W. R., Sponberg, S., Kenny, T. W., Fearing, R., Israelachvili, J. N. and Full, R. J. (2002b) Evidence for van der Waals adhesion in gecko setae. *Proc. Nat. Acad. Sci.*, 99: 12252–12256.

Bartholomew, G. A. (2005) Integrative biology, an organismic biologist's point of view. *Integr. Comp. Biol.*, 45: 330–332.

Bauer, A. M. (1999) Evolutionary scenarios in the *Pachydactylus* Group geckos of southern Africa: New hypotheses. *Afr. J. Herpetol.*, 48: 53–62.

Bauer, A. M. and Good, D. A. (1986) Scaling of scansorial surface area in the genus *Gekko*. In *Studies in Herpetology*, edited by, Rocek, Z. Prague: Charles University, pp. 363–366.

Bauer, A. M. and Good, D. A. (1996) Phylogenetic systematics of the day geckos, genus *Rhoptropus* (Reptilia: Gekkonidae), of south-western Africa. *J. Zool. Lond.*, 238: 635–663.

Bauer, A. M. and Lamb, T. (2002) Phylogenetic relationships among members of the *Pachydactylus capensis* group of southern African geckos. *Afr. Zool.*, 37: 209–220.

Bauer, A. M. and Lamb, T. (2005) Phylogenetic relationships of southern African geckos in the *Pachydactylus* Group (Squamata: Gekkonidae). *Afr. J. Herpetol.*, 54: 105–129.

Bauer, A. M., Böhme, W., and Weitschat, W. (2005) An early Eocene gecko from Baltic amber and its implications for the evolution of gecko adhesion. *J. Zool. Lond.*, 265: 327–332.

Bauer, A. M., Russell, A. P. and Powell, G. L. (1996) The evolution of locomotor morphology in *Rhoptropus* (Squamata: Gekkonidae): Functional and phylogenetic considerations. *Afr. J. Herpetol.*, 45: 8–30.

Bergmann, P. J. and Irschick, D. J. (2005) Effects of temperature on maximum clinging ability in a diurnal gecko: Evidence for a passive clinging mechanism? *J. Exp. Biol.*, 303A: 785–791.

Bock, W. J. and Von Wahlert, G. (1965) Adaptation and the form-function complex. *Evolution*, 19: 269–299.

Branch, B. (1988) *Field Guide to the Snakes and other Reptiles of Southern Africa*. Cape Town: Struik Publishers.

Briggs, G. A. D. and Briscoe, B. J. (1976) Effect of surface roughness on rolling friction between elastic solids. *Nature*, 260: 313–315.

Campolo, D., Jones, S. and Fearing, R. (2003) Fabrication of gecko foot-hair like nano structures and adhesion to random rough surfaces. *Nanotechnology*, 2: 856–859.

Cartmill, M. (1985) Climbing. In *Functional Vertebrate Morphology*, edited by Hildebrand, M., Bramble, D. M., Liem, K. F., and Wake, D. B. Cambridge, MA: The Belknap Press of Harvard University Press, pp 73–88.

Chow, T. S. (2003) Size-dependent adhesion of nanoparticles on rough substrates. *J. Phys. Condens. Matter*, 15: L83–L87.

Currey, J. (1984) *The Mechanical Adaptations of Bones*. Princeton: Princeton University Press.

Delannoy, S. (2006) *Subdigital setae of the Tokay gecko (Gekko gecko): Variation in form and implications for adhesion* . Department of Biological Sciences University of Calgary, MSc. Thesis, 227pp.

Gasc, J. P. and Renous, S. (1980) Les différentes formations piliformes de la surface épidermique sur la face palmaire chez *Coleodactylus amazonicus* (Anderson, 1918) (Sphaerodactylinae, Sauria), Lézarde de la litière dans les forêts de Guyane française. *C.R. Acad. Sci. Paris serie D*, 290: 675–678.

Gasc, J. P., Renous, S. and Diop, A. (1982) Structure microscopique de l'épiderme palmaire du saurien *Coleodactylus amazonicus* (Anderson, 1918) (Sphaerodactylinae), comparée à celle de l'épiderme des feuilles de la litière, substrat locomoteur de l'animal. *C.R. Acad. Sci., Paris serie III*, 294: 169–174.

Gay, C. (2002) Stickiness – some fundamentals of adhesion. *Integr. Comp. Biol.*, 42: 1123–1126.

Glassmaker, N. J., Jagota, A., Hui, C. Y. and Kim, J. (2004) Design of biomimetic fibrillar interfaces: 1, Making contact. *J. Roy. Soc. London Interfaces*, (doi:10.1098/rsif.2004.0004).

Grünert, N. (2000) *Namibia – Fascination of Geology*. Windhoek: Klaus Hess Publishers.

Hiller, U. (1968) Untersuchungen zum feinbau und zur funktion der haftborsten von reptilien. *Morph. Tiere*, 62: 307–362.

Hiller, U. (1969) Correlation between corona-discharge of polyethylene films and the adhering power of *Tarentola m. mauritanica* (Rept.). *Forma et Functio*, 1: 350–352.

Huber, G., Gorb, S. N., Hosada, N., Spolenak, R., and Arzt, E. (2007). Influence of Roughness on Gecko Adhesion. *Acta Biomaterialia* 3: 607–610.

Huber, G., Gorb, S. N., Spolenak, R., and Arzt, E. (2005b). Resolving the nanoscale adhesion of individual gecko spatulae by atomic force microscopy. *Biol. Lett.*, 1: 2–4.

Huber, G., Mantz, H., Spolenak, R., Mecke, K., Jacobs, K., Gorb, S. N. and Arzt, E. (2005a) Evidence for capillarity contributions to gecko adhesion from single spatula nanomechanical measurements. *Proc. Nat. Acad. Sci.*, 102: 16293–16296.

Hui, C. Y., Glassmaker, N. J., Tang, T. and Jagota, A. (2004) Design of biomimetic fibrillar interfaces: 2. Mechanics of enhanced adhesion. *J. Roy. Soc. London Interfaces* (doi10.1098/rsif.2004.005).

Irschick, D. J., Austin, C. C., Petren, K., Fisher, R. N., Losos, J. B. and Ellers, O. (1996) A comparative analysis of clinging ability among pad-bearing lizards. *Biol. J. Linn. Soc.*, 59: 21–35.

Jagota, A. and Bennison, S. J. (2002) Mechanics of adhesion through a fibrillar microstructure. *Integr. Comp. Biol.*, 42: 1140–1145.

Johnson, M. K., Russell, A. P. and Bauer, A. M. (2005) Locomotor morphometry of the *Pachydactylus* radiation of lizards (Gekkota: Gekkonidae): A phylogenetically and ecologically informed analysis. *Can. J. Zool.*, 83: 1511–1524.

Lamb, T. and Bauer, A. M. (2000) Relationships of the *Pachydactylus rugosus* group of geckos (Reptilia: Squamata: Gekkonidae). *African Zoology*, 35: 55–67.

Lamb, T. and Bauer, A. M. (2001) Mitochondrial phylogeny of Namib Day Geckos (*Rhoptropus*) based on cytochrome *b* and 16S rRNA sequences. *Copeia*, 2001: 775–780.

Lamb, T. and Bauer, A. M. (2002) Phylogenetic relationships of the large-bodied members of the African lizard genus *Pachydactylus* (Reptilia: Gekkonidae). *Copeia*, 2002: 586–596.

Maderson, P. F. A. (1970) Lizard hands and lizard glands: Models for evolutionary study. *Forma et Functio*, 3: 179–204.

Meine, K., Kloß, K., Schneider, T. and Spaltman, D. (2004) The influence of surface roughness on the adhesion force. *Surf. Interface Anal.*, 36: 694–697.

Peressadko, A. and Gorb, S. (2004) When less is more: Experimental evidence for tenacity enhancement by division of contact area. *Journal of Adhesion*, 80: 247–261.

Persson, B. N. J. (2003) On the mechanism of adhesion in biological systems. *Journal of Chemical Physics*, 118: 7614–7621.

Persson, B. N. J. and Gorb, S. (2003) The effect of surface roughness on the adhesion of elastic plates with application to biological systems. *Journal of Chemical Physics*, 119: 11437–11444.

Peterson, J. A., Benson, J. A., Ngai, M., Morin, J. and Ow, C. (1982) Scaling in tensile "skeletons": Structures with scale-independent length dimensions. *Science*, 217: 1267–1270.

Rimai, D. S. and Quesnel, D. J. (2001) *Fundamentals of Particle Adhesion*. Moorhead: Global Press.
Robinson, P. L. (1975) The function of the fifth metatarsal in lepidosauran reptiles. *Colloq. Int. C.N.R.S.*, 218: 461–483.
Ruibal, R. and Ernst, V. (1965) The structure of the digital setae of lizards. *J. Morphol.*, 117: 271–294.
Russell, A. P. (1975) A contribution to the functional analysis of the foot of the Tokay, *Gekko gecko* (Reptilia: Gekkonidae). *J. Zool. Lond.*, 176: 437–476.
Russell, A. P. (1976) Some comments concerning interrelationships amongst gekkonine geckos. In *Morphology and Biology of Reptiles, Linnean Society Symposium Series 3*, ed (Eds, Bellairs, A. d' A. and Cox, C. B.).London: Academic Press, pp. 217–244.
Russell, A. P. (1979) Parallelsim and integrated design in the foot structure of gekkonine and diplodactyline geckos. *Copeia*, 1: 1–21.
Russell, A. P. (1981) Descriptive and functional anatomy of the digital vascular system of the Tokay, *Gekko gecko*. *J. Morphol.*, 169: 293–323.
Russell, A. P. (1986) The morphological basis of weight-bearing in the scansors of the tokay gecko (*Gekko gecko*). *Can. J. Zool.*, 64: 948–955.
Russell, A. P. (2002) Integrative functional morphology of the Gekkotan adhesive system (Reptilia: Gekkota). *Integr. Comp. Biol.*, 42: 1154–1163.
Russell, A. P., Bauer, A. M. and Laroiya, R. (1997) Morphological correlates of the secondarily symmetrical pes of gekkotan lizards. *J. Zool. Lond.*, 241: 767–790.
Scherer, S. (2002) 3D surface analysis in scanning electron microscopy. *G.I.T. Imaging and Microscopy*, 3: 45–46.
Scherge, M. and Gorb, S. N. (2001) *Biological Micro- and Nanotribology: Nature's Solutions*. Berlin: Springer-Verlag.
Spolenak, R., Gorb, S. and Arzt, E. (2005) Adhesion design maps for bio-inspired attachment systems. *Acta Biomaterialia*, 1: 5–13.
Spolenak, R., Gorb, S., Gao, H. and Artz, E. (2004) Effects of contact shape on the scaling of biological attachments. *Proc. R. Soc. Lond. A*, 2004: 1–15. 10.1098/rspa2004.1326.
Sun, W., Neuzil, P., Kustandi, T. S., Oh, S. and Samper, V. D. (2005) The nature of the Gecko Lizard adhesive force. *Biophys. J.: Biophys. Lett.*, 89: L14–L17
Vanhooydonck, B., Andronescu, A., Herrel, A. and Irschick, D. J. (2005) Effects of substrate structure on speed and acceleration capacity in climbing geckos. *Biol. J. Linn. Soc.*, 85: 385–393.
Yamamoto, K., Tanuma, C. and Gemma, N. 1995. Competition between electrostatic and capillary forces acting on a single particle. *Jap. J. Appl. Physics* 34: 4176–4184.
Yu, N. and Polycarpou, A. A. (2004) Combining and contacting of two rough surfaces with asymmetric distribution of asperity heights. *J. Tribol.*, 126: 225–232.
Yurdumakan, B., Raravikar, N. R., Ajayan, P. M. and Dhinojwala, A. (2005) Synthetic gecko foot-hairs from multiwalled carbon nanotubes. *Chem. Commun.*, 2005: 3799–3801.
Zani, P. A. (2000) The comparative evolution of lizard claw and toe morphology and clinging performance. *J. Evol. Biol.*, 13: 316–325.

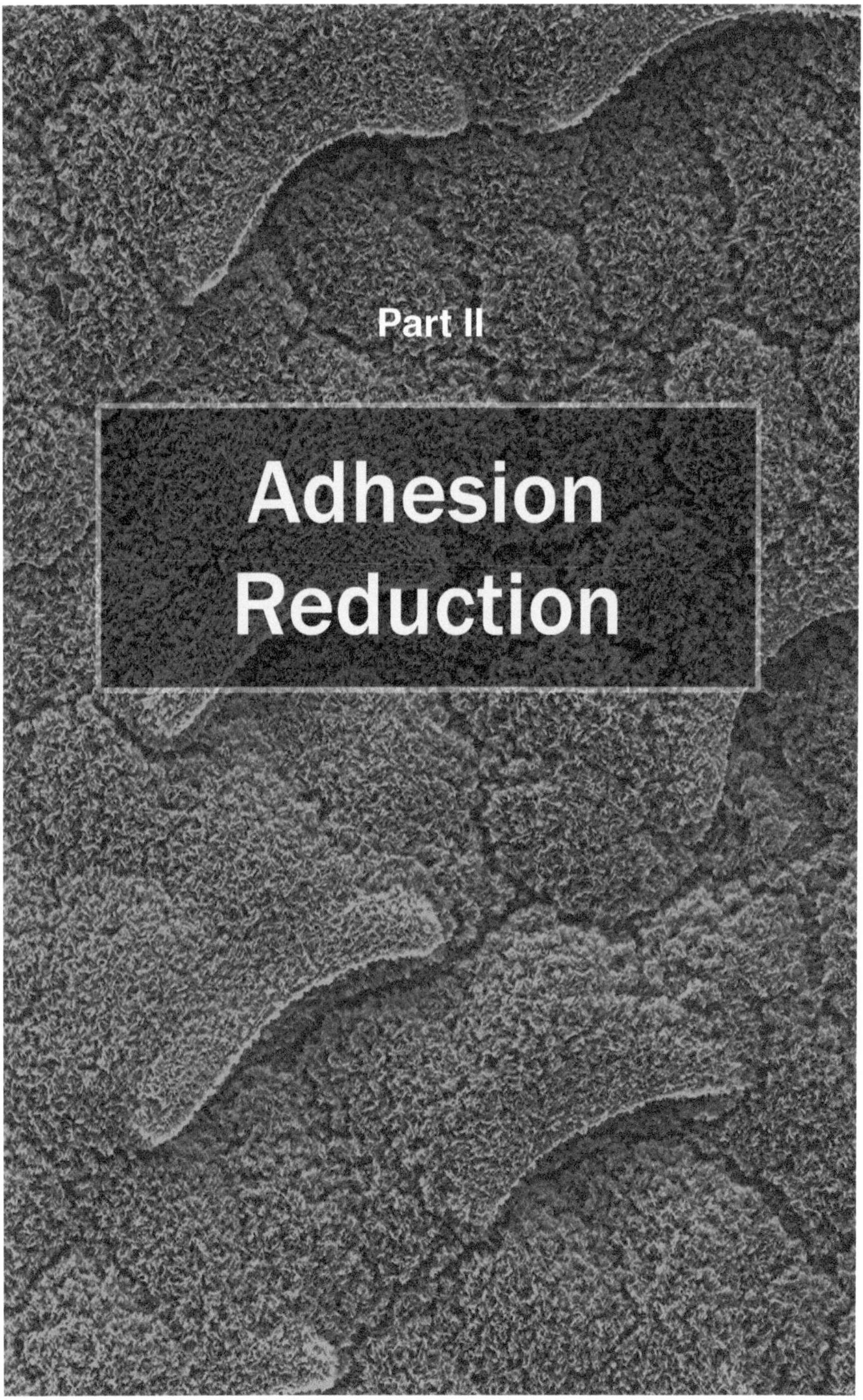

Part II

Adhesion Reduction

Chapter 6
Variable Attachment to Plant Surface Waxes by Predatory Insects

Sanford D. Eigenbrode, William E. Snyder, Garrett Clevenger, Hongjian Ding and Stanislav N. Gorb

In this chapter we will (1) provide an overview of evidence of the influence of habitat characteristics on predation, (2) provide specific evidence for the importance of the plant as habitat, and of variability in plant morphology's impact on the foraging of insect predators and parasitoids, (3) focus on the role of crystalline waxes on plant surfaces in mediating these types of interactions through their effects on insect attachment, (4) illustrate the implications of the variability of plant surface waxes and insect responses to surface waxes through a case study examining the attachment and performance of five species of predatory beetle on plants differing in surface wax, and (5) discuss implications for ecology of predation along with opportunities for further research.

6.1 Plant Traits and Carnivorous Insects

Habitat characteristics can affect the ability of predators to secure prey and therefore to impact prey populations. Habitat-dependent variation in predation has been observed in predatory taxa including snakes (Heinen, 1993), fish (Beukers and Jones, 1998; Nemeth, 1998), echinoderms (Dolmer, 1998), crabs (Bourgeois et al., 2006), insects and spiders (Rypstra et al., 1999; Finke and Denno, 2002; Farji-Brener, 2003; Denno et al., 2005), mites (Roda et al., 2001), and microbes (Ratnam et al., 1982; Rodriguezzaragoza, 1994). Mechanisms include variability in spatial refugia or opportunities for crypsis by prey, and habitat effects on predator and prey mobility or capacity to capture prey. Habitat dependent variability in foraging success of predators can influence habitat selection by altering the behavior of both predators (e.g., Webster and Hart, 2004) and prey (e.g., Sotka et al., 1999, Kicklighter and Hay, 2006). Variability in abilities of predators to forage effectively on different substrates potentially can influence their competitive interactions and their collective or community level impacts of prey populations (Byers, 2002).

S.D. Eigenbrode (✉)
Department of Plant, Soil and Entomological Sciences, University of Idaho, Moscow, ID, USA
e-mail: sanforde@uidaho.edu

S.N. Gorb (ed.), *Functional Surfaces in Biology*, Vol. 2,
DOI 10.1007/978-1-4020-6695-5_7, © Springer Science+Business Media B.V. 2009

6.1.1 Insect Predators and the Plant Surface

For many arthropod predators and parasitoids, the habitat in which they forage is predominantly the host plants of their prey species. Variability in both larger scale plant architectural features and fine morphological structure of the plants can affect these carnivores. The mechanisms range from differences in searching efficiency on plants with simple or complex architecture (Kareiva and Sahakian, 1990; Grevstad and Klepetka, 1992; Gingras et al., 2002, 2003; Andow and Olson, 2003; Legrand and Barbosa, 2003), effects of simple or glandular trichomes on searching, settling preferences or survival of predators (Agrawal and Karban, 1997; Roda et al., 2001; Hare, 2002; Kennedy, 2003; Romeis et al., 2005) and effects of crystalline waxy blooms on attachment and foraging behavior of predators and parasitoids (Eigenbrode, 2004). As a result of these effects and others mediated through the host plant, effectiveness of predators and parasitoids can differ substantially among plant species (Nordlund et al., 1988; Lill et al., 2002) and among genotypes within plant species (Pimentel, 1961; Pimentel and Wheeler, 1973; Hare, 1992; 2002). These effects are large enough to affect prey population dynamics and potentially influence the evolution of host plant specialization by insect herbivores species (Price et al., 1980; Bernays and Graham, 1988; Marquis and Whelan, 1996; Lill et al., 2002). Insofar as they influence prey populations and thereby host plant fitness, natural selection could select for plant traits that improve the effectiveness of predators and parasitoids (Price et al., 1980). From a practical standpoint, it may be possible to breed for crop plants with traits that enhance the effectiveness of predators and parasitoids (Hare, 1992; Bottrell et al., 1998; Cortesero et al., 2000).

Such possibilities are complicated, however, by the variable responses of carnivorous and herbivorous insects to different plant traits. Differences in plant form do not influence all species similarly and some effects are directly in opposition. For example Nordlund et al. (1988) reported that parasitism of eggs of *Helicoverpa zea* was greater on tomato than maize, but several predator species were more abundant on maize than tomato in the same study.

6.2 Attachment to Plant Surface Waxes and Predator–Prey Interactions

Many predators actively search for and subdue prey (Bell, 1991). Attachment to biological surfaces comes into play in these interactions in the devices employed by predators for securing prey (Betz and Kölsch, 2004) and for the devices employed by predators to attach to and run over the substrate while foraging (Eigenbrode et al., 1996). For arthropod predators of herbivores, the substrate is the plant surface, and attachment to plant surfaces typically requires attachment or adhesion to the surface waxes that overlay the cuticles of terrestrial plants (Jeffree, 1986).

6.2.1 Plant Surface Waxes

Plant surface waxes are mixtures of lipophilic compounds deposited within and on the exterior of the plant cuticle. Surface waxes typically are dominated by long-chain aliphatic hydrocarbons and secondary alcohols and ketones, with odd chain lengths predominating, and primary alcohols and fatty acids, with even chain lengths predominating. Other aliphatic compound classes and triterpenoids also can occur and predominate, depending upon the plant species (Baas and Figdor, 1978; Niemann, 1985; Walton 1990; Smith and Severson, 1992; Yang et al., 1993; Riederer and Markstädter, 1996). Although the primary function of plant surface waxes is to waterproof the cuticle (Schönherr 1976), they apparently have other ecological functions (Riederer and Müller, 2006) including providing some defense against pathogens (Blakeman, 1973; Carver et al., 1990; Nielsen et al., 2000). While clearly conferring benefits to plants, surface waxes are exploited by some herbivorous insects as cues for host recognition (Eigenbrode and Espelie, 1995; Udayagiri and Mason, 1995; 1997; Spencer, 1996; Eigenbrode and Pillai, 1998; Spencer et al., 1999; Morris et al., 2000; Cervantes et al., 2002; Müller and Riederer, 2005).

As pure materials, some plant wax constituents differ in surface energy or wettability (Holloway, 1969; Gorb; Gorb and Eigenbrode; unpublished data) and in properties that can affect the ability of insects to attach to them (Eigenbrode and Jetter, 2002, Gorb et al., unpublished data). As they occur on the plant surface, however, plant waxes are complex and variable mixtures that probably exert a range of possible effects on attachment by insects and other organisms. First, plant waxes are not deposited uniformly but occur in chemically distinct layers depending on their proximity to the cuticle or exterior of the plant. New methods have allowed this heterogeneity to be described (Jetter et al., 2000; Gorb et al., 2005). Second, the outermost layer is often elaborated into crystalline structures (Barthlott et al., 1998) and this crystalline epicuticular wax (CEW) interferes with attachment to the plants by insects and other organisms (Stork, 1980; Eigenbrode et al., 1998a,b; 1999; Duetting et al., 2003).

6.2.2 Functions of CEW on Plants

CEW where they occur are presumably adaptive for the plants that possess them. The waxy blooms are reflective and could reduce the amount of UV entering the plant tissues. The slipperiness of CEW could prevent the attachment of debris and provide plants with a self cleaning mechanism (Barthlott and Neinhuis, 1997). Lipophilic CEW could be more resistant to water penetrating or accumulating on the plant cuticle. The slipperiness of crystalline epicuticular waxes for insects is obviously adaptive for certain carnivorous plant species, notably *Nepenthes* spp. that trap insects in slippery-walled pitchers (Knoll, 1914; Juniper and Burras, 1962; Riedel et al., 2003; Gorb et al., 2005). Pitcher plant CEW are clearly selected for characteristics that render them slippery for insects. Although waxes on surfaces of other plant species are certainly constrained by other requirements and functions of waxes on stems, leaves, fruit and flowers, some are extremely slippery for insects and this effect could be adaptive. On the stems of certain plants such as *Salix alba*, CEW may serve to reduce depredation of pollen by ants (Harley, 1991; B. Juniper; personal communication). CEW also could protect plants from certain types of insect herbivory (Eigenbrode and Espelie, 1995; Juniper, 1995), if herbivores are unable to remain on plants to feed.

How CEW on plants disrupt insect attachment has yet to be resolved. There remain at least four possible mechanisms to explain it: 1) CEW are easily removed from the plant surface, reducing attachment, 2) waxes adsorb and render ineffective adhesive secretions of insect tarsi, 3) waxes present a rough surface that reduces the effective surface area available for adhesion to the points of maximum relief on that surface, 4) waxes can be dissolved by insect pad secretion and by this increase hydroplaning of the surface (Gorb, 2001; Gorb and Gorb, 2002). The distinctive crystals of *Nepenthes alata*, which are formed of mixtures of aldehydes, evidently impair insect attachment by minimizing potential attachment surface for the insect tarsal structures and possibly by reducing the wetting of the plant surface by whatever adhesive fluids are secreted by the insect tarsi (Riedel et al., 2003, see chapters by Gorb and Gorb and by Di Guisto et al., in this volume). The recent work with these plants has not supported prior conjecture that slipperiness of *Nepenthes* waxes is due to the wax crystals becoming easily dislodged from the pitcher wall (Knoll, 1914; Juniper and Burras, 1962). Rather, the wax crystals in conjunction with micromorphological features of the pitcher surface together represent a complex adaptive syndrome to reduce insect attachment (Gaume et al., 2002). The mechanisms by which CEW reduce insect attachment to other plant surfaces are evidently diverse. For example, some plants bear mechanically stable wax crystals, which, although not easily removed can reduce adhesion by increasing roughness of the surface, thereby minimizing contact area between the plant surface and insect attachment devices (Peressadko and Gorb, 2004; Gorb et al., 2005). On the other hand, scanning electron micrographs of the tarsi of insects after walking on plants with CEW show that the crystals from some plant species are indeed dislodged from the plant surface and become attached to and impair the adhesive structures of the insects (Stork, 1980; Eigenbrode et al., 1996; Gorb and Gorb, 2006).

Whatever the mechanism, the role of CEW in rendering plants slipperiness is apparent. Mechanical removal of the crystals from the slippery surfaces of carnivorous plants (Knoll, 1914), or from the leaves or stems of other plants with prominent waxy blooms, renders the surfaces less slippery (Bodnaryk, 1992; Federle et al., 1997). And the effects are large. Attachment forces by insects to *Brassica* spp. plants and *Pisum sativum* plants with normal waxy blooms are orders of magnitude greater than attachment to plants with waxy blooms removed mechanically or reduced due to mutations in wax biosynthesis or deposition (Stork, 1980; Eigenbrode et al., 1998a; Eigenbrode and Kabalo, 1999; Eigenbrode and Jetter, 2002). Among plant surfaces from 83 plant species tested for effects on attachment by the beetle *Chrysolina fastuosa*, stem of *Acer negundo*, with CEW, disrupted attachment substantially as compared with smooth surfaces (Gorb and Gorb, 2002). Finally, the importance of crystalline structure for disruption of insect attachment can be demonstrated experimentally. Plant wax mixtures or pure waxy components deposited on glass slides as crystals are more slippery (allow lower attachment forces by insects) than the same materials deposited as fused noncrystalline solids (Eigenbrode and Jetter, 2002).

6.2.3 Functional Significance of CEW for Insect Herbivores, Predators and Parasitoids

By reducing attachment to plants, CEW can provide protection against attack by some herbivores (Bodnaryk, 1992). Some reduced wax variants of crop plants become more susceptible to certain insect pests (Eigenbrode and Espelie, 1995). But CEW also can impair attachment by insect predators and parasitoids, interfering with the effectiveness of these insects at suppressing the insect herbivores they use as prey (Eigenbrode et al., 1996; 1999, 2000; Eigenbrode and Kabalo, 1999; White and Eigenbrode 2000b; Chang and Eigenbrode, 2004; Chang et al., 2004). Attachment ability (measured with a centrifuge or force transducer), efficacy, foraging behavior, or all of these have been found to be impaired by CEW for all predators and 4 of the 5 parasitoids examined so far (reviewed in Eigenbrode, 2004). Most of these studies have been carried out using plants expressing single gene mutations that reduce CEW in *Brassica oleracea, B. napus*, or *Pisum sativum*. The effect of CEW on attachment and predation appears quantitative. Using a range of genetic mutations that reduce CEW to varying degrees in *Brassica oleracea*, attachment forces generated by two predators, adult *Hippodamia convergens* (Coleoptera: Coccinellidae) and larval *Chrysoperla plorabunda* (Neuroptera: Chrysoperlidae), were found to be positively correlated with the rate of consumption of early instar caterpillars (*Plutella xylostella*) by each predator species on the plants (Eigenbrode and Kabalo, 1999; Eigenbrode et al., 1999; Eigenbrode, 2004).

This indirect effect of CEW on insect herbivores can help explain the preponderance of cases in which a reduced CEW confers apparent resistance to herbivores in the field (Eigenbrode et al., 1996; Eigenbrode, 2004) and suggests that reduced CEW could be beneficial in agricultural crops by promoting the effectiveness of important natural enemies (Eigenbrode et al., 2000). They also suggest that variation in CEW in natural systems can influence the capacity of natural enemies to suppress herbivore populations.

6.2.4 Variation in Insect Attachment to CEW

Insect herbivores and carnivores differ in their ability to attach to or compensate for poor attachment to CEW. For example, among 11 species of *Eucalyptus*-feeding chrysomelid beetle species studied by Edwards and Wanjura, (1991), *Chrysophtharta m-fuscum* and *Paropsis aegrota* predominated in the field on *Eucalyptus bicostata* and *E. bridgeiana*, each of which has a prominent CEW. Although attachment forces generated by these species to waxy leaves were not different from other *Eucalyptus* feeding chrysomelids, these two beetle species exhibit behaviors that facilitate their feeding on plants with the CEW, including straddling the leaf edge. Two crucifer-feeding aphid species, *Brevicoryne brassicae* and *Lipaphis erysimi* when placed on the upper surfaces of *Brassica napus* plants, move from upper to lower leaf surfaces within a few minutes. *B. brassicae* completes this maneuver equally rapidly on *Brassica napus* with a range of wax loads, whereas *L. erysimi* does so significantly more slowly on *B. napus* with CEW (Åhman, 1990).

Capacity to attach to CEW also differs among insect carnivores. The parasitoid wasp, *Eretmocerus* sp., studied by McAuslane et al. (2000) was equally effective at parasitizing *Bemisia argentifolii* on *Brassica oleracea* plants with either prominent or reduced CEW, whereas *Encarsia formosa* was more effective on plants with reduced CEW. Two other *Eretmocerus* spp. appear to be less hampered on *B. oleracea* CEW than does *Amitus bennetti*, and *Orius leavigatus* forages less effectively on *B. oleracea* with CEW than the mirid *Deraeocoris pallens* (D. Gerling, personal communication). Although attachment and rate of predation by *H. convergens* and *C. plorabunda* are reduced on *B. oleracea* with CEW, the relative effects are approximately twice as strong for *C. plorabunda* (Eigenbrode et al., 1999; Eigenbrode and Kabalo, 1999).

Variability in the capacity of insects to attach to CEW could be adaptive. For example, as part of a syndrome of host specialization for plants with waxy blooms (e.g., Brennan et al., 2001), or for predators and parasitoids that forage predominantly on plants with CEW.

The most striking example of variable adaptation to CEW among insects, and one in which adaptation appears to have played a role, occurs among ants associated with southeast Asian *Macaranga* spp. plants. Federle et al. (1997) showed that two ant species (*Crematogaster* [*Decacrema*] spp.) symbiotically associated with *Macaranga pruinosa* successfully climb the stems of this plant, which bear prominent CEW. Several ant species associated with another *Macaranga* species possessing CEW were also capable of climbing the stems of *M. pruinosa*. Eighteen other ant species tested could climb the stems of *M. pruinosa* with varying success, but none as well as those symbiotically associated with *Macaranga* species that have prominent CEW. Mechanical removal of CEW from *M. pruinosa* stems allowed all the tested species to climb these stems more successfully. The capacity for attachment to *Macaranga* CEW has apparently evolved in the ants as part of their coevolutionary association with *Macaranga* spp.

These differences suggest that certain predatory species may be more effective on plants with CEW and have a predominant role in reducing or regulating prey populations that occur on plants with CEW. Competitive interactions within guilds of carnivorous species may depend upon the relative abilities of predators to forage on surfaces with CEW and the occurrence of CEW in the habitat. Since the presence of CEW can cause as much as a 80% reduction in rates of prey consumption (Eigenbrode et al., 1999; Eigenbrode and Kabalo, 1999), these effects could be large. Not enough is known about variable attachment to CEW among insect carnivores and the implications of this variability for effectiveness of these insects.

6.3 Case Study: Responses of Five Coccinellidae Species to Crystalline Epicuticular Waxes

To make a first assessment of the variation in effects of CEW within a predatory guild, we compared the effects of CEW on attachment to plant surfaces and rate of prey consumption among five species of primarily aphidophagous Coccinellidae. The species selected for study were three native North American species, *Coccinella tranversoguttata* Brown, *Hippodamia convergens* Guerin Menéville and *Adalia bipunctata* (L.), and two introduced species *C. septempunctata* L. and *Harmonia axyridis* Pallas. All five species are now widely distributed in North America and occur in a variety of crop and other habitat types. *Adalia bipunctata* and *Ha. axyridis* occupy arboreal habitats more than the other species (Colunga-Garcia and Gage, 1998; Osawa et al., 2000; Sakuratani et al., 2000) but all species move among habitats to track resources.

6.3.1 Case Study: Experimental Methods

Two sister lines of peas, *Pisum sativum* L., differing in epicuticular waxes were developed from accession PI W6-15368 (Marx 406). As described in Eigenbrode et al. (1998b), the lines differ in expression of the wax mutation *wel* (Marx, 1969). One of the lines is homozygous for the wild-type allele *Wel* and has the prominent CEW typical of *P. sativum* on all photosynthetically active surfaces (flower petals have not been examined); the other line is homozygous for *wel* and has a very reduced CEW on all these surfaces. The mutation reduces wax coverage (mass/unit area) over the entire plant surface by approximately 80%, and changes wax composition as well as reduced CEW (Eigenbrode et al., 1998a,b). Both lines are also homozygous for the recessive mutation *tl* (acacia leaf), which expands tendrils to leaflets, providing greater surface area for foraging insects than do standard pea plants on which some leaflets are tendrils.

Peas were grown in 10 cm^2 pots in a 2:1 ratio of commercial potting mix (Sunshine Mix 1, SunGro Horticulture, Bellevue, WA) and sand. Plants were grown in a greenhouse at 25°C and a 16L:8D photoregime and were bottom-watered once a week. Seedlings at the 3 to 4 node stage were used in the experiments. Occasionally, it was necessary to use slightly older seedlings trimmed to 4 nodes.

Colonies of the coccinellids were established using individuals collected in eastern Washington. Founders of the *C. tranversoguttata, C. septempunctata* and *H. convergens* colonies were collected from agricultural fields and natural vegetation in the vicinity of the city of Pullman, WA, on the eastern edge of the state.

Although *Ha. axyridis* occurs near Pullman, WA, it is still rare there; therefore, we collected this species from agricultural fields in the vicinity of Prosser, WA, ca. 450 km southwest of Pullman, where this species has become relatively common. Coccinellids were maintained in 100 mm × 15 mm plastic Petri dishes on a mixed diet of pea aphid, *Acyrthosiphon pisum* (Harris), Russian wheat aphid, *Diuraphis noxia* (Kurdj.), and English grain aphid, *Sitobion avenae* (F.), at 22–25°C and a 16:8 L:D photoregime. Water was provided using a moistened dental wick. Larvae were separated at hatching and reared individually, on the same diet and under the same environmental conditions as adults.

Larvae were collected for use near the end of the penultimate (3rd) instar and allowed to molt to 4th instar immediately before being used in the experiments. Adults were collected from the colony soon after eclosion from pupae and sexed. Only female beetles were used in these experiments.

Acrythosiphon pisum used as prey in the predator efficiency experiments were reared on peas (cv. 'Columbia') and were from a laboratory colony developed from field collection from Whitman Co., WA.

Lateral attachment forces by adult females and 4th instar larvae of each species of coccinellid were measured with a centrifuge as described by Eigenbrode et al. (1999) and similar to that employed by Federle et al. (2000). To measure attachment, the test material (a *P. sativum* leaflet or glass) was fastened to a horizontal aluminum turntable (30 cm diameter). An individual insect was placed on the test material within a clear plastic canopy to eliminate effects of air resistance. The rotation speed of the turntable was increased at approximately Δ 4 rpm/second until the insect was separated from the leaf surface by centrifugal force. To permit observation of the insect during the test, a stroboscopic light directed at the insect was triggered to flash on each rotation of the turntable. The turntable revolutions per second (Hz) and the radius r [m] of the insect's location on the turntable when it was observed to detach were used to calculate the effective velocity v [ms^{-1}]. Together with the individual insect mass m [g] this gave the attachment force F [mN] according to:

$$F = m\,\frac{v^2}{r} = m\,\frac{(2\pi r \cdot Hz)^2}{r} = 4\pi^2\, m\, r\, Hz^2$$

Attachment by each insect was measured 2 to 3 times and the average of these values was used as a single observation. Attachment was measured to the upper leaf surface of each pea genotype (wild-type or reduced CEW). The insects of each developmental stage were tested over several days, taking care to test replicates of each pea CEW type and each insect species on each day. Each of the two experiments (adults and 4th instars) was analyzed with a factorial ANOVA for the effects of species, pea CEW type, and their interaction on attachment force achieved by the insect. A second pair of experiments was conducted following similar procedures to measure attachment by individuals of larvae of each species and adults of all species except *C. transversoguttata* to a clean glass coverslide. Each of these two experiments was analyzed as a single factor ANOVA with means separation performed using Tukeys HSD test. All analyses were conducted using SAS (Proc

GLM). The data conformed to the assumptions of ANOVA so no transformations were necessary.

To assess the rate of prey consumption by each species on plants of each CEW type, adult females and 4th instar larvae of each species of coccinellid were assessed for the number of pea aphids they removed from a caged pea plant during 24 h. Pea plants were grown under conditions described above until they reached the 4-node stage. The plants were then infested with pea aphids by placing a mass of approximately 50 aphids onto the plants with a small paint brush. After a 24-h settling period, the plants were inspected and aphids were removed to leave 40 healthy, feeding, pre-reproductive nymphs per plant. This number of aphids had been determined previously in a functional response assay to provide discriminating density for *H. convergens* on the same two pea genotypes (Rutledge and Eigenbrode, 2003). The infested pea plants were enclosed in 8-cm-diameter × 20-cm-tall acetate cylinders, covered on the top with organdy mesh and twisted into the soil in the pot to prevent aphids and predators from escaping. Prior to being added to the cages, the coccinellids to be tested were isolated from one another and given access only to water for 24 h. The predators were placed in the cage and allowed to feed on the aphids and the condition of the coccinellid was recorded after 24 h.

The experiments with larvae and adults were conducted in groups of replicates that included all five species of coccinellid on the two pea lines. This permitted us to analyze the results for each developmental stage of coccinellid as a single experiment with a factorial ANOVA for the response of aphids consumed (out of 40) and factors, coccinellid species, pea CEW type, and their interaction. Results for larvae and adults were analyzed as separate experiments. Analyses were conducted using SAS.

To compare the attachment structures of the insects, a subsample of tarsi from each stage of each species was examined with scanning electron microscopy. The legs of alcohol preserved specimens were removed under the dissection microscope, air dried, and mounted on aluminum stubs using carbon adhesive tape. Afterwards, the samples were sputter coated with gold-palladium (8–10 nm film thickness) and observed in Hitachi-S800 scanning electron microscope (SEM).

6.3.2 Case Study: Results

Adult females of each coccinellid species achieved lower attachment forces to pea plants with wild-type CEW than to those with reduced CEW (Fig. 6.1a), resulting in a significant effect of CEW on attachment. *Ha. axyridis* generated the greatest attachment forces, followed by *C. septempunctata*, and then the other species, which were similar to one another. This ranking of attachment forces among the coccinellid species was the same for wild-type and reduced CEW (Fig. 6.2a). The relative reduction in attachment force to leaves caused by the presence of CEW differed among the species based on the significant species × CEW interaction. A difference index (Krebs 1989) can be used to assess the effects of CEW on attachment for each

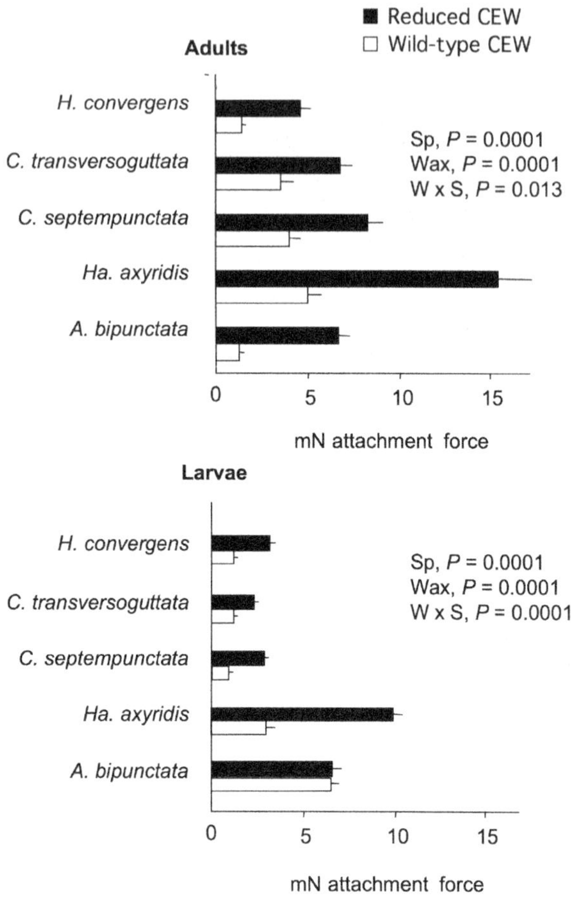

Fig. 6.1 Attachment by adult and larval Coccinellidae to pea plants with reduced crystalline leaf waxes (*filled bars*) and wild-type crystalline leaf waxes (*open bars*). Error bars are + standard errors of the mean. P values are for effects calculated using ANOVA: Species = coccinellid species, Wax = CEW or reduced CEW, W × S = CEW × species interaction

species, calculated as: (ARCEW − ACEW)/(ARCEW + ACEW) in which ARCEW and ACEW are attachment forces to peas with reduced and normal CEW, respectively. The difference indices are 0.67 for *A. bipunctata*, 0.61 for *H. convergens*, 0.46 for *Ha. axyridis*, and 0.29 for *C. septempunctata* and for *C. transversoguttata*, with a mean of 0.50.

Similar to adults, larvae of four of the coccinellid species achieved lower attachment forces to pea plants with wild-type CEW than to those with reduced CEW (Fig. 6.1b). Only the larvae of *A. bipunctata* achieved similar attachment forces to both CEW types, accounting for the significant species × CEW interaction term in the ANOVA. *Ha. axyridis* again achieved the greatest attachment forces, followed by *A. bipunctata*, and then the other species, which were similar. A similar pattern occurred on wild-type and reduced CEW peas. The difference indices calculated as for adults are 0.54 for *Ha. axyridis*, 0.51 for *C. septempunctata*, 0.46 for *H. convergens*, 0.32 for *C. transversoguttata*, and 0.01 for *A. bipunctata*, with a mean of 0.37.

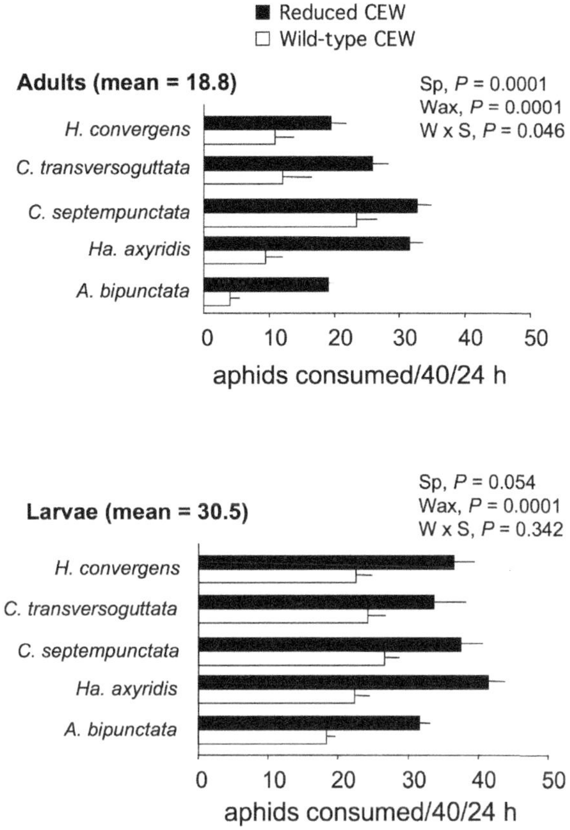

Fig. 6.2 Aphid consumption by adult and larval Coccinellidae to pea plants with reduced crystalline leaf waxes (*filled bars*) and wild-type crystalline leaf waxes (*open bars*). Error bars are + standard error of the mean. P values are for effects calculated using ANOVA: Species = coccinellid species, Wax = CEW or reduced CEW, W × S = CEW × species interaction

Attachment forces generated to glass differed significantly among the species, following patterns of relative attachment similar to those observed on reduced CEW plants (Table 6.1). The greatest attachment force to glass was generated by *Ha. axyridis* for both developmental stages. Force generated by *C. septempunctata* adults was similar to that of *H. axyridis* and adults of the other two species tested generated significantly lower forces. For larvae, *A. bipunctata* larvae generated smaller attachment forces than *Ha. axyridis*, but greater forces than the other three species (Table 6.1).

Adult females of all five species consumed significantly more aphids in 24 h when foraging on plants with reduced CEW than they did on plants with wild-type CEW ($P < 0.0001$) (Fig. 6.2a). In addition, consumption rates differed significantly among the species and the interaction between CEW and species was significant ($P < 0.05$). Consumption rate on reduced CEW plants followed the ranking: *C. septempunctata* = *Ha. axyridis* > *C. transversoguttata* > *H. convergens* = *A. bipunctata*; consumption rate on reduced EW plants followed the ranking: *C. septempunctata* > *Ha. axyridis* = *C. transversoguttata* = *H. convergens* > *A. bipunctata*. An index for the relative effect of CEW on consumption rates, calculated as for

Table 6.1 Attachment force (mN) to clean glass by larvae and adults of coccinellid species

Species	Adults	Larvae
Ha. axyridis	21.5a	11.37a
C. septempunctata	16.0a	2.8c
C. transversoguttata	–	2.6c
H. convergens	7.4b	2.6c
A. bipunctata	6.0b	6.3b
df	3,31	4,22
F	19.65	22.19
P	<0.0001	<0.0001

the attachment data above, indicates that CEW affected the species differently: *A. bipunctata*, 0.65 > *Ha. axyridis*, 0.54 > *C. transversoguttata*, 0.36 > *H. convergens*, 0.28 > *C. septempunctata*, 0.17. Average consumption rate for the adult coccinellids was 19.4 aphids/40 available/24 h.

Larvae of the five coccinellid species consumed significantly more aphids in 24 h when foraging on plants with reduced CEW than they did on plants with wild-type CEW ($P < 0.0001$) (Fig. 6.2b). Consumption rates did not differ significantly among the species, nor was the interaction between species and EW type significant, indicating that consumption rates by these species were improved similarly by reduced EW caused by the *wel* mutation. Although the interaction between CEW type and species was not significant, the relative effect of CEW type on consumption rates (estimated with difference indices) differed among the larvae of the five species: *Ha. axyridis*, 0.30 >y *A. bipunctata*, 0.27 > *H. convergens*, 0.24 > *C. septempunctata*, 0.17 > *C. transversoguttata*, 0.16. Average consumption rate for the adult coccinellids was 30.0 aphids/40 available/24 h.

The relationship between aphid consumption rate and attachment force was examined with a scatter plot of mean values for each species and by calculating Pearson's correlation coefficients for these measurements. For adult beetles these two variables were significantly and positively correlated (Pearson's $r = 0.800$, Spearman's $r = 0.793$, $P < 0.05$) (Fig. 6.3a). For larvae, the data were not correlated (Pearson's $r = 0.370$, Spearman's $r = 0.309$, $P > 0.05$) but followed a positive trend (Fig. 6.3b). Within each species, greater attachment on reduced CEW peas was associated with higher predation rate on those plants, as can be seen by inspecting the lines connecting the points for each species on Fig. 6.3.

Scanning electron micrographs revealed qualitatively similar tarsal morphology among adults and larvae of the five species (Fig. 6.4). These images do not permit quantification of differences in contact surface area but indicate that differences in attachment capabilities among the species is not related to obvious morphological differences.

Larval tarsi bear a single claw and an elongate pad covered with tenant setae. The pad appears to be inverted as an artifact of specimen preparation but may be everted and flexible in the living insects. The size, density, and distribution of the tenant setae appear to be similar among the five species. The form of the setae

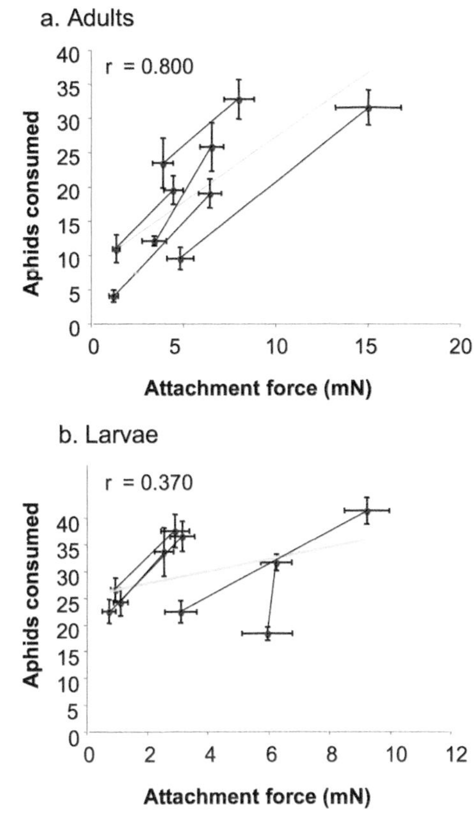

Fig. 6.3 Plot of mean attachment force as given in Fig. 6.1 and mean aphid consumption rate as given in Fig. 6.2 for the coccinellid species tested. *Open circles* are data for adult females and *closed circles* are data for fourth-instar larvae. Correlation coefficients were calculated for the data within each developmental stage. Points connected by lines are for the same species tested on wild-type CEW or reduced CEW pea plants. The point with the higher consumption rate in each connected pair of points is always for the reduced EW plants

differs slightly, but all are expanded at the tips into structures that evidently provide greater adhesive surface. In our preparations the expanded tips of the setae appear slightly concave. The tarsi of adult specimens bear two pads bearing tenent setae. The setae are of two types: spatulate and lanceolate. As for larval setae, the size and density are similar among the five species. *Adalia bipunctata* spatulate setae are less abruptly expanded than those of the other species and the lanceolate setae of this species are not acute as those of other species, terminating in a slightly grooved, expanded tip.

6.3.3 Case Study: Discussion

All five species of coccinellids generated significantly smaller attachment forces to leaflets from pea plants with wild-type CEW than to leaflets of a pea line with CEW reduced by the mutation *wel*. Attachment to reduced CEW peas was similar for all species to attachment to polished, clean glass, suggesting that the reduced CEW cuticle provides near optimal conditions for attachment/adhesion. Attachment

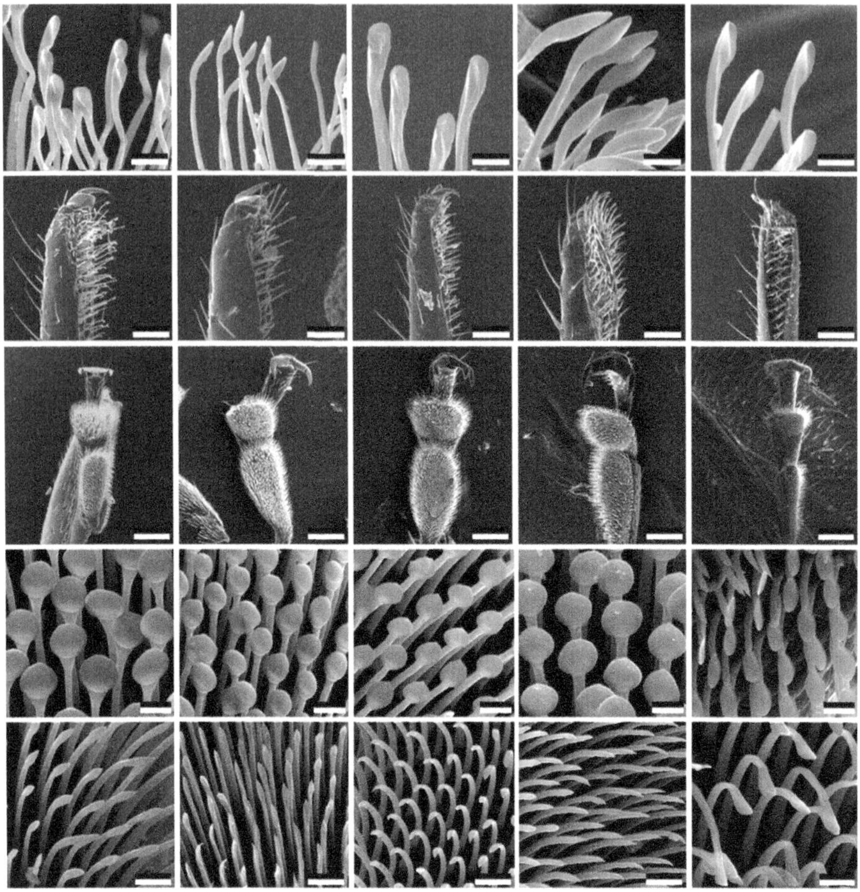

Fig. 6.4 Scanning electron micrographs of tarsi of the five Coccinellidae species tested in bioassays in the case study. Images in each column are from each species. Rows, from top to bottom (length of scale bar): 3rd instar larva, tarsal tenent setae (10 μm), 3rd instar larva tarsus (100 μm), adult tarsus (200 μm), adult tarsal tenent setae of males (discoid setae) (10 μm), adult tarsal tenant setae of females (spatulate and lanceolate). (10 μm, except for *A. bipunctata*, 5 μm)

forces differed significantly among the species and were not equally affected by CEW on pea. Attachment force generated by *Ha. axyridis* adults was substantially greater than for the other species, regardless of substrate (Fig. 6.3a). This might be expected because *Ha. axyridis*, although it exploits various habitats, is often encountered on trees (e.g., Colunga-Garcia and Gage, 1998; Osawa et al., 2000) where there may be selection for greater ability to attach strongly to leaf surfaces in order to avoid being dislodged by wind.

Despite the greater attachment forces, generated by *Ha. axyridis* on glass and reduced CEW pea, the CEW of pea reduced *Ha. axyridis* attachment force substantially. *A. bipunctata* is also sometimes arboreal (Sakuratani et al., 2000), but the adults did not attach more effectively to glass or to reduced EW pea leaves than the other species tested. Attachment by adults of *A. bipunctata* was also proportionally the most severely disrupted by CEWs. *C. septempunctata* appeared to be the least strongly affected by CEW, based on its smaller index of differential attachment.

Attachment force generated by adults of the coccinellid species was significantly and positively correlated with the 24-h rate of pea aphid consumption across all 5 species (Fig. 6.3). Within each species, greater attachment to reduced CEW pea plants was associated with greater rate of aphid consumption on these plants. This pattern is consistent with prior evidence that CEW disrupts attachment and effectiveness of active predatory insects, probably by impairing their mobility while foraging. Our result demonstrates that the effect can be significant and similar within a group of taxa that differs in foraging tendency and size. The average mass of adult *Ha. axyridis, C. transversoguttata* and *C. septempunctata* is 36 mg, the mass of *H. convergens* is 20 mg and *A. bipunctata* average mass is 13 mg. All forage widely and are aphid generalists, but tend to favor certain habitats.

Results for larvae were similar to those for adults in that four of the five species tested were able to generate stronger attachment forces to reduced CEW pea plants and, associated with this, were able to consume aphids at a greater rate on these plants. Larvae of *A. bipunctata* differed from the other species in its unusual ability to generate attachment forces on pea plants with CEW that were similar to those it could generate on glass or reduced CEW-bloom plants. No insect thus far tested has shown the ability to adhere equally effectively to plants with and without CEW (Stork, 1980, Edwards, 1982, Edwards and Wanjura, 1991, Gorb, 2001, Gorb and Gorb, 2002, Federle and Brüning, 2006).

Despite the better attachment ability of *A. bipunctata* larvae to pea CEW, aphid consumption by *A. bipunctata* larvae was substantially lower on pea with CEW. The mobility of *A. bipunctata* larvae appeared to be impeded by CEW, based on informal observations made during our experiments, but we did not quantify this effect. Shah reported that *A. bipunctata* fourth instar mobility was impaired on *Brassica oleracea* plants with CEW (Shah, 1982), which would be consistent with our observations here on pea CEW.

Our method for measuring attachment involves increasing the apparent force on the insect with a centrifuge, thereby mimicking gravitational or inertial forces that could occur in nature. It is possible that *A. bipunctata*, perhaps as part of its adaptive syndrome for arboreal foraging, is capable of anchoring itself to prevent being dislodged when challenged, but that this does not result in improved mobility during normal walking. Other insects appear to respond to the sensation of increasing apparent gravitational force by freezing and apparently gripping the substrate (Federle et al., 2000). Larval coccinellids possess an abdominal adhesive structure with which they can anchor to the substrate. In our bioassays, we did not observe *A. bipunctata* or any other species applying the terminal organ before being dislodged.

It remains possible, however, that *A. bipunctata* achieved stronger attachment to pea CEW by employing this organ.

Whether *A. bipunctata* is included or excluded from our data on larval coccinellids, there is no correlation across species between attachment force and aphid consumption rate (Fig. 6.3). This appears to be due primarily to the responses of *A. bipunctata* and *Ha. axyridis*, which differed in their attachment abilities from the other species. It is possible that these differences are associated with the adaptations each of these species possesses that allow them to exploit a wide range of plant habitats, including trees.

6.4 Implications of Variation in Attachment by Predatory Insects to Crystalline Epicuticular Plant Waxes

Our case study illustrates that CEW impairs attachment and predation by five different species of Coccinellidae, which is consistent with the majority of prior studies of the effects of CEW on carnivorous insects (Eigenbrode, 2004). It also shows that attachment to the plant surface and the effects of CEW on attachment vary considerably, even within a closely related guild of predators. Although we compared predation on two genotypes of *P. sativum* differing uniquely and strongly in the amount of CEW on their leaf surfaces, the data indicate these species and others could differ in their predatory efficiency on other plant surfaces on which they collectively forage and potentially compete. Although many Coccinellidae, especially the Coccinellini, are primarily aphidophagous but otherwise do not specialize in terms of diet (Hodek and Honek, 1996), some species exhibit habitat specialization (Carter and Dixon, 1982; Colunga-Garcia and Gage, 1998; Obrycki and Kring, 1998; Evans, 2003). The differing capacity of these predators to forage on CEW could contribute to this observed habitat specialization, with implications for species distributions and biological control.

Differences in effects of attachment to CEW that affect foraging potentially influence interactions among foliar predators competing indirectly for the same prey and directly through intraguild interference or predation. For coccinellids, both forms of competition can be important (Snyder et al., 2004; Snyder and Evans, 2006). These competitive interactions may or may not influence the effectiveness of the predatory fauna as a whole, but they can affect competitive displacement of particular species. Invasive coccinellids have displaced natives in some locations in North America (e.g., Michaud, 2002; Alyokhin and Sewell, 2004). Larvae and adults of *Ha. axyridis* and adults of *C. septempunctata* generated greater attachment forces to leaves than did the other coccinellid species, which might contribute to the greater competitive ability of these two species in parts of their invasive range. Since 1992 exotic *C. septempunctata* has been present in the Palouse region of northern Idaho and southeastern Washington, where it has reduced the relative abundance of the native *C. transversoguttata* (Elberson, 1992; White and Eigenbrode 2000a; Rutledge et al., 2003). *Ha. axyridis* was first collected in the Palouse region in

1998 (Youseff, unpublished), but continues to be rare, despite its predominance in agricultural regions close by. *Ha. axyridis* was relatively more strongly impaired by CEW than the other species, suggesting its competitiveness could be compromised in settings like the Palouse, where much of the aphid resource occurs on *P. sativum* and *Brassica* crops, which have prominent CEW.

We have shown that *Ha. convergens* and *C. septempunctata* are superior intraguild predators when paired with native Palouse coccinellids (Snyder et al., 2004). How the strength and direction of these competitive interactions might be influenced by the presence of CEW on the plant surface remains to be examined. Habitat variation, including adaptation to attach to various substrates can influence invasiveness of species. For example Cole et al. (2005) report that the introduced house gecko *Hemidactylus frenatus* is unable to adhere to certain substrates with loose particulate matter that afford apparent refugia from predation for the endemic *Nactus* species in the Mascarene islands. The presence of these refugia appears to prevent total displacement of *Nactus* by *H. frenatus* (Cole et al., 2005).

Within-species variation in ability to attach to CEW could also affect inter-predator interactions. Attachment to a smooth glass surface by two populations of *H. convergens*, one from the Palouse region and one from west of the Cascades, differs slightly, but significantly; attachment by the inland population is about 10% stronger than the western population (Eigenbrode, unpublished data). This variation may reflect adaptation by these populations for foraging on the dominant plants or other factors selecting for attachment attributes in their respective habitats. Genetic variation among other *H. convergens* populations has been documented (Obrycki et al., 2001). Within-species variability for adaptation to other specific plant characteristics also has been observed. For example, Andow and Olson, (2003) found evidence for genetic variability among families of *Trichogramma nubilale* for ability to forage effectively on complex plant surfaces. If selection alters attachment attributes of populations of generalist predators this could contribute to or reinforce regional resource or habitat partitioning among species of generalist predators.

Effects of larger scale plant architecture on foraging by predatory insects may be additionally influenced by CEW. Three studies comparing the effects of plant architecture on different coccinellid species (Frazer et al., 1981; Carter et al., 1984; Kareiva and Sahakian, 1990; Grevstad and Klepetka, 1992) did not consider possible contributions of differential attachment to CEW by the species examined. Grevstad and Klepetka, (1992) found that species of plants that differed in architecture caused different rates of falling and prey killing by coccinellids. The strongest effects occurred on those plants with a CEW in combination with larger scale features that seemed to force climbing or negotiating large smooth surfaces. Kareiva and Sahakian, (1990) showed that two species of coccinellid tended to fall more from pea plants with normal leaflets than those plants in which leaflets were converted to tendrils. But in their study, all the plants possessed a wild-type CEW and the attachment to CEW by the coccinellid species was not examined. In the absence of this CEW their result could have differed. Suggestive of this possibility, Chang et al. (2006) found a reduced number of coccinellids in field plots on those plants

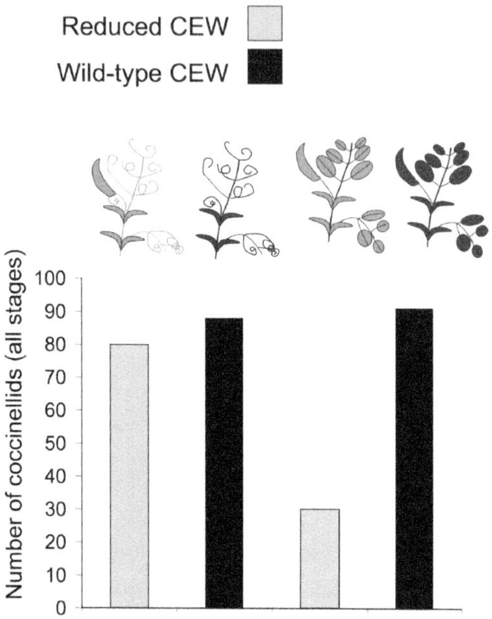

Fig. 6.5 Seasonal totals of coccinellids counted on *P. sativum* plants with different morphology and amounts of CEW. Replotted from Chang et al. (2006)

possessing leaflets but no tendrils and wild-type CEW, whereas other combinations (plants with leaflets converted to tendrils with wild-type or reduced amounts of CEW, and plants with leaflets without tendrils and reduced amounts of CEW) had similar numbers of these insects (Fig. 6.5). Chang et al.'s data pooled coccinellid species because they counted developmental stages difficult to distinguish to species in the field. Variation among the species in attachment to CEW could influence how these species responded to the combined effects of architecture and CEW.

Other implications of CEW on predator prey interactions remaining to be investigated include effects on predator defense behavior that depend on attachment or adhesion to the plant surface. The beetle *Hemisphaerota cyanea* (Coleoptera: Chrysomelidae: Cassidinae) defends itself against ant predation by attaching so strongly to the substrate with its expanded tarsi that the ants cannot remove it (Eisner and Aneshansley, 2000). Whether this strategy is affected by variable CEW on its hosts (all within Aracaceae) has not been investigated. Because many palms possess prominent CEW, *H. cyanea's* antipredator behavior may have emerged as an outgrowth of its extensively developed tarsi that originated as adaptations to exploit and attach to these waxy hosts. Other cassidine species may employ an attachment defense but that of *H. cyanea* has drawn the attention of naturalists because it is strikingly effective. Comparative study could reveal the role of CEW in facilitating or hampering such defenses.

Thus far, work with predators and parasitoids impacted by CEW has focused on generalist predators. An untested prediction is that specialist predators or parasitoids, because of their tight co-evolutionary relationship with particular prey and host plant species, are themselves better adapted than comparable species for attaching to particular CEW blends. The mechanisms to account for such adaptations if they are discovered, and the already documented variability in ability to attach to CEW that occurs among and within coccinellid species and that which occurs among ants associated with *Macaranga* spp., need to be examined and explained. Federle with colleagues (Federle et al., 1997, 2000; Federle and Brüning, 2006) has made the most progress towards elucidating these adaptations, but the details remain elusive. Our study illustrates the difficulties because the differences in attachment we have observed among coccinellids are not associated with obvious differences in the tarsal morphology of these insects. It has been shown that adhesive fluids excreted onto insect tarsi can be complicated mixtures of lipophilic and lipophobic materials (Gorb, 2001; Federle et al., 2002; Vötsch et al., 2002), and differences in these fluids could account for interspecific variability in attachment to CEW. Attachment structures such as the abdominal adhesive pads *Chrysoperla* spp., which produce an abundant and chemically complex secretion, appear to function extremely well on CEW (Eigenbrode, unpublished data). Morphological adaptations for walking on CEW potentially include changes in body proportions other than the tarsi; e.g., relatively longer legs are associated with better walking on CEW for aphids (Åhman, 1990; Eigenbrode et al., 1996) and for *Macaranga* specialist ants (Federle and Brüning, 2006). Finally, has been previously noted, attachment to specific types of plants surfaces, including CEW could be accomplished through behavioral rather than, or in addition to, morphological adaptations (Southwood 1986; Kennedy, 1986; Edwards and Wanjura, 1991; Federle and Brüning, 2006).

6.5 Conclusions

Habitat characteristics affect carnivores and their capacity to take prey. For many carnivorous insects, the habitat is largely or exclusively the surfaces of plants on which their herbivorous prey are feeding. CEW on these surfaces reduce the attachment and therefore the mobility of most predator and parasitoid species that have been studied. The severity of these effects of CEW evidently varies among the insect species based on a few published studies. In this chapter, we have shown that variation in effects of CEW can differ among species, including closely related species within a single predatory guild. This variation has implications for the effects of predatory guilds on prey suppression and on competitive interactions among predatory species. Through such mechanisms structure of arthropod communities and outcomes of applied biological control can be affected in complex ways by CEW on plants.

Acknowledgments This work was supported by Federal Ministry of Education, Science and Technology, Germany to SG (project BIONA 01 RB 0802A) and by National Research Initiative of the USDA Cooperative Research, Education and Extension Service (CSREES), grant #2004-01215 to WES and grant #2000-35302 to SDE.

References

Agrawal, A.A. and Karban, R. (1997) Domatia mediate plant-arthropod mutualism. *Nature* 387: 562–563.

Åhman, I. (1990) Plant-surface characteristics and movements of two *Brassica*-feeding aphids, *Lipaphis erysimi* and *Brevicoryne brassicae*. *Symposium Biologica Hungarica* 39: 119–125.

Alyokhin, A. and Sewell, G. (2004) Changes in a lady beetle community following the establishment of three alien species. *Biological Invasions* 6: 463–471.

Andow, D.A. and Olson, D.M. (2003) Inheritance of host finding ability on structurally complex surfaces. *Oecologia* 136: 324–328.

Baas, W.J. and Figdor, C.G. (1978) Triterpene composition of *Hoya australis* cuticular wax in relation to leaf age. *Zeitschrift fur Pflanzenphysiologie* 87: 243–253.

Barthlott, W. and Neinhuis, C. (1997) Purity of the sacred lotus, or escape from contamination in biological surfaces. *Planta* 202: 1–8.

Barthlott, W., Neinhuis, C., Cutler, D., Ditsch, F., Muesel, I., Theisen, I. and Wilhelm, H. (1998) Classification and terminology of plant epicuticular waxes. *Botanical Journal of the Linnean Society* 126: 237–260.

Bell, W.J. (1991) *Searching behavior: The behavioral ecology of finding resources*, Chapman and Hall, London.

Bernays, E. and Graham, M. (1988) On the evolution of host specificity in phytophagous arthropods. *Ecology* 69: 886–892.

Betz, O. and Kölsch, G. (2004) The role of adhesion in prey capture and predator defence in arthropods. *Arthropod Structure & Development* 33: 3–30.

Beukers, J.S. and Jones, G.P. (1998) Habitat complexity modifies the impact of piscivores on a coral reef fish population. *Oecologia* 114: 50–59.

Blakeman, J.P. (1973) The chemical environment of the leaf surface with special reference to spore germination of pathogenic fungi. *Pesticide Science* 4: 575–588.

Bodnaryk, R.P. (1992) Leaf epicuticular wax, an antixenotic factor in Brassicaceae that affects the rate and pattern of feeding in flea beetles, *Phyllotreta cruciferae* (Goeze). *Canadian Journal of Plant Science* 72: 1295–1303.

Bottrell, D.G., Barbosa, P. and Gould, F. (1998) Manipulating natural enemies by plant variety selection and modification: a realistic strategy? *Annual Review of Entomology* 43: 347–367.

Bourgeois, M., Brethes, J.C. and Nadeau, M. (2006) Substrate effects on survival, growth and dispersal of juvenile sea scallop, Placopecten magellanicus (Gmelin 1791). *Journal of Shellfish Research* 25: 43–49.

Brennan, E.B., Weinbaum, S.A., Rosenheim, J.A. and Karban, R. (2001) Heteroblasty in *Eucalyptus globulus* (Myricales: Myricaceae) affects ovipositonal and settling preferences of *Ctenarytaina eucalypti* and *C. spatulata* (Homoptera: Psyllidae). *Environmental Entomology* 30: 1144–1149.

Byers, J.E. (2002) Physical habitat attribute mediates biotic resistance to non-indigenous species invasion. *Oecologia* 130: 146–156.

Carter, M.C. and Dixon, A.F.G. (1982) Habitat quality and the foraging behavior of coccinellid larvae. *Journal of Animal Ecology* 51: 865–878.

Carter, M.C., Sutherland, D. and Dixon, A.F.G. (1984) Plant structure and the searching efficiency of coccinellid larvae. *Oecologia* 63: 394–397.

Carver, T.L.W., Thomas, B.J., Ingerson-Morris, S.M. and Roderick, H.W. (1990) The role of the abaxial leaf surface waxes of *Lolium* spp. in resistance to *Erysiphe graminis*. *Plant Pathology* 39: 573–583.

Cervantes, D., Eigenbrode, S.D., Ding, H. and Bosque-Perez, N. (2002) Oviposition preference of Hessian fly, *Mayetiola destructor*, on winter wheats varying in surface waxes. *Journal of Chemical Ecology* 28: 193–210.

Chang, G.C. and Eigenbrode, S.D. (2004) Delineating the effects of a plant trait on interactions among associated insects. *Oecologia* 139: 123–130.

Chang, G.C., Neufeld, J. and Eigenbrode, S.D. (2006) Leaf surface wax and plant morphology of peas influences insect density. *submitted to Entomologia Experimentalis et Applicata* 119: 197–205.

Chang, G.C., Neufeld, J., Duetting, P.S. and Eigenbrode, S.D. (2004) Waxy bloom in peas influences the performance and behavior of *Aphidius ervi*, a parasitoid of the pea aphid. *Entomologia Experimentalis et Applicata* 110: 257–265.

Cole, N.C., Jones, C.G. and Harris, S. (2005) The need for enemy-free space: The impact of an invasive gecko on island endemics. *Biological Conservation* 125: 467–474.

Colunga-Garcia, M. and Gage, S.H. (1998) Arrival, establishment, and habitat use of the multicolored Asian lady beetle (Coleoptera: Coccinellidae) in a Michigan landscape. *Environmental Entomology* 27: 1574–1580.

Cortesero, A.M., Stapel, J.O. and Lewis, W.J. (2000) Understanding and manipulating plant attributes to enhance biological control. *Biological Control* 17: 35–49.

Denno, R.F., Lewis, D. and Gratton, C. (2005) Spatial variation in the relative strength of top-down and bottom-up forces: causes and consequences for phytophagous insect populations. *Annales Zoologici Fennici* 42: 295–311.

Dolmer, P. (1998) The interactions between bed structure of Mytilus edulis L. and the predator *Asterias rubens* L. *Journal of Experimental Marine Biology and Ecology* 228: 137–150.

Duetting, P.S., Ding, H., Neufeld, J. and Eigenbrode, S.D. (2003) Plant waxy bloom on peas affects infection of pea aphids by *Pandora neoaphidis*. *Journal of Invertebrate Pathology* 84: 149–158.

Edwards, P.B. (1982) Do waxes on juvenile *Eucalyptus* leaves provide protection from grazing insects? *Australian Journal of Ecology* 7: 347–352.

Edwards, P.B. and Wanjura, W.J. (1991) Physical attributes of eucalypt leaves and the host range of chrysomelid beetles. In: *Insects-plants '89*, ed. by Szentesi, A. and Jermy, T. Budapest: Akadémiai Kiadó, pp. 227–236.

Eigenbrode, S.D. (2004) The effects of plant epicuticular waxy blooms on attachment and effectiveness of predatory insects. *Arthropod Structure and Development* 33: 91–102.

Eigenbrode, S.D. and Espelie, K.E. (1995) Effects of plant epicuticular lipids on insect herbivores. *Annual Review of Entomology* 40: 171–194.

Eigenbrode, S.D. and Jetter, R. (2002) Attachment to plant surface waxes by an insect predator. *Integrative and Comparative Biology* 42: 1091–1099.

Eigenbrode, S.D. and Kabalo, N.N. (1999) Effects of *Brassica oleracea* waxblooms on predation and attachment by *Hippodamia convergens*. *Entomologia Experimentalis et Applicata* 91: 125–130.

Eigenbrode, S.D. and Pillai, S.K. (1998) Neonate *Plutella xylostella* L. responses to surface wax components of a resistant cabbage (*Brassica oleracea* L.). *Journal of Chemical Ecology* 24: 1611–1627.

Eigenbrode, S.D., Castagnola, T., Roux, M.-B. and Steljes, L. (1996) Mobility of three generalist predators is greater on cabbage with glossy leaf wax than on cabbage with a wax bloom. *Entomologia Experimentalis et Applicata* 81: 335–343.

Eigenbrode, S.D., Kabalo, N.N. and Stoner, K.A. (1999) Predation, behavior, and attachment by *Chrysoperla plorabunda* larvae on *Brassica oleracea* with different surface waxblooms. *Entomologia Experimentalis et Applicata* 90: 225–235.

Eigenbrode, S.D., Rayor, L., Chow, J. and Latty, P. (2000) Effects of waxbloom variation in *Brassica oleracea* on foraging by a vespid wasp. *Entomologia Experimentalis et Applicata* 97: 161–166.

Eigenbrode, S.D., White, C., Rohde, M. and Simon, C.J. (1998a) Behavior and effectiveness of adult *Hippodamia convergens* (Coleoptera: Coccinellidae) as a predator of *Acyrthosiphon pisum* on a glossy-wax mutant of *Pisum sativum*. *Environmental Entomology* 91: 902–909.

Eigenbrode, S.D., White, C., Rohde, M. and Simon, C.J. (1998b) Epicuticular wax phenotype of the *wel* mutation and its effect on pea aphid populations in the greenhouse and in the field. *Pisum Genetics* 29: 13–17.

Eisner, T. and Aneshansley, D.J. (2000) Defense by foot adhesion in a beetle (*Hemisphaerota cyanea*). *Proceedings of the National Academy of Sciences* 97: 6568–6573.

Elberson, L. (1992) In *Division of entomology*, University of Idaho, Moscow.

Evans, E.W. (2003) Searching and reproductive behaviour of female aphidophagous ladybirds (Coleoptera: Coccinellidae): a review. *European Journal of Entomology* 100: 1–10.

Farji-Brener, A.G. (2003) Microhabitat selection by antlion larvae, *Myrmeleon crudelis*: Effect of soil particle size on pit-trap design and prey capture. *Journal of Insect Behavior* 16: 783–796.

Federle, W., Brüning, T. (2006) Ecology and biomechanics of slippery wax barriers and waxrunning in *Macaranga*-ant mutualisms. In: Ecology and biomechanics: A mechanical approach to the ecology of animals and plants, ed. by Herrel, A., Speck, T. and Rowe, N. Boca Raton: CRC Press, pp. 163–183.

Federle, W., Maschwitz, U., Fiala, B., Riederer, M. and Hölldobler, B. (1997) Slippery ant-plants and skillful climbers: Selection and protection of specific ant partners by epicuticular wax blooms in *Macaranga* (Euporbiaceae). *Oecologia* 112: 217–224.

Federle, W., Riehle, M., Curtis, A.S.G. and Full, R.J. (2002) An integrative study of insect adhesion: Mechanics and wet adhesion of pretarsal pads in ants. *Integrative and Comparative Biology* 42: 1100–1106.

Federle, W., Rohrseitz, K. and Hölldobler, B. (2000) Attachment forces of ants measured with a centrifuge: better 'wax-runners' have a poorer attachment to a smooth surface. *The Journal of Experimental Biology* 203: 505–512.

Finke, D.L. and Denno, R.F. (2002) Intraguild predation diminished in complex-structured vegetation: implications for prey suppression. *Ecology* 83: 643–652.

Frazer, B.D., Gilbert, N., Ives, P.M. and Raworth, D.A. (1981) Predation of aphids by coccinellid larvae. *Canadian Entomologist* 113: 1043–1046.

Gaume, L., Gorb, S. and Rowe, N. (2002) Function of epidermal surfaces in the trapping efficiency of *Nepenthes alata* pitchers. *New Phytologist* 156: 479–489.

Gingras, D., Dutilleul, P. and Boivin, G. (2002) Modeling the impact of plant structure on host-finding behavior of parasitoids. *Oecologia* 130: 396–402.

Gingras, D., Dutilleul, P. and Boivin, G. (2003) Effect of plant structure on host finding capacity of lepidopterous pests of crucifers by two Trichogramma parasitoids. *Biological Control* 27: 25–31.

Gorb, E., Haas, K., Henrich, A., Enders, S., Barbakadze, N. and Gorb, S. (2005) Composite structure of the crystalline epicuticular wax layer of the slippery zone in the pitchers of the carnivorous plant *Nepenthes alata* and its effect on insect attachment. *The Journal of Experimental Biology* 208: 4651–4662.

Gorb, E.V. and Gorb, S.N. (2002) Attachment ability of the beetle *Chrysolina fastuosa* on various plant surfaces. *Entomologia Experimentalis et Applicata* 105: 13–28.

Gorb, E.V. and Gorb, S.N. (2006) Do plant waxes make insect attachment structures dirty? Experimental evidence for the contamination-hypothesis. In: *Ecology and biomechanics: A mechanical approach to the ecology of animals and plants*, ed. by Herrel, A., Speck, T. and Rowe, N. Boca Raton: CRC Press, pp. 147–162.

Gorb, S.N. (2001) *Attachment devices of the insect cuticle*, Kluwer, Dordrecht.

Grevstad, F.S. and Klepetka, B.W. (1992) The influence of plant architecture on the foraging efficiencies of a suite of ladybird beetles feeding on aphids. *Oecologia* 92: 399–404.

Hare, J.D. (1992) Effects of plant variation on herbivore-natural enemy interactions. In: *Plant resistance to herbivores and pathogens*, ed. by Fritz, R.S. and Simms, E.L. Chicago: University of Chicago Press, pp. 278–300.

Hare, J.D. (2002) Plant genetic variation in tritrophic interactions. In: *Multitrophic level interactions*, ed. by Tscharntke, T. and Hawkins, B.A. Cambridge: Cambridge University Press, pp. 8–43.

Harley, R. (1991) The greasy pole syndrome. In: *Ant – plant interactions*, ed. by Huxley, C.R. and Cutler, D.F. Oxford: Oxford University Press, pp. 430–433.

Heinen, J.T. (1993) Substrate choice and predation risk in newly metamorphosed American toads *Bufo americanus* – an experimental analysis. *American Midland Naturalist* 130: 184–192.

Hodek, I. and Honek, A. (1996) *Ecology of Coccinellidae*, Kluwer Academic, Dordrecht.

Holloway, P.J. (1969) Chemistry of leaf waxes in relation to wetting. *Journal of Science and Food Agriculture* 20: 124–128.

Jeffree, C.E. (1986) The cuticle, epicuticular waxes and trichomes of plants, with reference to their structure, functions and evolution. In: *Insects and the plant surface*, ed. by Juniper, B. and Southwood, T.R.E. London: Edward Arnold, pp. 23–64.

Jetter, R., Schaffer, S. and Riederer, M. (2000) Leaf cuticular waxes are arranged in chemically and mechanically distinct layers: evidence from Prunus laurocerasus L. *Plant Cell and Environment* 23: 619–628.

Juniper, B.E. (1995) Waxes on plant surfaces and their interactions with insects. In: *Waxes: chemistry, molecular biology and functions*, ed. by Hamilton, R.J. Dundee, U.K.: The Oily Press, pp. 157–174.

Juniper, B.E. and Burras, J.K. (1962) How pitcher plants trap insects. *New Scientist* 13: 75–77.

Kareiva, P. and Sahakian, R. (1990) Tritrophic effects of a simple architectural mutation in pea plants. *Nature* 345: 433–434.

Kennedy, C.E.J. (1986) Attachment may be a basis for specialization in oak aphids. *Ecological Entomology* 11: 291–300.

Kennedy, G.G. (2003) Tomato, pests, parasitoids, and predators: tritrophic interactions involving the genus *Lycopersicon. Annual Review of Entomology* 48: 51–72.

Kicklighter, C.E. and Hay, M.E. (2006) Integrating prey defensive traits: Contrasts of marine worms from temperate and tropical habitats. *Ecological Monographs* 76: 195–215.

Knoll, F. (1914) Über die Ursache des Ausgleitens der Insectenbeine an wachsbedekten Pflanzentheilen. *Jahrbuch für wissenschaftliche Botanik* 54: 448–497.

Krebs, C. J. (1989) Ecological methodology. Harper & Row, New York, 652p.

Legrand, A. and Barbosa, P. (2003) Plant morphological complexity impacts foraging efficiency of adult *Coccinella septempunctata* L. (Coleoptera : Coccinellidae). *Environmental Entomology* 32: 1219–1226.

Lill, J.T., Marquis, R.J. and Ricklefs, R.E. (2002) Host plants influence parasitism of forest caterpillars. *Nature* 417: 170–173.

Marquis, R.J. and Whelan, C. (1996) Plant morphology and recruitment of the third trophic level: subtle and little-recognized defenses? *Oikos* 75: 330–334.

Marx, G.A. (1969) Two additional genes conditioning wax formation. *Pisum Newsletter* 1: 10–11.

McAuslane, H.J., Simmons, A.M. and Jackson, D.M. (2000) Parasitism of silverleaf whitefly, *Bemisia argentifolii*, on collard with reduced or normal leaf wax. *Florida Entomologist* 83: 428–437.

Michaud, J.P. (2002) Invasion of the Florida citrus ecosystem by *Harmonia axyridis* (Coleoptera: Coccinellidae) and asymmetric competition with a native species, *Cycloneda sanguinea. Enviromental Entomology* 31: 827–835.

Morris, B.D., Foster, S.P. and Harris, M.O. (2000) Identification of 1-octacosanal and 6-methoxy-2-benzoxazolinone from wheat as ovipositional stimulants for Hessian fly, *Mayetiola destructor. Journal of Chemical Ecology* 26: 859–867.

Müller, C., and Riederer, M. (2005) Plant surface properties in chemical ecology. *Journal of Chemical Ecology* 31: 2621–2651.

Nemeth, R.S. (1998) The effect of natural variation in substrate architecture on the survival of juvenile bicolor damselfish. *Environmental Biology of Fishes* 53: 129–141.

Nielsen, K.A., Nicholson, R.L., Carver, T.L.W., Kunoh, H. and Oliver, R.P. (2000) First touch: an immediate response to surface recognition in conidia of *Blumeria graminis. Physiological and Molecular Plant Pathology* 56: 63–70.

Niemann, G.J. (1985) Biosynthesis of pentacyclic triterpenoids in leaves of *Ilex aquifolium* L. *Planta* 166: 51–56.

Nordlund, D.A., Lewis, W.J. and Altieri, M.A. (1988) Influences of plant-produced allelochemicals on the host/prey selection behavior of entomophagous insects. In: *Novel aspects of insect-plant interactions*, ed. by Barbosa, P. and Letourneau, D. New York: John Wiley & Son, pp. 65–83.
Obrycki, J.J. and Kring, T. (1998) Predacious coccinellids in biological control. *Annual Review of Entomology* 43: 295–321.
Obrycki, J.J., Krafsur, E.S., Bogran, C.E., Gomez, L.E. and Cave, R.E. (2001) Comparative studies of three populations of the lady beetle predator *Hippodamia convergens* (Coleoptera: Coccinellidae). *Florida Entomologist* 84: 55–62.
Osawa, N. (2000) Population field studies on the aphidophagous ladybird beetle *Harmonia axyridis* (Coleoptera: Coccinellidae): resource tracking and population characteristics. *Population Ecology* 42: 115–127.
Peressadko, A.G. and Gorb, S.N. (2004) Surface profile and friction force generated by insects. *Fortschritt-berichte VDI, First International Industrial Conference Bionik* 15 (249): 257–263.
Pimentel, D. (1961) An evaluation of insect resistance in broccoli, Brussels sprouts, cabbage, collards, and kale. *Journal of Economic Entomology* 54: 156–158.
Pimentel, D. and Wheeler, A. (1973) Influence of alfalfa resistance on a pea aphid population and its associated parasites, predators, and competitors. *Environmental Entomology* 2: 1–11.
Price, P.W., Bouton, C.E., Gross, P., McPheron, B.A., Thompson, J.N. and Weis, A.E. (1980) Interactions among three trophic levels: influence of plants on interactions between insect herbivores and natural enemies. *Annual Review of Ecology and Systematics* 11: 41–63.
Ratnam, D.A., Pavlou, S. and Fredrickson, A.G. (1982) Effects of attachment of bacteria to chemostat walls in a microbial predator-prey relationship. *Biotechnology and Bioengineering* 24: 2675–2694.
Riedel, M., Eichner, A. and Jetter, R. (2003) Slippery surfaces of carnivorous plants: composition of epicuticular wax crystals in *Nepenthes alata* Blanco pitchers. *Planta* 218: 87–97.
Riederer, M. and Markstädter, C. (1996) Cuticular waxes: a critical assessment of current knowledge. In: *Plant cuticles, an integrated functional approach*, ed. by Kerstiens, G. Oxford: BIOS Scientific Publishers, pp. 189–200.
Riederer, M. and Müller, C. (Eds.) (2006) *Biology of the plant cuticle*, Blackwell Publishing.
Roda, A., Nyrop, J., English-Loeb, G. and Dicke, M. (2001) Leaf pubescence and two-spotted spider mite webbing influence phytoseiid behavior and population density. *Oecologia* 129: 551–560.
Rodriguezzaragoza, S. (1994) Ecology of free-living Amebas. *Critical Reviews in Microbiology* 20: 225–241.
Romeis, J., Babendreier, D., Wackers, F.L. and Shanower, T.G. (2005) Habitat and plant specificity of Trichogramma egg parasitoids – underlying mechanisms and implications. *Basic and Applied Ecology* 6: 215–236.
Rutledge, C.E. and Eigenbrode, S.D. (2003) Epicuticular wax on pea plants decreases instantaneous search rate of *Hippodamia convergens* larvae and reduces their attachment to leaf surfaces. *The Canadian Entomologist* 135: 93–101.
Rutledge, C.E., Robinson, A. and Eigenbrode, S.D. (2003) Effects of a simple plant morphological mutation on the arthropod community and the impacts of predators on a principal insect herbivore. *Oecologia* 135: 39–50.
Rypstra, A.L., Carter, P.E., Balfour, R.A. and Marshall, S.D. (1999) Architectural features of agricultural habitats and their impact on the spider inhabitants. *Journal of Arachnology* 27: 371–377.
Sakuratani, Y., Matsumoto, Y., Oka, M., Kubo, T., Fujii, A., Uotani, M. and Teraguchi, T. (2000) Life history of *Adalia bipunctata* (Coleoptera: Coccinellidae) in Japan. *European Journal of Entomology* 97: 555–558.
Schönherr, J. (1976) Water permeability of isolated cuticular membranes: the effects of cuticular waxes on diffusion of water. *Planta* 131: 159–164.
Shah, M.A. (1982) The influence of plant surfaces on the searching behavior of *Coccinellid* larvae. *Entomologia Experimentalis et Applicata* 31: 377–380.

Smith, M.T. and Severson, R.F. (1992) Host recognition by the blackmargined aphid (Homoptera: Aphididae) on pecan. *Journal of Entomological Science* 27: 93–112.
Snyder, W.E. and Evans, E.W. (2006) Ecological effects of invasive, arthropod generalist predators. *Annual Review of Ecology Evolution and Systematics* 37: 95–122.
Snyder, W.E., Clevenger, G.M. and Eigenbrode, S.D. (2004) Intraguild predation and the replacement of native ladybird beetles by exotics. *Oecologia* 140: 559–565.
Sotka, E.E., Hay, M.E. and Thomas, J.D. (1999) Host-plant specialization by a non-herbivorous amphipod: advantages for the amphipod and costs for the seaweed. *Oecologia* 118: 471–482.
Spencer, J.L. (1996) Waxes enhance *Plutella xylostella* oviposition in response to sinigrin and cabbage homogenates. *Entomologia Experimentalis et Applicata* 81: 165–173.
Spencer, J.L., Pillai, S. and Bernays, E.A. (1999) Synergism in the oviposition behavior of *Plutella xylostella*: sinigrin and wax compounds. *Journal of Insect Behavior* 12: 483–500.
Stork, N.E. (1980) Role of waxblooms in preventing attachment to brassicas by the mustard beetle, *Phaedon cochleariae. Entomologia Experimentalis et Applicata* 28: 100–107.
Udayagiri, S. and Mason, C.E. (1995) Host plant constituents as oviposition stimulants for a generalist herbivore: European corn borer. *Entomologia Experimentalis et Applicata* 76: 59–95.
Udayagiri, S. and Mason, C.E. (1997) Epicuticular wax chemicals in *Zea mays* influence oviposition in *Ostrinia nubilalis. Journal of Chemical Ecology* 23: 1675–1687.
Vötsch, W., Nicholson, G., Müller, R., Stierhof, Y.D., Gorb, S. and Schwarz, U. (2002) Chemical composition of the attachment pad secretion of the locust *Locusta migratoria. Insect Biochemistry and Molecular Biology* 32: 1605–1613.
Walton, T.J. (1990) Waxes, cutin and suberin. In: *Methods in plant biochemistry*, ed. by Dey, P.M. and Harborne, J.B. San Diego, CA: Academic Press Inc., pp. 105–158.
Webster, M.M. and Hart, P.J.B. (2004) Substrate discrimination and preference in foraging fish. *Animal Behaviour* 68: 1071–1077.
White, C. and Eigenbrode, S.D. (2000a) Effects of surface wax variation in *Pisum sativum* L. on herbivorous and entomophagous insects in the field. *Environmental Entomology* 29: 776–780.
White, C. and Eigenbrode, S.D. (2000b) Leaf surface waxbloom in *Pisum sativum* influences predation and intra-guild interactions involving two predator species. *Oecologia* 124: 252–259.
Yang, G., Espelie, K.E., Todd, J.W., Culbreath, A.K., Pittman, R.N. and Demeski, J.W. (1993) Cuticular lipids from wild and cultivated peanuts and the relative resistance of these peanut species to fall armyworm and thrips. Journal of Agricultural and Food Chemistry 41: 814–818.

Chapter 7
The Waxy Surface in *Nepenthes* Pitcher Plants: Variability, Adaptive Significance and Developmental Evolution

Bruno Di Giusto, Michaël Guéroult, Nick Rowe and Laurence Gaume

7.1 Introduction

The *Nepenthes* pitcher plants are represented by at least 100 species, most of which are found in South-East Asia (Cheek and Jebb, 2001). The pitchers of these carnivorous plants are epiascidate leaves that have evolved in response to habitat nutrient stress (Juniper et al., 1989). They have developed morphological and physico-chemical adaptations for attraction, capture, retention and digestion of arthropods (Darwin, 1875; Lloyd, 1942; Juniper et al., 1989) from which they derive most of their nitrogen supplies (Schultze et al., 1997; Moran et al., 2001; Ellison et al., 2003; but see Moran et al., 2003). The morphological diversity of the pitchers has for long fascinated collectors and naturalists, as well as scientists (Danser, 1928; Clarke, 1997, 2001; Cheek and Jebb, 2001) and could provide relevant information for phylogenetic analyses (Meimberg et al., 2006). The different surfaces of the pitcher present diverse textures (Adams and Smith, 1977; Owen and Lennon, 1999), which have been shown to contribute, in complementary ways, to insect fall (Bohn and Federle, 2004) and retention (Gaume et al., 2002; Gorb et al., 2004). Until these recent studies, most of the work aimed at elucidating trapping mechanisms of *Nepenthes* pitcher plants focused on the capture and retentive function of the slippery waxy surface that covers the upper inner pitcher wall (Lloyd, 1942; Juniper et al., 1989; Gaume et al., 2002; Riedel et al., 2003; Gaume et al., 2004; Gorb et al., 2005). However, several species are polymorphic regarding the presence of a waxy layer (Lloyd, 1942), and other species are even monomorphic for the absence of waxy layer (Cheek and Jebb, 2001; Bohn and Federle, 2004). These observations (illustrated by Fig. 7.1) therefore raise questions whether this layer, generally supposed to be a "key trapping device" in *Nepenthes* pitcher plants, is always of adaptive significance and whether other surfaces could play more crucial roles in insect capture for some of these carnivorous plants.

Assessing the fitness effects of the waxy layer on pitcher plants is not an easy task and cannot be done directly in field experiments aimed at comparing wax-bearing

L. Gaume (✉)
UMR CNRS 5120 AMAP, Botanique et Bioinformatique de l'architecture des plantes, Montpellier, France

S.N. Gorb (ed.), *Functional Surfaces in Biology*, Vol. 2,
DOI 10.1007/978-1-4020-6695-5_8, © Springer Science+Business Media B.V. 2009

Fig. 7.1 Some of the lowland species of *Nepenthes* distributed in Brunei Darussalam, showing the diversity of pitcher types within the genus. The pitchers framed in blue possess a waxy zone (*indicated by the arrow*) while the pitchers framed in red lack such a zone. The knife (9 cm long) gives the scale. The pitcher of *N. gracilis* has the same length as the knife

phenotypes with ones in which the wax layer has been experimentally removed. Complete removal of the waxy layer can indeed be destructive to the pitcher itself (e.g. the use of hot chloroform according to Riedel et al., 2003; Gorb et al., 2005). During our field studies of *Nepenthes rafflesiana* Jack in Borneo, we noticed two levels of polymorphism for the presence of a waxy surface in this species. The first level is found among species, the second level is found among varieties of the same species. *Nepenthes rafflesiana* is reported to be extremely variable displaying at least 13 varieties (Phillipps and Lamb, 1996; Clarke, 1997; Cheek and Jebb, 2001). In Borneo, four varieties are present: *N. rafflesiana* var. *typica* Beck, the most common one; *N. rafflesiana* var. *elongata* Hort. whose pitchers are softer, more cylindrical and longer than pitchers of the typical one; *N. rafflesiana* var. *gigantea* whose pitcher size is enormous and spectacular (but whose taxonomic status is not recognized by Cheek and Jebb [2001]), and *N. rafflesiana* var. *alata* Adam and Wilcock, which has pronounced wings on its tendril (Clarke, 1997). All these discriminant characteristics are stable in cultivation (Clarke, 1997; Gaume, pers. observ.), meaning that these variants are not simply dependent on variations of the environment, but should at least represent "ecotypes" (locally adapted variants) or even genotypes with a taxonomic significance such as sub-species. We noticed from field studies that pitchers of the variety *elongata* always bear a waxy layer and that the plant never flowers at the same time as the variety *typica* when it grows in sympatry. This means that a temporal barrier prevents the two varieties or sub-species from fertilizing each other. We also noticed that pitchers of the giant form never bear a waxy layer. By contrast within populations of the typical variety, some pitchers bear a waxy layer while others do not. These observations mean that intra-specific comparisons are possible and make *N. rafflesiana* a suitable candidate for investigating the functional role and the selective advantage conferred by the waxy layer.

This study is aimed at characterizing the polymorphism of the waxy layer in *N. rafflesiana,* testing whether this layer plays a preponderant role in insect trapping in the species; and if not, investigating what other characteristics of the species may contribute to its trapping success. The first part of the study compares the length of pitchers of the two varieties throughout the development of the plant in order to characterize pitcher differentiation. The second part concerns *N. rafflesiana* var. *typica,* which is polymorphic for the waxy layer, and intends to address the two following questions. (1) Does the waxy layer vary with plant ontogeny? (2) Does the presence of the waxy layer provide a substantial benefit to the plant? For the latter question, waxy and non-waxy phenotypes are compared for their retention effect on ants and flies and for prey abundance. The third part of the study compares functional and ecological attributes of the waxy variety, *N. rafflesiana* var. *elongata,* and the typical one. Field experiments were carried out to assess (1) attraction, (2) retention and (3) trapping efficiency of both varieties. The ultimate purpose is to investigate how and why some *Nepenthes* taxa could have lost their waxy layer during evolution.

7.2 Material and Methods

7.2.1 Studied Taxa

Nepenthes rafflesiana is a lowland *Nepenthes* vine characterized, like most *Nepenthes* species, by an ontogenetic pitcher-dimorphism. Lower terrestrial pitchers are usually ovoid with wings and borne by rosette plants. They are attached to the leaf lamina by a straight tendril which joins the pitcher at its base. The second type of pitchers is the "upper" or aerial one and is found on plants that have begun to climb. They lack "wings"and are attached to the plant by a coiled tendril. These upper pitchers are infundibular (funnel-shaped) and more slender at the base than lower pitchers (Fig. 7.2). The pitcher lifespan is 2–3 months. The plant is characterised by numerous extrafloral nectaries, scattered on the tendril, leaf blade, external surface of the pitcher, lid and between the curved teeth of the peristome (the upper grooved rim of the pitcher) (Adam, 1997; Merbach et al., 2001). The plant captures arthropods from different orders with ants being the most commonly trapped (Adam, 1997; Moran, 1996; Moran et al., 1999). The variety *typica* is the most common of the Bornean varieties of *N. rafflesiana* and it grows in northern Borneo, northern Sumatra and the Peninsular Malaysia (Clarke, 1997). A combination of visual and olfactory cues accounts for prey attraction in this variety (Moran, 1996). It is common in "kerangas" or heath forest usually formed on white acidic sands,

Fig. 7.2 The three varieties of *Nepenthes rafflesiana* found in Brunei with upper and lower pitchers shown in the upper and lower parts of the figure, respectively. Note the presence of a waxy layer (pale area) in the *variety elongata* (*long arrow*), whose basal limit can be assessed by the presence of a hip (*short arrow*) on the outer part of the pitcher

or in degraded forests on such formations. The variety *elongata* has taller and more slender pitchers and is sparsely distributed throughout the heath and peat swamp forests of western Borneo (Clarke, 1997) in more closed habitats. This plant possesses neither UV patterns, nor fragrance and catches significantly less prey than the typical form (Moran, 1996). The variety *gigantea* is threatened with extinction, being confined to rare sites in heath forest of northern Borneo (According to S. Nyawa from the Brunei Museum and CITES commission, it will be put in Annexe 1 of CITES). The above three varieties are the sole varieties of *N. rafflesiana* reported so far from Brunei (Phillipps and Lamb, 1996; Clarke, 1997, Fig. 7.2).

7.2.2 Study Sites

The study on *N. rafflesiana* var. *typica* was mainly carried out at a site located in a zone of degraded "kerangas" or heath forest in Brunei (site 1: 4°38 N, 114°30 E) in July 2003 during the dry season. Typical vegetation of such open "kerangas" includes shrubs from the genera *Melastoma* and *Syzygium* and by *Gleichenia* ferns. Three other species of *Nepenthes* were also found at the study site: *N. gracilis*, *N. mirabilis* var. *echinostoma* and *N. ampullaria*. The analysis of insect attraction in the variety *typica* was carried out in July 2004 at the second site (site 2: 4°34 N, 114°25 E) in the Badas forest reserve (in a border zone of heath forest and peat swamp forest) and compared with results for the variety *elongata*, found in sympatry at this second site. All experiments focused on *N. rafflesiana* var. *elongata* were carried out at the latter site. Three other species of *Nepenthes* were found at this study site: *N. gracilis*, *N. bicalcarata* and *N. ampullaria*. The length measurements of pitchers of the variety *elongata* were conducted at two further sites (site 3: 4°35 N, 114°30 E and site 4: 4°29 N, 114°27 E) in wet and semi-closed areas of heath forest in June 2006. The elevations of all these sites were less than 50 m.

7.2.3 Pitcher Measurements

For these analyses, we used pitchers that had completed their entire development, i.e. mature pitchers of maximal size, whose lid was open and which had already trapped insects. 51 lower and 50 upper pitchers of *N. rafflesiana var. typica* were selected at site 1 from 101 plants displaying an extended range of different developmental stages. 29 lower and 29 upper pitchers of *N. rafflesiana var. elongata* were selected at sites 3 and 4 from 58 plants of a similar range of developmental stages. The presence of a waxy surface was easily detected by the lighter glossy surface appearance characterising the waxy layer in the inner upper part of the pitcher. If the glossy appearance was weak, the presence of a "hip" or rib on the outer part of the pitcher that marks the separation between the digestive zone of the basal and flared part of the pitcher and the adjacent conductive waxy zone forming the more cylindrical and slender upper part of the pitcher (see Fig. 7.2, small arrows) was used. For each plant, the height of both plant and pitcher were measured as well as the length of the waxy zone in each pitcher. The main objectives were to plot and compare the length of pitchers over plant development between the two varieties as well as the relative length of the waxy zone, expressed as the length of the waxy zone divided by the length of the pitcher. To test whether the relative length of the waxy zone in a pitcher undergoes a logarithmic decrease with plant age, a regression analysis was performed using the relative length of the waxy zone as the dependent variable and the plant height (ln-transformed) as continuous covariate.

7.2.4 Retention Experiments

The first experiment was aimed at testing whether the presence of the waxy zone provides a benefit to *N. rafflesiana* var. *typica* in terms of insect retention. In June 2006, we tested and compared the retentive ability of lower pitchers (whose peristome had been removed) with and without a waxy zone on ants (*Oecophylla smaragdina* [*ca.* 7 mm long] captured in the field) and flies (*Drosophila melanogaster* [*ca.* 3.5 mm long] reared in the laboratory). 10 pitchers with wax and 10 pitchers without wax were thus selected, each from different plants. One fly was drawn into a soft tube and blown onto the digestive pitcher liquid without direct manipulation. Observations of fly behaviour including whether the fly escaped or was trapped were made for 5 minutes. A second trial was then conducted on the same pitcher. For each of the 10 pitchers (plants), the number of escapes could be 0, 1 or 2. The same experiment and analysis were performed with ants. The pitchers were then emptied of their digestive liquids and rinsed with water and the experiment was conducted again with both ants and flies dropped into the empty pitchers. This second experiment was performed to separate the retentive effect of the pitcher liquid from that of the pitcher wall. As the total number of trials per pitcher was constant (= 2), we used a poisson regression model to test the effect of wax (presence or absence), the effect of liquid (presence or absence) and the effect of insect identity (ant *O. smaragdina* or fly *D. melanogaster*) on the number of insect escapes.

Two sets of experiments were designed to compare the retentive ability of *N. rafflesiana* var. *typica* and *N. rafflesiana* var. *elongata* in 5 min. - observation sessions. Seven lower and seven upper pitchers of the variety *typica* were selected on different plants at site 1. Two ant species, which are common prey of *N. rafflesiana* were selected: *Anoplolepis gracilipes* (*ca.* 5 mm long) and *Polyrhachis* sp1. (*ca.* 1 cm long). The experiment consisted of dropping one ant worker into the pitcher liquid of each type of pitcher and recording its behaviour and success of escape or whether it sank in the pitcher liquid. The experiment was repeated for each insect species so that there were two trials with *A. gracilipes* and two trials with *Polyrhachis* sp. per pitcher. A similar experiment but with only one trial per pitcher was conducted on 16 lower pitchers and 18 upper pitchers of different plants of *N. rafflesiana* var. *elongata* at site 2 with *Polyrhachis* sp. and 15 lower and 15 upper pitchers of different plants of the same variety with *A. gracilipes*. As the total number of trials per pitcher varies according to the plant variety (4 for *typica* and 2 for *elongata*), we used a logistic regression to test the effect of plant variety (*typica* or *elongata*), type of pitcher (lower or upper) and ant species (*Polyrhachis* sp. or *A. gracilipes*) on the frequency of ant escapes by pitchers (plants).

7.2.5 Analyses of Attraction

We gathered empirical data on attraction in site 2 by observing for 10 minutes 10 lower and 11 upper pitchers belonging to 21 distinct plants of *Nepenthes rafflesiana* var. *typica* and counting the number of arthropod visitors and the number of visitor species. We conducted the same type of observation on 20 lower and 18 upper pitchers belonging to 38 distinct plants of *N. rafflesiana* var. *elongata* in the same site. Using Poisson regression models, we analysed the effects of the variety and pitcher type on the number of insects or species visiting each pitcher. The aim was to compare the attractiveness of the two varieties and of the two types of pitcher in terms of number of arthropod visitors and species diversity.

7.2.6 Analyses of Prey Spectra

At site 1, we collected and preserved (in 75% ethanol), the contents of 6 waxy pitchers and 6 non-waxy pitchers of *Nepenthes rafflesiana* var. *typica* whose height was previously measured. The total number of prey items was counted. Using an analysis of covariance, we tested whether there was an effect of pitcher size and wax presence on the number of prey items, the number of terrestrial arthropods and the number of flying insects.

The contents of 17 upper pitchers from 17 randomly selected *N. rafflesiana* var. *typica* plants at site 1 and the contents of 20 upper pitchers from 20 randomly selected plants of the waxy variety *elongata* at site 2 were analysed. The pitchers were approximately of the same age (1 month). Using a binocular microscope, we sorted, counted, and identified the prey items to at least genus level for the ants, and to family level for other arthropods. Sometimes, the cuticular remains of the arthropods did not permit complete identification (especially in the case of the Lepidoptera, whose soft wings were always completely digested). In most cases, the prey could be identified to order level. Ten orders were defined: Hymenoptera, Coleoptera, Diptera, Lepidoptera, Dictyoptera, Orthoptera, Isoptera, Thysanoptera, Heteroptera and Arachnida. A MANOVA was performed to test for an overall difference in prey abundance for the different orders between the two varieties of *N. raflesiana* and to know which prey order variable most discriminates these two varieties.

7.2.7 Statistical Analyses

Statistical analyses were carried out using the software package SAS v.9 (SAS Institute Inc., Cary, NC, USA.). For model selection, backward procedures were adopted, starting with the removal of the non-significant highest order interactions. The analyses of variance, the analyses of covariance and the multiple analysis of variance were carried out using the procedure GLM. The normal distribution of the residuals was checked using the Shapiro test. The poisson and logistic regressions were carried out using the procedure GENMOD, with a poisson and a binomial error distribution, respectively. Correction for over-dispersion was applied when necessary using the square root of the ratio of Pearson's χ^2 to the associated number of degrees of freedom.

7.3 Results

7.3.1 Differentiations of the Two Varieties During Ontogeny

The two varieties of *Nepenthes rafflesiana* differentiate from each other during plant ontogeny according to their pitchers. As shown by the Fig. 7.3, the size of pitchers undergoes a logarithmic increase with plant ontogeny (significant and positive slope

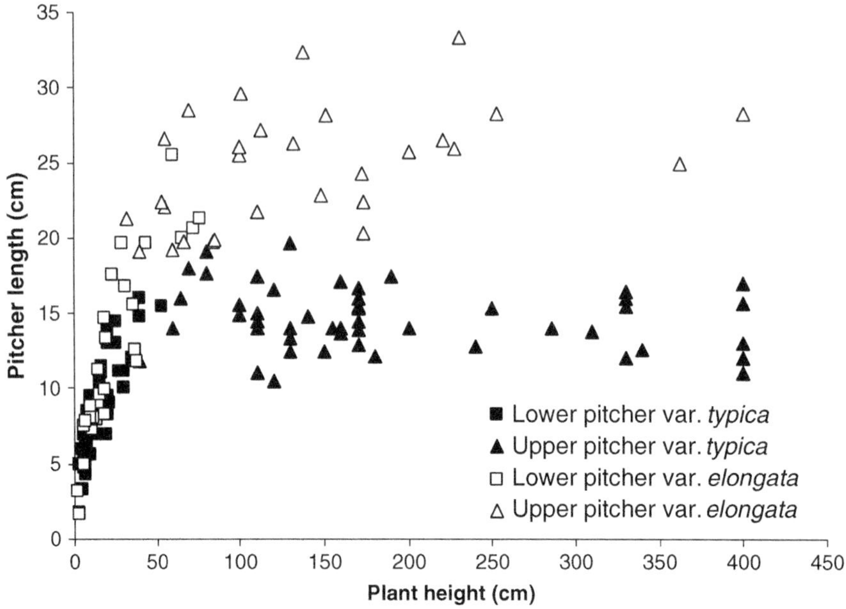

Fig. 7.3 Pitcher height in *N. rafflesiana* over plant development

in the analysis of covariance performed on ln-transformed data, Table 7.1). The increase is much more pronounced for lower pitchers than for upper ones (comparison of the slopes in Table 7.1a and Table 7.1b). This increase, though similar for the two varieties in the beginning (common intercept) is more rapid for the variety *elongata* than the variety *typica* at later stages (significant interaction, two different slopes, Table 7.1a).

Pitchers of the variety *typica* reach a plateau, where upper pitchers no longer increase in size with plant ontogeny, early (slope not significantly different from zero, Table 7.1b). By contrast, upper pitchers of the variety *elongata* continue

Table 7.1 Analyses of covariance testing for the effect of plant height (ln-transformed) and variety identity (*typica* vs. *elongata*) on pitcher length for (a) lower pitchers ($R^2 = 0.81$, residuals normally distributed: Shapiro statistic $W = 0.99$, $p = 0.78$) and (b) upper pitchers ($R^2 = 0.79$, residuals normally distributed: $W = 0.98$, $p = 0.40$).

Dependent variable:	(a) height of lower pitcher				(b) height of upper pitcher			
Covariate	*df*	SS	*F*	*P*	*df*	SS	*F*	*P*
Ln(plant height)	1	1234	273.4	<0.0001	1	34.91	4.94	0.0292
Variety	1	10.4	2.3	0.1333	1	15	2.12	0.1491
Ln(plant height)*Variety	1	29.8	6.6	0.0121	1	87.13	12.34	0.0008
Parameter	estimate	S.E.	*T*	*P*	estimate	S.E.	*T*	*P*
Common intercept	−3.24	0.86	−3.8	0.0003	15.54	2.59	5.61	<0.0001
Slope for *N. r. typica*	4.74	0.33	14.31	<0.0001	0.01	0.51	0.02	0.9834
Slope for *N. r. elongata*	5.42	0.31	17.69	<0.0001	2.16	0.54	3.97	<0.0001

to undergo a logarithmic increase and their size reaches a plateau later in plant ontogeny (Fig. 7.3, weak slope but significantly different from zero, Table 7.1b). Hence, upper pitchers of *N. rafflesiana* var. *elongata* (mean height = 24.8 cm, SD = 3.9, n = 50) are on average 1.7 times longer than upper pitchers of the variety *typica* (m = 14.6 cm, SD = 2.1, n = 29).

7.3.2 Ontogenetic Loss of Wax in the Variety typica

The waxy zone in the *elongata* form is thicker and greater in relative and absolute length than in the *typica* form and this provides part of explanation why pitchers are longer in the first variety. Indeed, the waxy zone here occupies on average 60% of the pitcher length for lower pitchers and 47% for upper pitchers but for the *typica* form only 20% and 0% of lower and upper pitchers, respectively (Fig. 7.4a). There is a significant tendency for the relative length of the waxy zone to be higher in lower than in upper pitchers and to be higher in var. *elongata* than in var. *typica* (Two-way

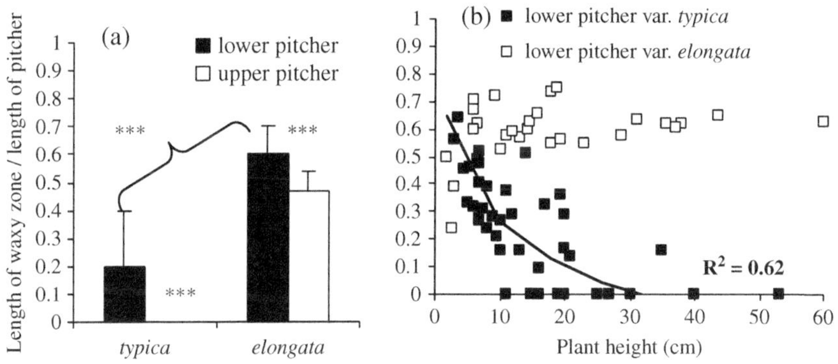

Fig. 7.4 (**a**) Relative length of the waxy zone compared between varieties and types of pitchers in *N. rafflesiana*. ***: differences significant at $\alpha = 0.001$. The error bars refer to standard deviations. (**b**) Ontogenetic changes of the waxy zone in lower pitchers. The different points correspond to distinct pitchers on distinct plants

Table 7.2 Two-way analysis of variance testing for the effect of variety identity (*typica* vs. *elongata*) and pitcher type (lower *vs.* upper) on the relative length of the waxy zone (arcsin(sqrt)-transformed to approach a normal distribution), $R^2 = 0.77$

Dependent variable: arcsin ([wax length/pitcher length]$^{0.5}$)					Parameter variety*pitcher	Estimate	S.E.	*T*	*P*
Covariate	*df*	SS	*F*	*P*	*typica* - UP	0	0.03	0	1
Variety	1	15.1	427.6	<0.0001	*typica* - LP	0.36	0.03	13.6	<0.0001
Pitcher	1	3.0	84.8	<0.0001	*elongata* - UP	0.76	0.03	21.6	<0.0001
Variety*pitcher	1	0.5	14.5	0.0002	*elongata* - LP	0.88	0.03	25.7	<0.0001

ANOVA: significant effect of variety and pitcher type, Table 7.2), but the difference between pitcher types is far more pronounced for the *typica* form than for the *elongata* form (significant interaction) since in the *typica* form, upper pitchers never bear a waxy zone. As for lower pitchers of the *typica* form, there is a significant trend in the relative length of the waxy layer to decrease as a logarithmic function of the plant height (regression using ln[plant height]): $F_{1,48} = 78.46$, $P < 0.0001$, Fig. 7.4b). The two last observations mean that for the variety *typica*, there is a trend of the waxy zone to be lost with plant ontogeny.

7.3.3 Absence of a Significant Effect of the Wax on Insect Retention and Prey Abundance in the Variety typica

The lower pitchers of *N. rafflesiana* var. *typica* are efficient in retaining insects (trapping 100% of ants and more than 50% of flies) but this efficiency cannot be ascribed to a wax effect. Indeed, the insect bioassays showed only a weak effect (statistically non-significant) of the pitcher wax on insect retention. The effect of wax presence ($\chi^2 = 1.72$, $p = 0.19$, Table 7.3, Fig. 7.5) was negligible compared to the effect of the presence of the digestive liquid ($\chi^2 = 46.18$, $p < 0.0001$); the trend of waxy pitchers

Table 7.3 Poisson regression model testing for the effect of wax (presence *vs.* absence), digestive liquid (presence *vs.* absence) and insect identity (ant *O. smaragdina vs.* fly *D. melanogaster*) on the retention of insects by lower pitchers of *N. rafflesiana* var. *typica*.

Dependent variable: number of insect escapes			
Covariate	*df*	χ^2	*P*
Wax	1	1.72	0.19
Liquid	1	46.18	< 0.0001
Insect	1	0.24	0.6224
Liquid*insect	1	22.85	< 0.0001

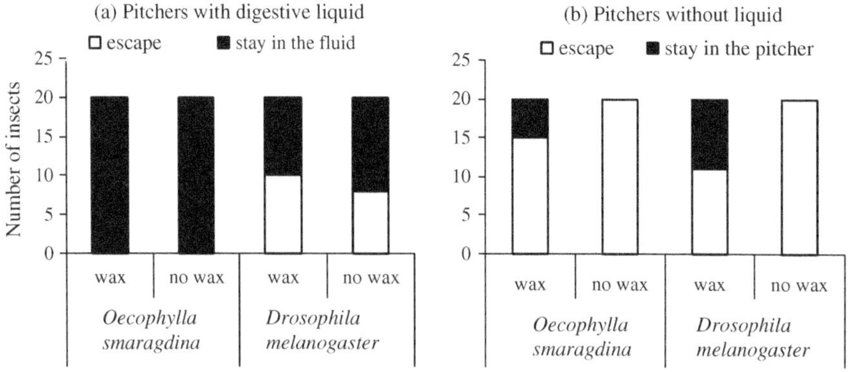

Fig. 7.5 Insect retention in *N. rafflesiana* var. *typica* compared for pitcher types (with and without wax) and insect types (ants and flies) for lower pitchers (**a**) with the digestive liquid and (**b**) emptied

to more efficiently retain insects compared to non-waxy ones was only detectable when the liquid was removed from pitchers (Fig. 7.5b). Moreover, retention was greatly dependent on insect type but with contrasting effects according to presence or absence of the liquid (highly significant interaction insect*liquid): retention was more efficient for ants than for flies when the liquid was present (Fig. 7.5a) but ants were slightly better at escaping than flies when the liquid was absent from pitchers (Fig. 7.5b).

The observations of insect behaviour provide part of the mechanistic explanation of the retention process. In pitchers with the liquid present, ants could clearly not generate sufficient force to move their body out of the liquid. When by chance they succeded in reaching the pitcher wall, whether wax was present or not, they could not move their legs out of the liquid and were retained in the liquid without possibility of escape. As for the flies, they could never escape directly from the liquid by flight because they were unable to free their wings from the liquid. When they succeded in reaching the wall, their legs were retained under the level of the liquid. However, as they could generate more force than ants by moving their wings, they sometimes succeded in extracting their forelegs and their body from the liquid. When this occurred, they escaped from the pitcher either by walking (mostly in non-waxy pitchers) or by taking off and flying (mostly in waxy pitchers). But prior to being able to achieve this, flies were observed to groom themselves for several minutes. In emptied pitchers, the main difficulty to escape comes from the slippery waxy layer, which often makes insects fall several times preventing them from either climbing out (ants) or positioning for take off (flies). Even if they succeded in escaping from the emptied pitchers, the time of escape process was longer in waxy pitchers than in non-waxy pitchers (significant for flies: $m_{wax} = 117.5 \pm 149.2$ sec., $m_{no-wax} = 3.6 \pm 3.5$ sec., *t*-test for unequal variances: $t = -2.41$, $p = 0.04$; not significant for ants: $m_{wax} = 75.3 \pm 91.3$ sec., $m_{no-wax} = 27 \pm 38.1$ sec., *t*-test for unequal variances: $t = -1.54$, $p = 0.14$).

The analyses of prey abundance did not show any effect of wax presence on the quantity of insect prey. In order to separate the effect of pitcher height on prey abundance from the effect of wax, whose presence in the pitchers was also dependant on pitcher height (Figs. 7.3, 7.4b), we used an analysis of covariance. The total number of prey items increased as an exponential function of pitcher height ($R^2 = 0.73$, $F_{1,10} = 26.9$, $p = 0.0004$, Fig. 7.6) independently of its wax phenotype. There was no effect of wax in the ANCOVA performed on ln[number of prey items] (same intercept: effect of wax, $F_{2,9} = 0.52$, $p = 0.49$, same slope: wax $*$ pitcher $F_{3,8} = 0.18$, $p = 0.68$ for the two regression lines estimated for the waxy pitchers and the non-waxy ones, respectively). This means that for a given pitcher height, the number of captured arthropods was similar whether the pitcher has a waxy zone or not. In terms of taxon diversity, the lower pitchers of *N. rafflesiana* var. *typica* were found to trap mostly ants (m = 73.5%, SD = 33.5%, n = 12). The other terrestrial arthropods (Orthoptera, Arachnida, Myriapoda, Crustacea) represented 9.2% of the total prey items while the flying insects (Diptera, Coleoptera, Lepidoptera, Dictyoptera, Hemiptera) represented 15% of the prey. There was no difference on the patterns of prey spectra between waxy pitchers and non-waxy ones.

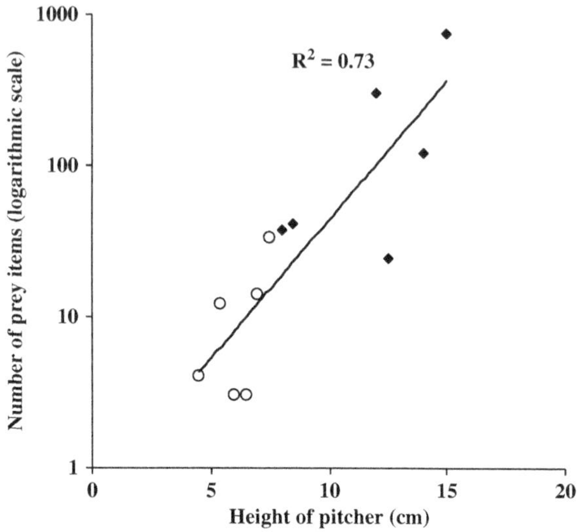

Fig. 7.6 Number of prey items as an exponential function of pitcher height. There was no significant difference between waxy pitchers (*white circles*) and non-waxy ones (*black diamonds*), for which a common regression line was estimated

7.3.4 Better Retention in the Waxy Variety elongata

The insect bioassay carried out on ant species compared retention between varieties of *N. rafflesiana* (*typica* vs. *elongata*), between pitcher types (lower *vs.* upper) and between ant species (the large ant *Polyrhachis* sp. *vs.* the smaller one *Anoplolepis gracilipes*). Retention greatly depended on ant species since 100% of *A. gracilipes* were retained within *N. rafflesiana* pitchers (for both varieties) against only 53% of *Polyrhachis* ants (Fig. 7.7). Interestingly, 73.5% of the *Polyrhachis* ants were retained within pitchers of the variety *elongata* against only 28.5% in the case of the variety *typica*. According to the logistic regression performed on the data subset

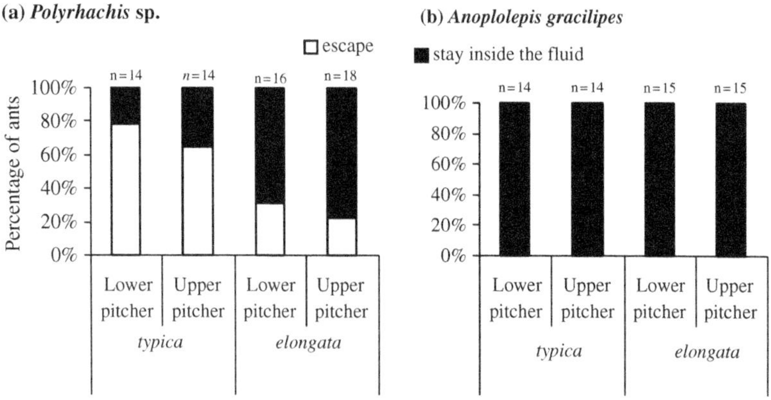

Fig. 7.7 Results of the retention experiment carried out on the two varieties of *N. rafflesiana* for (**a**) the ant *Polyrhachis sp.* and (**b**) the ant *Anoplolepis gracilipes*

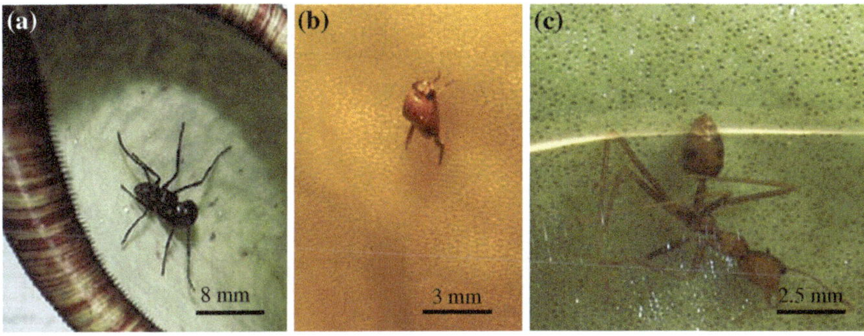

Fig. 7.8 Insect behaviour within the pitchers of *N. rafflesiana*. (**a**) A worker of *Polyrhachis* sp. trying with great difficulty to climb up the waxy surface of a lower pitcher of the variety *elongata*. Note that it applies entire tarsa against the waxy wall to increase the contact area (**b**) A *Drosophila melanogaster* fly that had succeeded in moving its body from the digestive liquid of the variety *typica* and that is slowly climbing up the glandular surface by walking. Its wings have been wetted by the digestive liquid and are adhering to its body. (**c**) An *Oecophylla smaragdina* ant, which has begun to sink inside a pitcher of the variety *typica* after several unsuccessful attempts to emerge from the liquid

corresponding to the *Polyrhachis* ants, there was an effect of the *Nepenthes* variety on ant retention ($\chi^2 = 12.9$, $p = 0.0003$) but no significant effect of either pitcher type ($\chi^2 = 1.01$, $p = 0.31$) or interaction between variety and pitcher type ($\chi^2 = 0.05$, $p = 0.83$). Although the trend was not significant, for each variety, the successes of escape were slightly higher in the lower pitchers than in the upper ones (Fig. 7.7).

Behavioural observations (Fig. 7.8) showed that the small ants *Anoplolepis gracilipes* and *Oecophylla smaragdina* could not at all free their bodies from the digestive liquid of the variety *typica*. As for the bigger *Polyrhachis* ants, they experienced the same difficulty in extracting their forelegs and body out of the liquid but they could apparently generate more muscular force than *A. gracilipes* in order to haul themselves up onto the pitcher wall. Once out of the liquid, walking on the wall was easier on variety *typica* than on variety *elongata*, because on variety *elongata* as ants slipped repeatedly along the waxy walls of the latter and succeeded in escaping from the pitchers only after a mean of 12 attempts (min = 1, max = 30). In the upper pitchers of var. *elongata*, the ants seemed to be more entrapped in the liquid than in the lower pitchers. Moreover in these pitchers, they were observed on five occasions to be directly drawn down and retained deep within the base of the pitcher.

7.3.5 Lower Attraction of the Waxy Variety elongata

The poisson regression performed on the number of insects recorded during the ten minutes - observation sessions showed significant effects of both *Nepenthes* variety and pitcher type on the number of insect visitors (data slightly over-dispersed, $\hat{c} = 1.29$, therefore F instead of χ^2 are provided; effect of variety: $F_{1,56} = 4.29$,

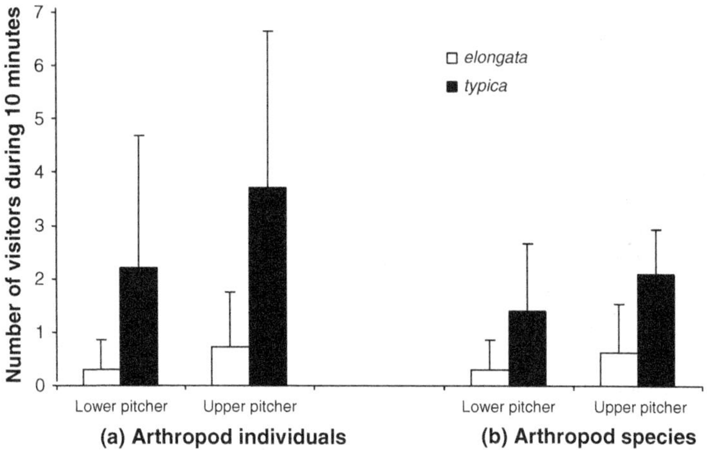

Fig. 7.9 Attractiveness of pitchers in terms of (**a**) number of individual visitors and (**b**) number of species visitors compared for the two *N. rafflesiana* varieties and for the two pitcher types. The error bars refer to standard deviations

$p = 0.04$, effect of pitcher type: $F_{1,56} = 33.48$, $p < 0.0001$, no significant interaction: $F_{1,55} = 0.24$, $p = 0.62$). The number of insect visitors was significantly higher on the variety *typica* than on the variety *elongata* and on upper rather than lower pitchers for both of the two varieties (Fig. 7.9). The same trend was observed for the number of visitor species, although the pitcher effect was only marginally significant in this case (effect of variety: $\chi^2 = 23.21$, $p < 0.0001$, effect of pitcher type: $\chi^2 = 3.24$, $p = 0.07$, no significant interaction: $\chi^2 = 0.26$, $p = 0.61$). This means that pitchers of *N. rafflesiana* var. *typica* are more attractive than those of *N. rafflesiana* var. *elongata* and that, for a given variety, upper pitchers are more attractive than lower ones.

In both varieties, ants were the main visitors (several ants were common, such as Formicinae of the genera *Polyrhachis*, *Camponotus*, *Anoplolepis*, Myrmicinae of the genus *Crematogaster* and Dolichoderinae of the genus *Tapinoma*). The variety *elongata* discriminates from the variety *typica* in attracting few insects belonging to generalist pollinator orders such as flying Hymenoptera, Coleoptera, Lepidoptera and differs from the variety *typica* in attracting hemipterans, for example a reduviid bug (assassin bug). As a matter of fact, the Hemiptera was the only order, for which the number of prey items was more abundant in the upper pitchers of the elongate form than in the upper pitchers of the typical one (Fig. 7.10).

7.3.6 Lower Prey Abundance and Poorer Prey Spectrum in the Waxy Variety elongata

Indeed, the MANOVA procedure carried out to compare the two *Nepenthes* varieties in terms of their arthropod abundance within each prey order (Fig. 7.10a) showed a global and significant tendency (overall variety effect: $F_{10,26} = 2.27$, $p = 0.04$) of

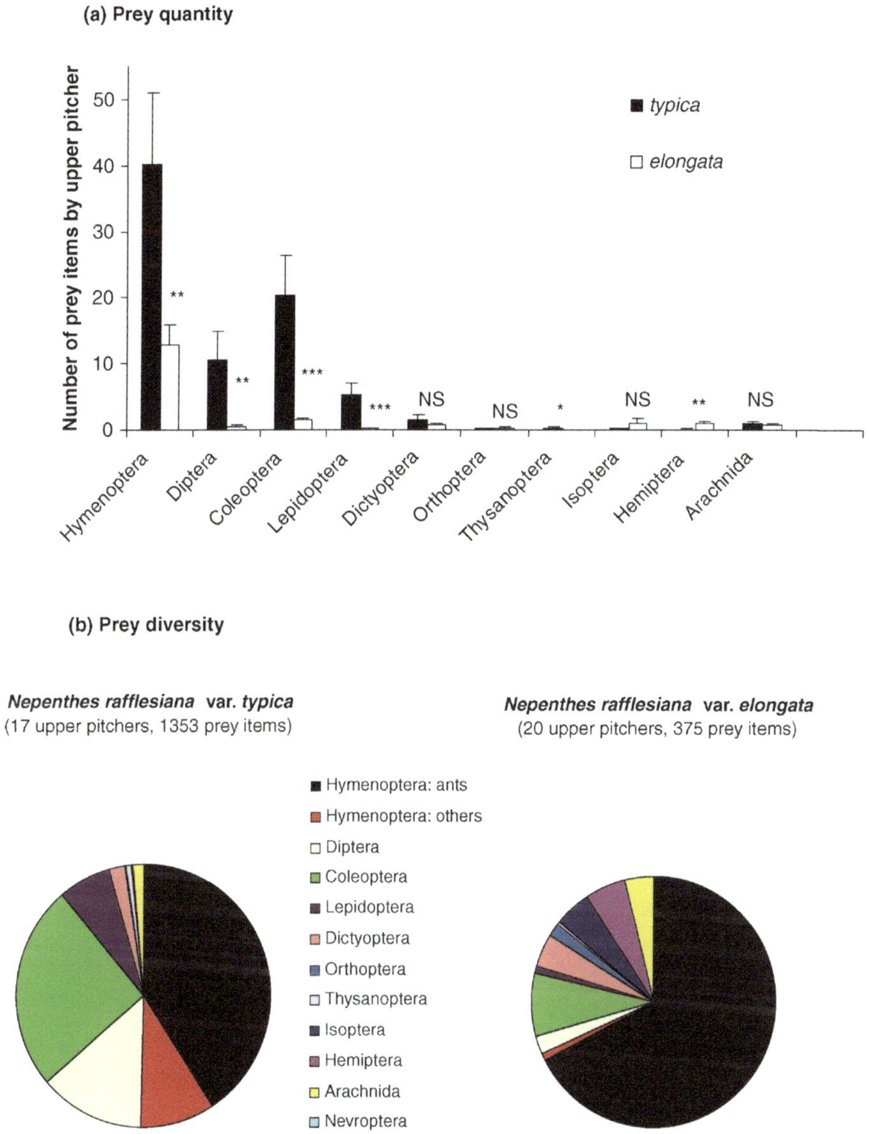

Fig. 7.10 Prey spectra of the two studied *Nepenthes* varieties for their (**a**) quantity and (**b**) diversity. NS, means not significant from each others; *, $p < 0.1$; **, $p < 0.001$; ***, $p < 0.0001$. The errors bars refer to standard deviations

upper pitchers of the typical variety to trap more arthropods than the elongate one with an average of 4.2 times more ($m_{typica} = 79.7 \pm 77.6$, $m_{elongata} = 18.8 \pm 14.8$). The most discriminant orders were the Hymenoptera (especially the flying insects other than ants), Lepidoptera, Coleoptera, Diptera and Thysanoptera (orders particularly rich in pollinating species), which were trapped in much higher quantities by var. *typica* than by var. *elongata* (Fig. 7.10a). The hymenopterans other than ants, lepidopterans and thysanopterans were almost absent from the prey spectrum of the variety *elongata*. The percentage of flying insects was 52.7± 25.3 for the variety *typica* and only 29.5% ± 25.2 for the variety *elongata* whose prey repartition was shown to be less scattered than for the variety *typica*, being composed mostly of ants (Fig. 7.10b).

7.4 Discussion

Nepenthes rafflesiana displays several forms that have been classified as varieties by taxonomists (Danser, 1928; Cheek and Jebb, 2001) but their status has been tested neither by genetic analysis nor in ecological studies. This work compared the two varieties *typica* and *elongata* in an ontogenetic and ecological context. It focused on the waxy trapping surface, which appeared to be the most discriminant character between the two taxa. The adaptive significance of the waxy trapping surface has been tested and shown to be only weak at least in the variety *typica*.

7.4.1 The Waxy Zone: A Labile Element of Pitcher Development in Nepenthes

The waxy zone in *Nepenthes* pitcher plants has been defined as a conductive zone by Hooker (1859) because of the perceived importance of the structure for conducting insects down to the digestive liquid. It is located above the glandular zone, which has been defined functionally as the digestive zone. The waxy zone has for long being considered as the primary surface for insect trapping (Juniper and Burras, 1962) and this is still the general belief today (but see Bohn and Federle, 2004) possibly because functional studies investigating interactions between plant and insect surfaces have often been carried out on horticultural hybrids, in which this layer is often well-expressed. However quite early, some authors (Lloyd, 1942) already noticed that the waxy zone could be present or absent in some 'natural' species; and within a given species, could be present in some pitchers but absent in others. However, no further work has been carried out to understand this kind of polymorphism. In this study, we show that the polymorphism concerning the waxy zone can be readily explained in the light of plant's development. In *N. rafflesiana* var. *typica*, when the plant advances in ontogeny, its produces pitchers with a reduced waxy zone, and plants higher than 30–40 cm produce pitchers with no waxy zone. This observation is linked to the change in form of pitchers. The small lower pitchers of young plants are rounded at the base and elongate in their distal part (corresponding to the waxy zone) with a rib or "hip" separating the two parts. Lower pitchers produced by older plants are ovoid with the hip only visible at the top of the back face of the pitcher just beneath the peristome. Upper pitchers characterising climbing stages of the plant never bear a waxy zone and are infundibular or funnel-shaped as if they even lack the upper part of the digestive zone. Such changes in pitcher form during ontogeny have been reported on several *Nepenthes* species (Clarke, 2001).

Thus, the differences observed in form and length of pitchers between the two varieties of *N. rafflesiana* are probably partly linked to developmental differences of the waxy zone. The waxy zone is always present in the elongate variety and its length represented 2/3 of the total pitcher length but in *N. rafflesiana* var. *typica*, the development of the waxy zone seems to have aborted. Why has such a slippery surface, which is usually considered to be the main insect trapping zone in *Nepenthes* pitcher plants, been lost during ontogeny?

7.4.2 Weak Adaptive Significance of the Waxy Surface in the Variety typica

There was no effect of wax presence on insect retention in the variety *typica*, nor was there an effect of the wax on the number of prey items trapped by this variety. Insects are known to provide many essential nutrients needed by the carnivorous plant (Schultze et al., 1997). They were estimated to provide 70% of the nitrogen used by this variety (Moran and Moran, 1998; Moran et al., 2001). Therefore the number of prey items should correlate with the fitness (survival and reproductive capacities) of the plant. The presence of wax in *N. rafflesiana* var. *typica* thus probably has no significant effect on fitness of the plant. The wax has therefore only a weak, if any, adaptive significance for plants of the variety *typica*. It is therefore not surprising to observe a developmental loss of the waxy layer in this variety.

Furthermore, the wax layer is composed mainly of aliphatic compounds dominated by very-long-chain aldehydes, such as triacontanal being the main constituent, which includes 30 atoms of carbon (Riedel et al., 2003). Hence, elaboration of the waxy layer must be costly to the plant, and if it does not provide any substantial benefit, it might not be maintained by natural selection. Why does the waxy zone in *N. rafflesiana* var. *typica* provide no (or no more) selective advantage to the plant? We suggest that the net benefits provided by the waxy layer may have become negligible in comparison to those provided by other more adaptive traits of the pitcher in this taxon.

The ecological success of the variety *typica* is of little doubt. In Borneo at least, it is the most abundant variety of *N. rafflesiana*, being found in many areas on white sand formations in heath forests and degraded heath forests (Clarke, 2001). As a matter of fact, upper pitchers trapped on average four times more arthropods than upper pitchers of the variety *elongata* and their prey spectra were much richer than those of the variety *elongata*. If wax does not explain this success, what trait(s) have contributed to such a trapping efficiency and ecological success?

First, our insect bioassays revealed the importance of the digestive liquid in insect retention. Its physical rather than chemical properties might explain the high efficiency of insect capture. In both varieties, insects were maintained in the liquid with part of their body tending to sink. Surface properties and viscosity of the fluid could be important in insect retention but further investigations are needed to clarify the mechanisms involved. However, the liquid of the waxy variety *elongata* appeared to show similar retentive properties. Hence, the liquid alone cannot replace the slippery waxy surface and explain the higher ecological success of the variety *typica*.

According to our results, the variety *typica* displays high insect attractive properties. It was shown to attract more insects in total and more flying insects, especially pollinator categories than the variety *elongata*. Higher attraction was shown by upper pitchers, which were observed in the field to produce a sweet odour (Di Giusto and Gaume, pers. observ.) supposed to be part of the attractive cue in this variety (Moran, 1996). Carnivorous plants mimic flowers in a number of ways (Joel, 1988), and we suggest that upper pitchers of the variety *typica* are also capable of mimicking flowers biochemically and attracting generalist pollinator species. Most of the lower pitchers of the variety *typica* were not fragrant. As for the waxy and elongate variety, no odour was detected in the field for either of the two pitcher types. The greater number of insect visits observed in upper pitchers compared to lower ones might be the result of a larger abundance of insects at higher strata. This might be the case for the variety *elongata* but it cannot explain the far greater difference between upper and lower pitchers observed for the variety *typica*. Moreover, the comparison of the number of prey items caught in the variety *typica* by lower pitchers and upper pitchers put on the soil showed that the latter were still more attractive (Moran, 1996).

Is it possible that waxy phenotypes are not compatible with odour-attractive ones? This would explain the loss of the waxy zone in *N. rafflesiana* var. *typica*. We observed in the field that none of the waxy lower pitchers of the variety *typica* were fragrant; the only lower pitchers that were found to deliver a sweet odour were a few non-waxy pitchers. We also observed in Brunei that none of the waxy *Nepenthes* species were fragrant (neither *N. gracilis*, nor *N. albomarginata*, or *N. mirabilis*, for example). Moreover, the only *Nepenthes* species that are reported to deliver a sweet odour have non-waxy pitchers, such as *N. sumatrana* (Clarke, 2001). Could the lipid composition of the wax be a barrier to odour emission? This is a hypothesis that warrants further investigation.

The higher insect attracting properties of the variety *typica* are probably responsible for their higher prey captures. The better retention provided by the slippery and waxy pitchers of the variety *elongata* was not sufficient to compensate for their weaker power of attraction. The observed difference in prey abundance between the two varieties can not be attributed to a site effect. First, at the site where the two species were found together, simple visual inspection of the pitchers permitted us to establish lower prey contents of the variety *elongata*. Second, Moran obtained similar results on this point though on a much smaller sample size (Moran, 1996). Finally, the recordings of insect visitors for both varieties were done at the same precise site and showed that the pitchers of the variety *typica* attracted more insects than the pitchers of the variety *elongata*. As a matter of fact, the variety *elongata* is more sparsely distributed than the variety *typica*, being confined to more restricted and closed areas. Therefore the advantage provided by the waxy layer in terms of insect retention might be under-exploited, if the plant does not display sufficient cues (physical ones, such as colour or UV-patterns or chemical ones, such as sweet odours) that attract insects from long distance. In these cases, there is no clear correlation between wax phenotype and plant fitness, and the adaptive significance of the waxy layer might therefore also be questionable for the variety *elongata*. Is it conceivable that the typical form is on the way to exclude the elongate form through a competitive process? Niche differentiation led by competition may occur for the two taxa whose habitat and prey spectra seem to differ. Added to the fact that a reproductive barrier seems to prevent cross-fertilization between the two taxa (they never flower at the same time in sympatry), we would not be surprised to discover that they are in the process of speciation.

7.4.3 Heterochrony and Evolutionary Loss of the Waxy Zone in some Nepenthes Species?

Heterochrony refers to changes in the rate and timing of growth and development events or patterns. It is believed to play a key role in the evolution of biodiversity since it often leads to changes in size and shape during ontogeny and may sometimes result in speciation events if they favor the reproductive isolation of the individual (McNamara, 1997; Cronk et al., 2002). We may have observed evidence of a heterochronic process in the development of pitchers of *N. rafflesiana*. If we assume that the waxy trait, which is shared by most of the *Nepenthes* species, is an ancestral character, *N. rafflesiana* var. *elongata* would be closer to the common ancestor than *N. rafflesiana* var. *typica*. Ontogenetic loss of the waxy zone in *Nepenthes rafflesiana* var. *typica* would represent evidence of peramorphosis (McNamara, 1997) via acceleration of the developmental process for example. According to this process, only young plants or juveniles of the derived forms (such as *N. rafflesiana* var. *typica*) express the waxy phenotypes, which are typical of the ancestral forms (such as *N. rafflesiana* var. *elongata*). Similar events might have occurred several times in the evolution of *Nepenthes*, whose pitchers display considerable diversity in sizes and shapes (Danser, 1928; Cheek and Jebb, 2001; Clarke, 2001), and may have been involved in many instances of speciation. Tests of such hypotheses will become possible when a fully resolved molecular phylogeny of *Nepenthes* is available.

Acknowledgments We thank L. Lim for her kind cooperation, and D. Marshall, D. Lane, and D. Edwards for their administrative help in the Universiti Brunei Darussalam. We are grateful to the Forestry Department of Brunei, which supplied permits to carry out this research in the field. We also greatly acknowledge the help of S. Nyawa from the Brunei Museum and CITES commission and of M. Idris from The National Herbarium of Brunei. D. McKey is thanked for his helpful comments on the manuscript.

References

Adam, J. H. (1997) Prey spectra of Bornean *Nepenthes* species (Nepenthaceae) in relation to their habitat. *Pertanika Journal of Tropical Agriculture Science* 20: 121–134.

Adams, R.M., and Smith, G.W. (1977) A S.E.M. survey of the five carnivorous pitcher plant genera. *American Journal of Botany* 64: 265–272.

Bohn, H.F., and Federle, W. (2004) Insect aquaplanning: *Nepenthes* pitcher plants capture prey with the peristome, a fully wettable water-lubricated anisotropic surface. *Proceedings of the National Academy of Sciences USA* 101: 14138–14143.

Cheek, M., and Jebb, M. (2001) Nepenthaceae. In: *Flora Malesiana*, ed. by H.P. Nooteboom. Publications Department of the Nationaal Herbarium Nederland, Leiden, S1, 15; pp. 1–164.

Clarke, C. (1997) Nepenthes*of Borneo*. Kota Kinabalu, Malaysia: Natural History Publications.

Clarke, C. (2001) Nepenthes*of Sumatra and Peninsular Malaysia*. Kota Kinabalu, Malaysia: Natural History Publications.

Cronk, Q.C.B., Bateman, R.M., and Hawkins, J.A. (2002) *Developmental genetics and plant evolution*. London, UK: Taylor and Francis.

Danser, B.H. (1928) The Nepenthaceae of the Netherlands Indies. *Bulletin du Jardin Botanique de Buitenzorg* 9: 249–438.

Darwin, C. (1875) *Insectivorous plants*. London, UK: John Murray.

Ellison, A.M., Gotelli, N.J., Brewer J.S., Cochran-Stafira, D.L., Kneitel, J.M., Miller, T.M., Worley, A.C., and Zamora, R. (2003) The evolutionary ecology of carnivorous plants. *Advances in Ecological Research* 33: 1–74.

Gaume, L., Gorb, S., and Rowe, N. (2002) Function of epidermal surfaces in the trapping efficiency of *Nepenthes alata* pitchers. *New Phytologist* 156: 479–489.

Gaume, L., Perret, P., Gorb, E., Gorb, S., Labat, J-J., and Rowe, N. (2004) How do plant waxes cause flies to slide? Experimental tests of wax-based trapping mechanisms in three pitfall carnivorous plants. *Arthropod Structure and Development* 33: 103–111.

Gorb, E., Haas, K., Henrich A., Enders S., Barbakadze, N., and Gorb S. (2005) Composite structure of the crystalline epicuticular wax layer of the slippery zone in the pitchers of the carnivorous plant *Nepenthes alata* and its effect on insect attachment. *The Journal of Experimental Biology* 208: 4651–4662.

Gorb, E., Kastner, V., Peressadko, A., Arzt, E., Gaume, L., Rowe, N., and Gorb, S. (2004) Structure and properties of the glandular surface in the digestive zone of the pitcher in the carnivorous pitcher plant *Nepenthes ventrata* and its role in insect trapping and retention. *Journal of Experimental Biology* 207: 2947–2963.

Hooker, J.D. (1859) On the origin and the development of the pitcher of *Nepenthes*, with an account of some new Bornean plants of the genus. *Trans.Linnean Society* 22: 415–424.

Joel, D. M. (1988) Mimicry and mutualism in carnivorous pitcher plants (Sarraceniaceae, Nepenthaceae, Cephalotaceae, Bromeliaceae). *Biological Journal of the Linnean Society* 35: 185–197.

Juniper, B.E., and Burras, J. (1962) How pitcher plants trap insects. *New Scientist* 13:75–77.

Juniper, B.E., Robins, R.J., and Joel, D. (1989) *The carnivorous plants*. London, UK: Academic press.

Lloyd, F.E. (1942) *The carnivorous plants*. Waltham, Mass., US: Chronica Botanica Co.

McNamara, K.J. (1997) *Shapes of time: The evolution of growth and development*. Baltimore: John Hopkins University Press.

Meimberg, H., Thalammer, S., Brachmann, A., and Heub, G. (2006) Comparative analysis of a translocated copy of the *trn*K intron in carnivorous family Nepenthaceae. *Molecular Phylogenetics and Evolution* 39: 478–490.

Merbach, M.A., Zizka, G., Fiala, B., Maschwitz, U., and Booth, W.A. (2001) Patterns of nectar secretion in five *Nepenthes* species from Brunei Darussalam, Northwest Borneo, and implications for ant-plant relationships. *Flora* 196: 153–160.

Moran, J.A. (1996) Pitcher dimorphism, prey composition and the mechanisms of prey attraction in the pitcher plant *Nepenthes rafflesiana* in Borneo. *Journal of Ecology* 84: 515–525.

Moran, J.A., and Moran, A.J. (1998) Foliar reflectance and vector analysis reveal nutrient stress in prey-deprived pitcher plants (*Nepenthes rafflesiana*). *International Journal of Plant Science* 159: 996–1001.

Moran, J.A., Booth, W.E., and Charles, J.K. (1999) Aspects of pitcher morphology and spectral characteristics of six Bornean *Nepenthes* pitcher plant species: implications for prey capture. *Annals of Botany* 83: 521–528.

Moran, J.A., Clarke, C.M. and Hawkins, B.J. (2003) From carnivore to detritivore? Isotopic evidence for leaf litter utilization by the tropical pitcher plant *Nepenthes ampullaria*. *International Journal of Plant Science* 164: 635–639.

Moran, J.A., Merbach, M.A., Livingston, N.J., Clarke, C.M., and Booth, W.E. (2001) Termite prey specialization in the pitcher plant *Nepenthes albomarginata* – Evidence from stable isotope analysis. *Annals of Botany* 88: 307–311.

Owen, T.P., and Lennon, K.A. (1999) Structure and development of the pitchers from the carnivorous plant *Nepenthes alata* (Nepenthaceae). *American Journal of Botany* 86: 1382–1390.

Phillipps, A., and Lamb, A. (1996) *Pitcher plants of Borneo*. Kota Kinabalu, Malaysia: Natural History Publications.

Riedel, M., Eichner, A., and Jetter, R. (2003) Slippery surfaces of carnivorous plants: composition of epicuticular wax crystals in *Nepenthes alata* Blanco pitchers. *Planta* 218: 87–97.

Schultze, W., Schultze, E.D., Pate, J.S., and Gillison, A.N. (1997) The nitrogen supply from soils and insects during growth of the pitcher plants *Nepenthes mirabilis, Cephalotus follicularis* and *Darlingtonia californica*. *Oecologia* 112: 464–471.

Chapter 8
Functional Surfaces in the Pitcher of the Carnivorous Plant *Nepenthes alata*: A Cryo-Sem Approach

Elena V. Gorb and Stanislav N. Gorb

8.1 Introduction

Since carnivorous plants grow in habitats deprived of such nutrients as nitrogen and phosphorus, they use insects and other arthropods as an important nutrient source (Thum, 1988; Juniper et al., 1989; Schulze and Schulze, 1990; Schulze et al., 1997; Ellison and Gotelli, 2001). Carnivorous plants from the genus *Nepenthes* have evolved specialized trapping organs called pitchers, adapted to prey attracting, capturing and digestion (Juniper, 1986; Juniper et al., 1989). These passive "pitfall" traps originated from modified leaves (Troll, 1932; Goebel, 1923; Lloyd, 1942; Pant and Bhatnagar, 1977; Juniper et al., 1989) and have a rather complex structure (Fig. 8.1A). They are composed of several distinct zones, which differ in macromorphology, surface microsculpture and chemistry, and serve different functions (Hooker, 1859; Lloyd, 1942; Juniper et al., 1989). In most *Nepenthes* species, such structural and functional pitcher zones as a leaf-like lid, ribbed rim called peristome, slippery (waxy) zone, and digestive (glandular) zone are clearly distinguished (Fig. 8.2B) (Adams and Smith, 1977; Owen and Lennon, 1999; Clarke, 1997, 2001; Gaume et al., 2002). Recently, a smooth transitional zone, separating the waxy surface from the glandular one, has been described (Gaume et al., 2002).

Numerous previous studies on *Nepenthes* traps were focused mainly on the pitcher morphology, using various microscopy methods, including conventional scanning electron microscopy (SEM) (Adams and Smith, 1977; Pant and Bhatnagar, 1977; Owen and Lennon, 1999; Riedel et al., 2003; Gorb et al., 2004), and on functions of different zones (Fenner, 1904; Lloyd, 1942; Juniper and Burras, 1962; Luettge, 1964, 1965; Schulze et al., 1999; Owen et al., 1999; Gaume et al., 2002; Bohn and Federle, 2004; Gorb et al., 2005). In a few studies, surface properties of some zones were estimated (Bohn and Federle, 2004; Gorb et al., 2004; Gorb and Gorb, 2006).

The aim of this study is to reexamine the structure and microtopography of epidermal surfaces in functional zones of *N. alata* pitchers, using a cryo-SEM

E.V. Gorb (✉)
Evolutionary Biomaterials Group, Max Planck Institute, Stuttgart, Germany

S.N. Gorb (ed.), *Functional Surfaces in Biology*, Vol. 2,
DOI 10.1007/978-1-4020-6695-5_9,

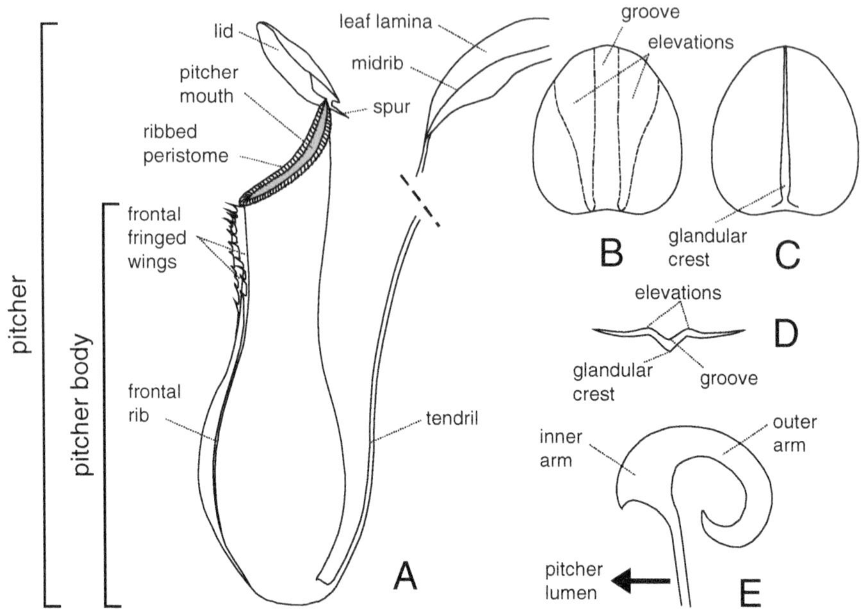

Fig. 8.1 Diagram of the upper pitcher in *Nepenthes alata*. **A**. Side view of the entire pitcher. **B**. Upper lid face. **C**. Inner lid face. **D**. In plane view of the lid. E. Transverse section of the peristome

method allowing high-resolution imaging of frozen and fractured samples. The existing electron microscopy studies of *Nepenthes* pitchers have used the conventional technique of sample preparation, including treatment in strong solvents. Since surfaces are covered with layers of waxes or fluids, it was previously impossible to combine high-resolution imaging with keeping samples in native condition. Using the cryo-SEM technique, we obtained a new insight into the ultrastructure of superficial layers in different pitcher zones under native conditions at high resolution. Based on results we obtained and literature data, the role of different functional pitcher surfaces in prey capturing and retention will be discussed.

8.2 Material and Methods

8.2.1 Plant Material

Leaf and pitcher material of the mountainous Philippine plant *N. alata* Blanco (Nepenthaceae) was obtained from plants cultivated in the greenhouse of the Botanical Garden at the University of Hohenheim (Stuttgart, Germany).

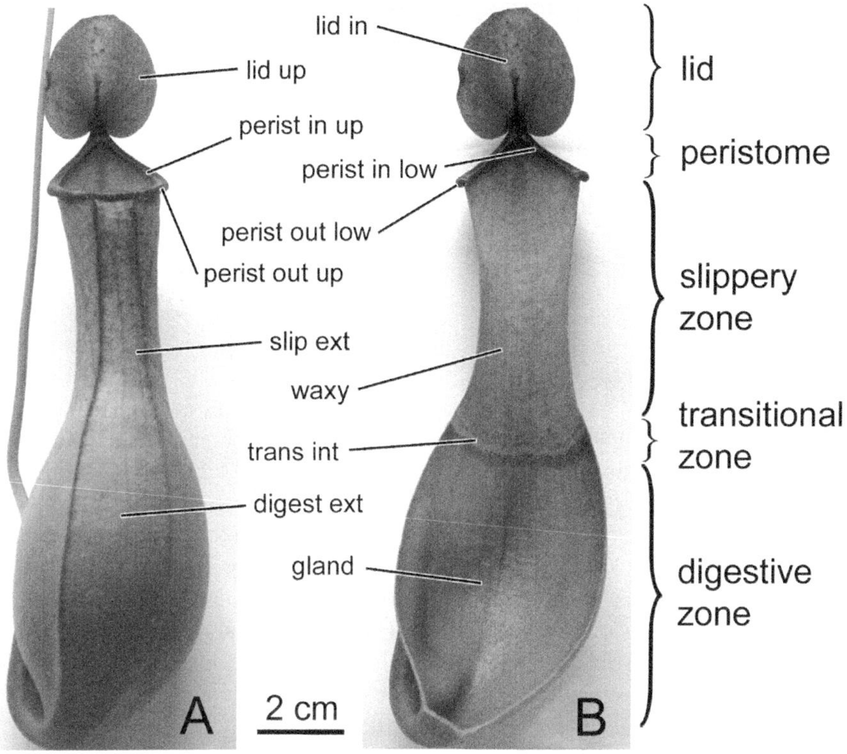

Fig. 8.2 Pitcher zones and functional epidermal surfaces, studied in *Nepenthes alata*. **A**. Intact pitcher. **B**. Longitudinally dissected pitcher, internal surfaces are visible. *gland*, glandular surface of the digestive zone; *digest ext*, external surface of the digestive zone; *lid in*, inner face of the lid; *lid up*, upper face of the lid; *perist in low*, lower surface of the inner arm of the peristome; *perist in up*, upper surface of the inner arm of the peristome; *perist out low*, lower surface of the outer arm of the peristome; *perist out up*, upper surface of the outer arm of the peristome; *slip ext*, external surface of the slippery zone; *trans int*, internal surface of the transitional zone; *waxy*, waxy surface of the slippery zone

Following pitcher surfaces were examined (Fig. 8.2):

a) the lid, upper and inner faces;
b) the peristome, upper and lower surfaces of outer and inner arms;
c) the slippery zone, external and waxy surfaces;
d) the digestive zone, external and glandular surfaces.

Since upper and lower pitchers do not differ very much morphologically in *N. alata* (Danser, 1928; Slack, 1979; Jebb and Cheek, 1997), we examined only upper pitchers.

Pitchers were previously considered to be evolved from leaves, in which the adaxial surface of the epiascidiate (container) leaf forms the inner wall of the pitcher tube (Martin and Juniper, 1970; Juniper et al., 1989). Therefore, the lamina of the pitcher-bearing leaf, both adaxial and abaxial surfaces, was studied here as a reference structure.

8.2.2 Cryo-SEM Method

General morphology of the leaf, pitcher and different pitcher zones was examined on fresh, untreated samples in a binocular microscope Leica MZ 12.5 with a built-in video chip Leica IC A (Leica Microsystems GmbH, Wetzlar, Germany). For cryo-SEM, small pieces ($\sim$ 1 cm long for the peristome and $\sim$ 2 cm^2 for other samples) were cut out with a razor blade from the leaf/pitcher, glued with polyvinyl alcohol Tissue-Tek® O.C.T.™ Compound (Sakura Finetek Europe B.V., Zoeterwoude, The Netherlands) to or mechanically gripped in a small vice on holders, and frozen in a cryo-stage preparation chamber at $-140°C$ (Gatan ALTO 2500 cryo-preparation system, Gatan Inc., Abingdon, UK). Frozen samples, either entire or fractured with a cold metal fracture knife in the cryo-stage preparation chamber, were sputter-coated with gold-palladium (6 nm) and studied in frozen condition in a cryo-SEM Hitachi S-4800 (Hitachi High-Technologies Corporation, Tokyo, Japan) at 3 kV accelerating voltage and $-120°C$ temperature. This method allowed us, for the first time, to visualize cuticle structure together with waxes and fluids, located on surfaces. Variables of surface structures and thickness of cuticle layers were measured from digital images, using SigmaScan software (SPSS Inc., Chicago, USA).

We used the terminology, proposed by Payne (1978), to characterize trichome morphology. For the cuticular membrane, or cuticle, we employed the nomenclature according to Holloway (1982), Martin and Juniper (1970), and Jeffree (1986, 1996). Cuticle structure was determined and described according to Holloway (1982) and Jeffree (1986, 1996). Type of epicuticular wax coverage was identified according to the classification, introduced by Barthlott et al. (1998).

8.3 Results

8.3.1 General Morphology of the Pitcher

Pitcher-bearing leaves, as well as other leaves, are epitiolate and have rather thick lanceolate-ovate lamina with a quite narrow base and acute apex. A midrib extends in a tendril with a hanging pitcher at the end; the twining tendril is attached to the pitcher at the back (Fig. 8.1A). Pitchers are up to 15 cm long and 4 cm wide, with an elongated cylindrical body, becoming slightly bulbous in its lower part. The basal part is abruptly attenuate to the tendril and does not differ in texture from the rest of the pitcher. Two front wings, being rather narrow and fringed in the upper part of the pitcher body, are gradually reduced to ribs, running in the direction towards the pitcher base, where they merge with the tendril. A pitcher mouth is slightly oval and oblique, with a relatively narrow cylindrical ribbed peristome. The inner arm of the peristome has indistinct teeth and is much shorter than the outer arm (Fig. 8.1E). The lid is almost round, elliptic or slightly cordate and possesses, on the inner face, a glandular crest, being especially prominent in the basal lid part (Fig. 8.1C, D). The upper lid face is uneven because of the groove, surrounded with two elevations, all running from the proximal to the distal part of the lid (Fig. 8.1B, D). The spur is non-branched and acutely-pointed (Fig. 8.1A). The pitcher body is light green and becomes flecked with red in the upper part, peristome and upper face of the lid.

8.3.2 Leaf Lamina

Leaf surfaces have a shiny appearance, which is expressed less on the abaxial side. In the binocular microscope, the surfaces are smooth, almost even (abaxial side) or slightly uneven because of somewhat prominent epidermal cells (adaxial side). At high magnifications in the cryo-SEM, surfaces show some microsculpturing that is rather irregular on the adaxial side (Fig. 8.3A) and has scale- or knob-like, flat, randomly distributed projections on the abaxial side (Fig. 8.3D).

Both leaf surfaces bear regularly scattered, flattened peltate trichomes (approx. 50–60 μm in diameter or length), situated in small shallow depressions (Fig. 8.3B, C, E). On the adaxial side, the trichomes are usually clearly four-lobed, regularly-shaped, with rather equal lobes in most cases (Fig. 8.3B, C) or rarely, asymmetrical

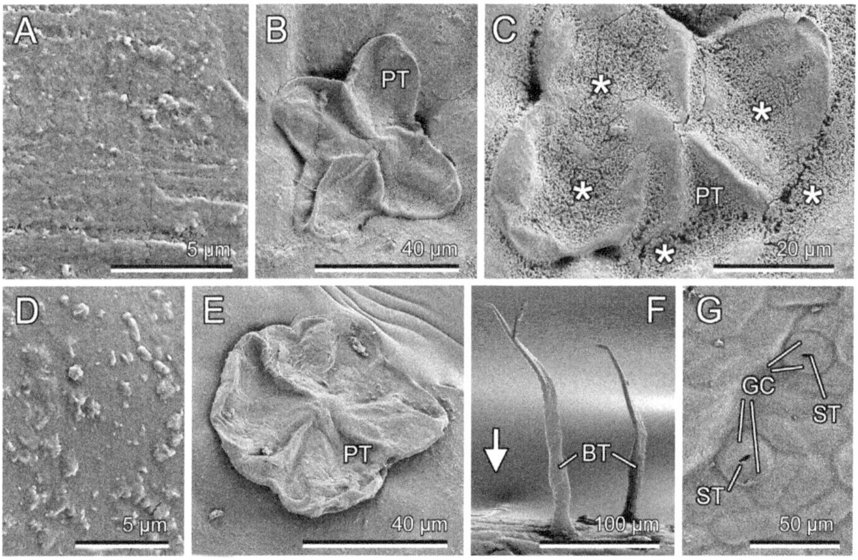

Fig. 8.3 Cryo-SEM micrographs of the leaf lamina surfaces. **A–C**. Adaxial surface. **D**–**G**. Abaxial surface. *BT*, bootjack trichomes; *GC*, guard cells; *PT*, peltate trichomes; *ST*, stomata. *Asterisks* in **C** show a foam-like coverage on the trichome surface; *arrow* in **F** indicates the direction towards the midrib of the lamina

with unequal lobes. On the abaxial side, round-shaped trichomes with less distinguishable lobes (Fig. 8.3E) occur much more often than those observed on the adaxial side. These trichomes are slightly larger and distributed more densely compared to those on the adaxial side. In some samples, the surface of the trichomes and in surrounding areas, is covered with a foam-like coverage (Fig. 8.3C), which may be interpreted as residuals of a frozen fluid. This fluid is probably a sugar-containing water solution, secreted by peltate trichomes, serving as water-excreting glands, hydathodes (Haberlandt, 1894; Stern, 1916).

On both leaf sides, long ($121.20\pm51.30\,\mu m$, n=15) erect bootjack trichomes are densely situated on the very leaf lamina margins (Fig. 8.3F). These trichomes are either non-branched or have few (up to 4) flattened branches with acute tips originating at shallow angles, with one branch, being usually notably longer than the others. Both leaf lamina surfaces bears regularly distributed stomata, each having two guard cells (Fig. 8.3G). On the abaxial leaf side, stomata are far more frequent than on the adaxial surface.

The leaf lamina is about 350 – 400 μm thick and has a layered inner structure. It is composed of the abaxial and adaxial epidermal cell layers, and palisade and spongy mesophyll in between (Fig. 8.4A). The outer epidermal wall (the external cellulose wall of the epidermal cells together with the cuticle, terminology after Lyshede (1982)) is much thicker (adaxial: $2.35\pm0.15\,\mu m$, n=8; abaxial: $2.52\pm0.25\,\mu m$, n=10) than inner and radial cell walls in both leaf lamina surfaces (Fig. 8.4B, D).

On the adaxial side, the thickness of the external cellulose wall ($0.37\pm0.10\,\mu m$, n=8) considerably exceeds the cuticle thickness ($1.95\pm0.16\,\mu m$, n=8) (Fig. 8.4C). The cuticle shows clear reticulate structure, which is coarse in inner portions, finer in the middle part, and disappears entirely in distal regions. The outermost layer constitutes an amorphous epicuticular wax, which forms a thin ($0.08\pm0.02\,\mu m$, n=8) continuous smooth layer. It is easily detectable because of artificial cracks, located close to the fracture plane. On the abaxial side, the outer epidermal wall is nearly of the same thickness, with a thicker ($1.03\pm0.20\,\mu m$, n=10) external cellulose wall and a thinner ($1.51\pm0.22\,\mu m$, n=10) cuticle than those of the adaxial side (Fig. 8.4D, E). The cuticle has reticulate texture in deeply located sites (weaker compared to the adaxial side) and becomes amorphous close to the surface (Fig. 8.4E). The epicuticular wax layer appears much thicker ($0.42\pm0.10\,\mu m$, n=10) than in the adaxial surface, consists of slightly inhomogeneous material, and forms regular wavy surface sculpturing (Fig. 8.4E, F).

8.3.3 Lid

The inner lid face has a rather plain surface appearance, whereas in the outer face of the lid, the epidermal cell topography causes a slightly uneven surface relief. A superficial thin layer, covering the surface of the inner lid face, appears visible at high magnifications (Fig. 8.5E, H). This coverage is either relatively smooth (Fig. 8.5H) or more sculptured (Fig. 8.5E).

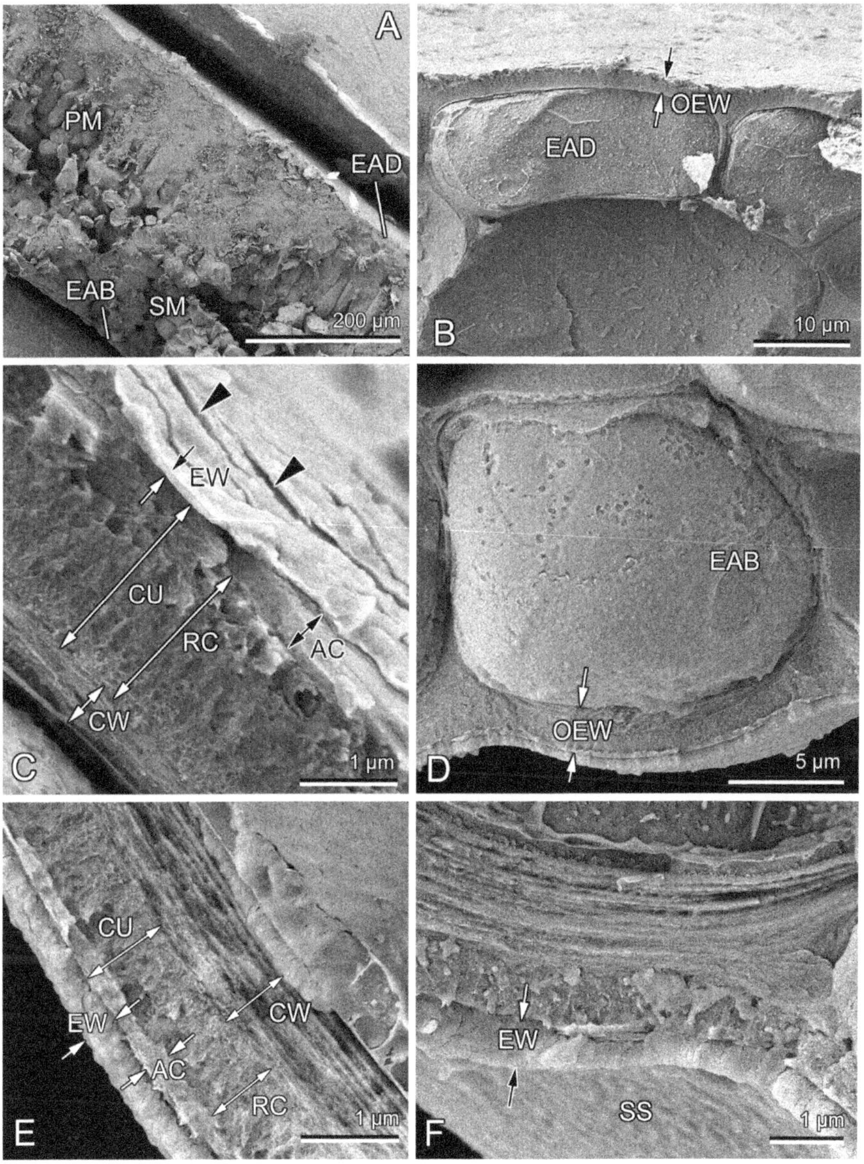

Fig. 8.4 Cryo-SEM micrographs of the inner structure of the leaf (**A**) and outer epidermal wall structure (**B**–**F**), fractured samples. **B**, **C**. Adaxial surface. **D**–**F**. Abaxial surface. *AC*, amorphous cuticle; *CU*, cuticle; *CW*, external cellulose wall; *EAB*, abaxial epidermal cell layer; *EAD*, adaxial epidermal cell layer; *EW*, epicuticular wax layer; *OEW*, outer epidermal wall; *PM*, palisade mesophyll; *RC*, reticulate cuticle; *SM*, spongy mesophyll; *SS*, surface sculpturing. *Arrowheads* in **C** point to artificial cracks of the epicuticular wax layer

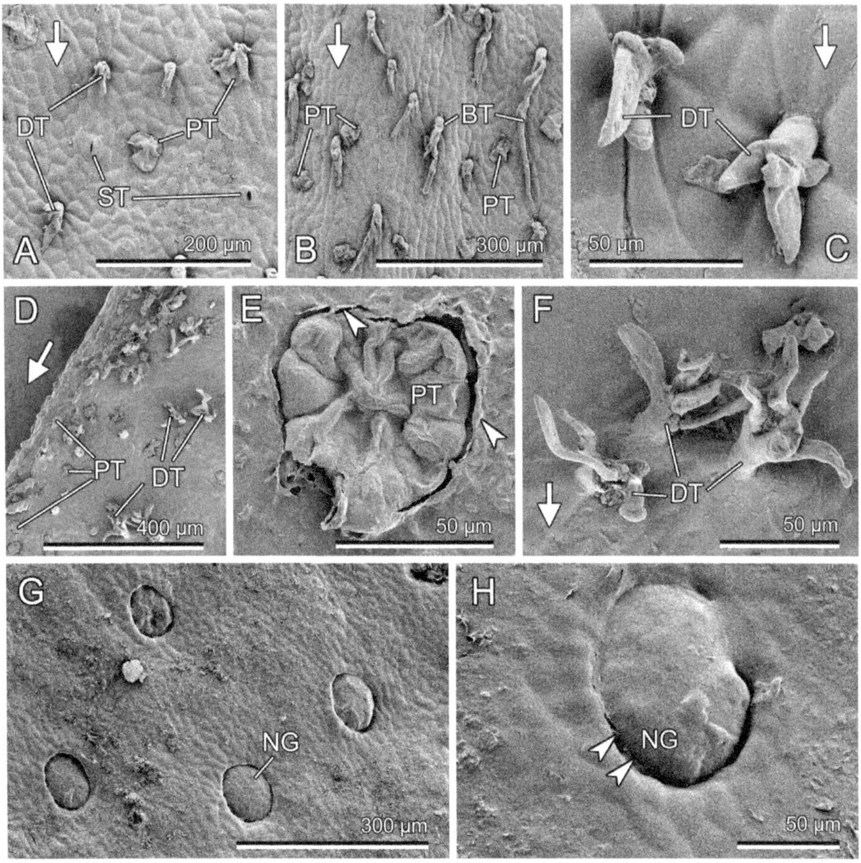

Fig. 8.5 Cryo-SEM micrographs of the upper (**A–C**) and inner (**D–H**) faces of the lid. *BT*, bootjack trichomes; *DT*, dendroid trichomes; *NG*, nectar glands; *PT*, peltate trichomes; *ST*, stomata with guard cells. *Arrows* in **A–D** and **F** show the direction to the base of the lid; *arrowheads* in **E** and **H** indicate a superficial surface coverage

Both faces bear numerous macroscopic surface structures. Stomata occur sporadically on both lid faces (Fig. 8.5A). On the upper face, round-shaped, peltate trichomes (approx. 40–60 μm in diameter), similar to those found on the abaxial leaf surface, are regularly distributed (Fig. 8.5A, B). Other types of trichomes (bootjack and dendroid), densely covering the basal part of the lid and especially the surface of two elevations (Fig. 8.5A–C), vary essentially in dimensions, shape and branching pattern.

Bootjack trichomes of greatly variable size (length: 121.20±51.30 μm, n=15) are the most abundant ones. They have few (2–5) flattened branches with acute tips, and one branch is always much longer than the others (Fig. 8.5B). These trichomes are usually crowded to the surface and proximally-pointed. Dendroid trichomes are shorter (length: 46.86±8.66 μm, n=16), and their flattened branches, with acute tips, have nearly the same length (Fig. 8.5C). These trichomes are erect, and their branches originate at shallow angles. They are distributed more sparsely over the surface and prevail at sites situated close to lid margins.

In the inner lid face, flattened peltate trichomes and dendroid trichomes occur (Fig. 8.5D), while no bootjack ones are present. Peltate trichomes are similar to those observed at the upper lid face (Fig. 8.5E). Dendroid trichomes usually have more branches of smaller dimensions compared to those on the upper face; tips are rounded, not acute (Fig. 8.5F). Peltate trichomes, although being much more numerous at lid margins and on the crest, are widespread throughout the whole lid surface, whereas dendroid ones are concentrated close to the margins (Fig. 8.5D). The basal and central parts of the inner lid face bear regularly distributed, extrafloral nectar glands (Fig. 8.5G). These are large (diameter or length: 85.95±9.12 μm, n=8) round- or oval-shaped structures, sunken in quite deep epidermal depressions. Some depressions are completely filled with secretion so that glands appear to be completely hidden (Fig. 8.5H). A superficial coverage, detected on the inner lid surface (Fig. 8.5E, H), probably constitutes a layer of gland secretion.

The lid is thinner (up to 250 μm) than the leaf lamina; its mesophyll consists of closely packed, undifferentiated parenchyma cells (Fig. 8.6A). Outer epidermal walls in both lid faces are thicker than inner and radial cell walls (Fig. 8.6B, E), being, however, much thinner than those of the leaf lamina. In the upper face, the outer epidermal wall (thickness: 1.40±0.17 μm, n=10) is reinforced by a rather heavy external cellulose wall (thickness: 0.71±0.10 μm, n=10), having almost the same thickness as the cuticle (0.67±0.11 μm, n=10) (Fig. 8.6B, C). The cuticle has a structure similar to that in the adaxial leaf surface; however, reticular texture is less pronounced, and distal amorphous regions often prevail (Fig. 8.6C). A smooth thin continuous layer of the epicuticular wax of variable thickness (0.08±0.04 μm, n=23) overlays the cuticle (Fig. 8.6D). It creates irregular microrelief of the surface and is cracked in some places.

The outer epidermal wall and the cuticle of the inner face of the lid, although slightly thinner, are very similar in structure and thickness ratios (outer wall: 1.01±0.15 μm; cellulose wall: 0.44±0.06 μm; cuticle: 0.53±0.15 μm, n=10) to those of the outer lid face (Fig. 8.6E, F). The only essential difference lies in the epicuticular coverage. A thin wax film is hardly visible in a few sites (Fig. 8.6F). The superficial layer, varying considerably in thickness (0.17±0.11 μm, n=11) and showing distinct lamellate structure, is clearly detectable, since it flakes away easily as a result of the fracture procedure (Fig. 8.6E, F). This layer may be interpreted as a secretion, produced by nectar glands.

8.3.4 Peristome

The upper peristome surface, although having a shiny appearance, shows a regularly ribbed topography, created by a hierarchical system of radial ridges and grooves. Ridges of the first order (Fig. 8.7A) emerge close to the margin of the outer peristome arm (Fig. 8.7H, I) and end in tooth-like outgrowths, located at the margin

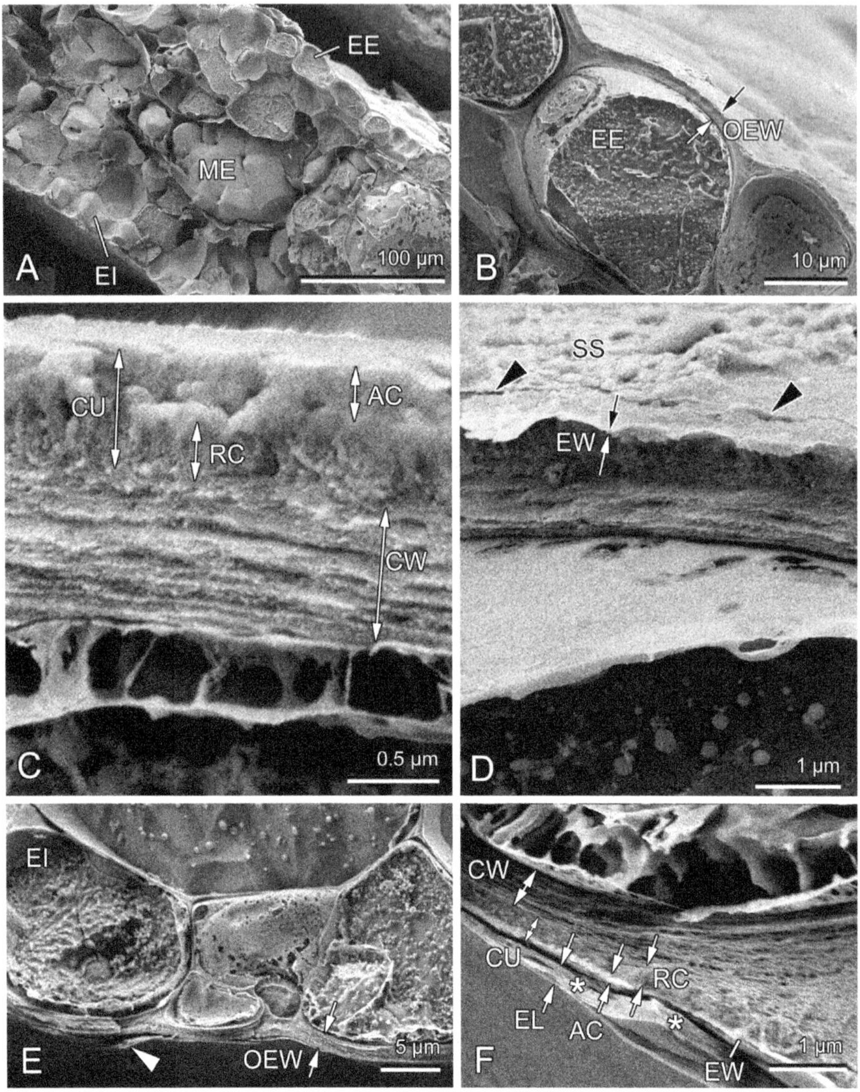

Fig. 8.6 Cryo-SEM micrographs of the inner structure of the lid, fractured samples. **A**. Fracture of the lid. **B–F**. Outer epidermal wall structure of the upper (**B–D**) and inner (**E, F**) lid faces. *AC*, amorphous cuticle; *CU*, cuticle; *CW*, external cellulose wall; *EE*, external epidermal cell layer; *EI*, inner epidermal cell layer; *EL*, epicuticular layer; *EW*, epicuticular wax layer; *ME*, mesophyll; *OEW*, outer epidermal wall; *RC*, reticulate cuticle; *SS*, surface sculpturing. *Black arrowheads* in **D** point to cracks of the epicuticular wax layer; *white arrowhead* in **E** indicates a flaked superficial layer; *asterisks* in **F** show lamellate structure of the superficial layer

in the inner arm of the peristome (Fig. 8.7E). Teeth hang over the pitcher wall and are pointed towards the pitcher cavity. Close to the inner arm margin, there are numerous nectar glands, each situated in a deep depression between two teeth (Fig. 8.7E, F). The depression turns into a deep stria, running in the direction of the

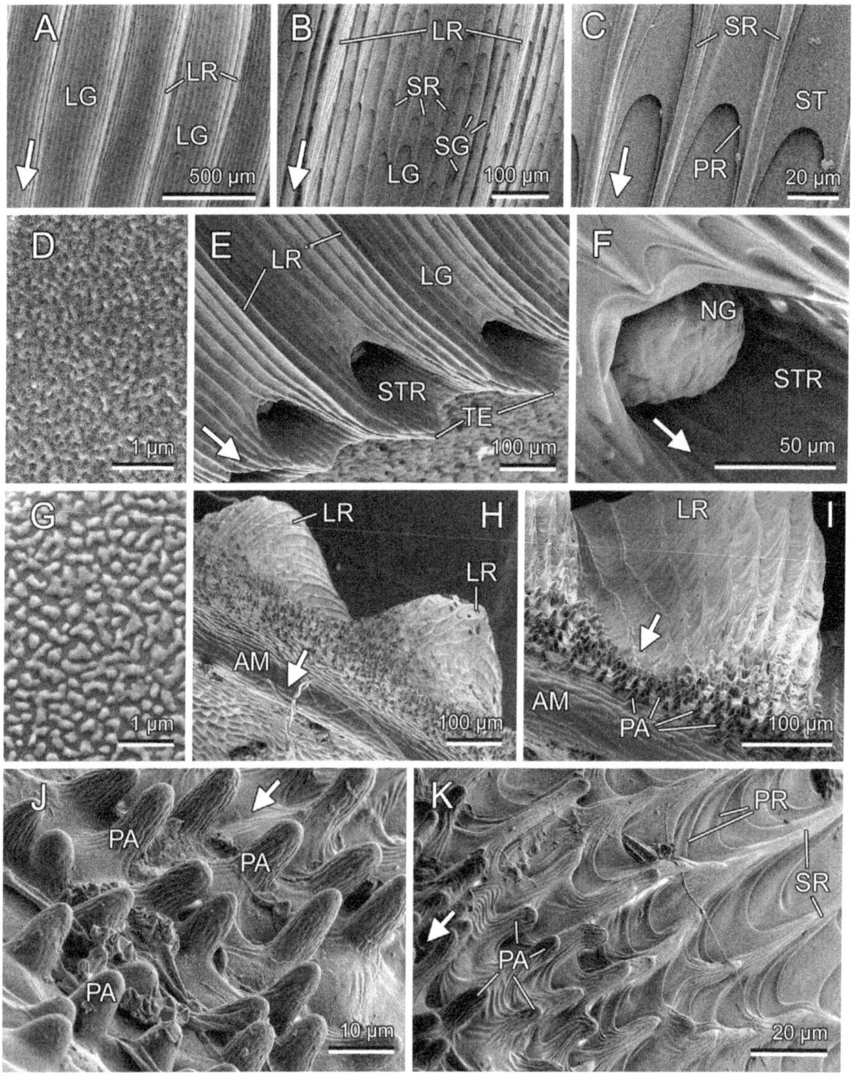

Fig. 8.7 Cryo-SEM micrographs of the upper surface of the peristome. **A–D**. Central part. **E–G**. Inner arm. **H–K**. Outer arm. *AM*, margin of the outer arm; *LG*, large grooves between ridges of the first order; *LR*, large ridges of the first order; *NG*, nectar glands; *PA*, papillae; *PR*, parabolic ledges; *SG*, small grooves between ridges of the second order; *SR*, small ridges of the second order; *ST*, steps; *STR*, striae; *TE*, teeth of the inner arm margin. *Arrows* in **A–C**, **E**, **F** and **H–K** indicate the direction towards the interior of the pitcher body

pitcher cavity. Grooves between large first-order ridges (width: 325.77±25.50 µm, n=9) are additionally sculptured with smaller ridges of the second order (Fig. 8.7B).

The second-order grooves (width: 30.65±2.89 µm, n=15) bear rows of parabolic ledges, forming step-like surface structures of greatly variable size (width: 58.89± 21.07 µm, n=18) (Fig. 8.7C). Steps are pointed towards the interior of the pitcher body, and this causes evident surface anisotropy of the peristome. Both second-order grooves and steps derive from papillae, occuring close to the margin of the outer arm (Fig. 8.7I–K). The margin has smooth uneven surface architecture, caused by epidermal cell topography, lacking any outgrowths (Fig. 8.7H, I). The erect papillae, having distinct pitted ornamentation, emerge at the margin, increase gradually in size, get additionally rough rugose microsculpturing (Fig. 8.7I, J), and become more and more crowded towards the surface (Fig. 8.7K). Finally, they create second-order ridges and steps that are characteristic features in the upper side of the peristome. The surface of steps demonstrates regular smoothed pitted ornamentation (Fig. 8.7D), and bears densely distributed droplet-like prominences of an irregular shape and variable size close to the inner arm margin (Fig. 8.7G).

In a separate experiment, we sprayed water over the upper peristome surface. Water readily wetted the surface and was present as a layer, entirely covering the areas between large first-order ridges (Fig. 8.8A, B).

The lower surface shows an essential difference between the outer and inner arms of the peristome. The uneven surface of the outer arm bears densely distributed, slightly sunken symmetrical peltate trichomes (length: 51.91±4.88 μm, n=15)

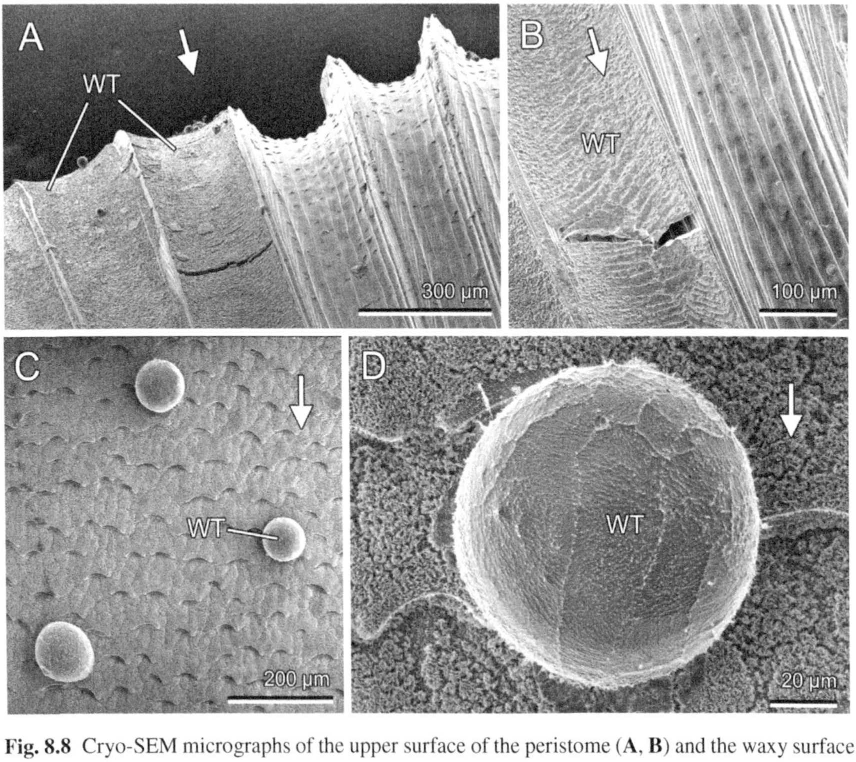

Fig. 8.8 Cryo-SEM micrographs of the upper surface of the peristome (**A**, **B**) and the waxy surface of the slippery zone (**C**, **D**) after spraying water over the surfaces. *WT*, water. *Arrows* in **A** and **B** show the direction towards the interior of the pitcher body, in **C** and **D** – to the bottom of the pitcher

similar to those described above in the lid (Fig. 8.9B, C). As in the abaxial leaf side, the surface is partly covered with a foam-like coverage (Fig. 8.9D). Several morphological regions are distinguished in the outer arm (Fig. 8.9A). Close to the margin, only peltate trichomes occur. Moving away from the margin, solitary dendroid trichomes (length: 82.29±14.13 μm, n=14), having the same structure as those in the inner face of the lid, appear (Fig. 8.9E). Close to the boundary

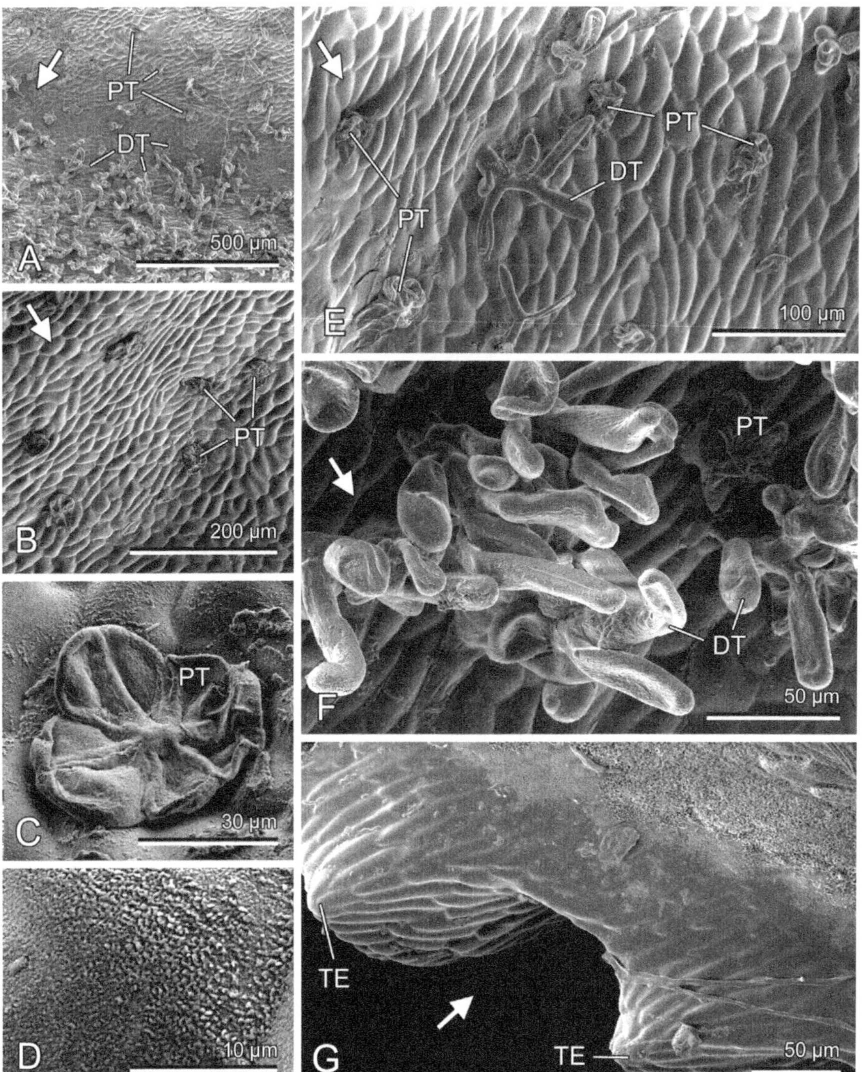

Fig. 8.9 Cryo-SEM micrographs of the lower surface of the peristome. **A–F**. Outer arm. **G**. Inner arm. *DT*, dendroid trichomes; *PT*, peltate trichomes; *TE*, teeth of the inner arm margin. *Arrows* in **A**, **B** and **E–G** show the direction towards the bottom of the pitcher

between the peristome and the external surface of the pitcher, they are numerous and situated very densely (Fig. 8.9F). In the lower side of the inner arm, teeth surface has an uneven relief, formed by relatively flat, protuberant, overlapping cells (Fig. 8.9G). Towards the boundary between the peristome and the waxy zone, surface sculpturing becomes less expressed and finally disappears.

The peristome is nearly as thick (approx. 350 μm) as the leaf lamina. Being much thicker than the lid, it shows, however, similar inner structure, being composed of two epidermal layers and undifferentiated parenchyma cells (Fig. 8.10A). The outer epidermal wall of the upper side is very massive (minimal thickness: 4.14±0.95 μm, n=14) (Fig. 8.10B) because of an essentially thickened (3.53±0.74 μm, n=14) cuticle; the external cellulose wall is relatively thin (0.55±0.36 μm, n=10) (Fig. 8.10C, D). It is relevant to note that characteristic surface features of the peristome, at least at the level of second-order ridges and steps, originated from the cuticle thickening and formation of folds. Thus, the ridges arise as considerable cuticle protrusions (Fig. 8.10C), whereas parabolic ledges represent either an incrassate cuticle in sites close to the ridges and in low steps (Fig. 8.10D), or cuticle folds in the case of deep overhanging steps (Fig. 8.10E). The cuticle shows layered structure, where the upper amorphous part, having a rather homogeneous appearance, and the course reticulate region beneath are distinguished (Fig. 8.10C, D). Overall cuticle thickness, as well as the ratio of thicknesses of these two parts, varies extremely, depending on site location. Smooth thin wax layer (thickness: 0.08±0.03 μm, n=15), covering the amorphous cuticle region, is visible in a few fractures. It cracks near the fracture plane, either forming series of small parallel clefts or loosing broken flinders when thicker (Fig. 8.10F, G). This layer creates a regular smoothed, pitted microrelief of the surface (Fig. 8.10F), observed in most preparations.

In the lower side, the outer epidermal wall is much thinner (0.78±0.12 μm, n=11) compared to the upper side, with the external cellulose wall being half as thick (0.24±0.10 μm, n=10) as the cuticle (thickness: 0.52±0.08 μm, n=10) (Fig. 8.11A, B). Reticulate region in deeply located sites, having course texture, and amorphous region in upper sites of the cuticle are of almost the same thickness (Fig. 8.11B). Smooth thin wax layer on the top of the cuticle, although showing considerable variability in thickness (0.23±0.06 μm, n=12), is heavier compared to that of the upper surface (Fig. 8.11C). It cracks readily near the fracture plane (Fig. 8.11D, E). This layer creates irregular pitted sculpturing of the surface (Fig. 8.11B, D).

8.3.5 External and Waxy Surfaces of the Slippery Zone

The external surface of the pitcher at the level of the slippery zone has a slightly uneven profile, caused by convex epidermal cells (Fig. 8.12A), and shows a weak, irregular surface microrelief at higher magnifications (Fig. 8.12B). It bears randomly scattered stomata and regularly distributed trichomes of several types: flat peltate, short branched (dendroid), and long non-branched (attenuate) (Fig. 8.12A). Symmetrical rounded or four-lobed peltate trichomes (diameter or length: 67.74±

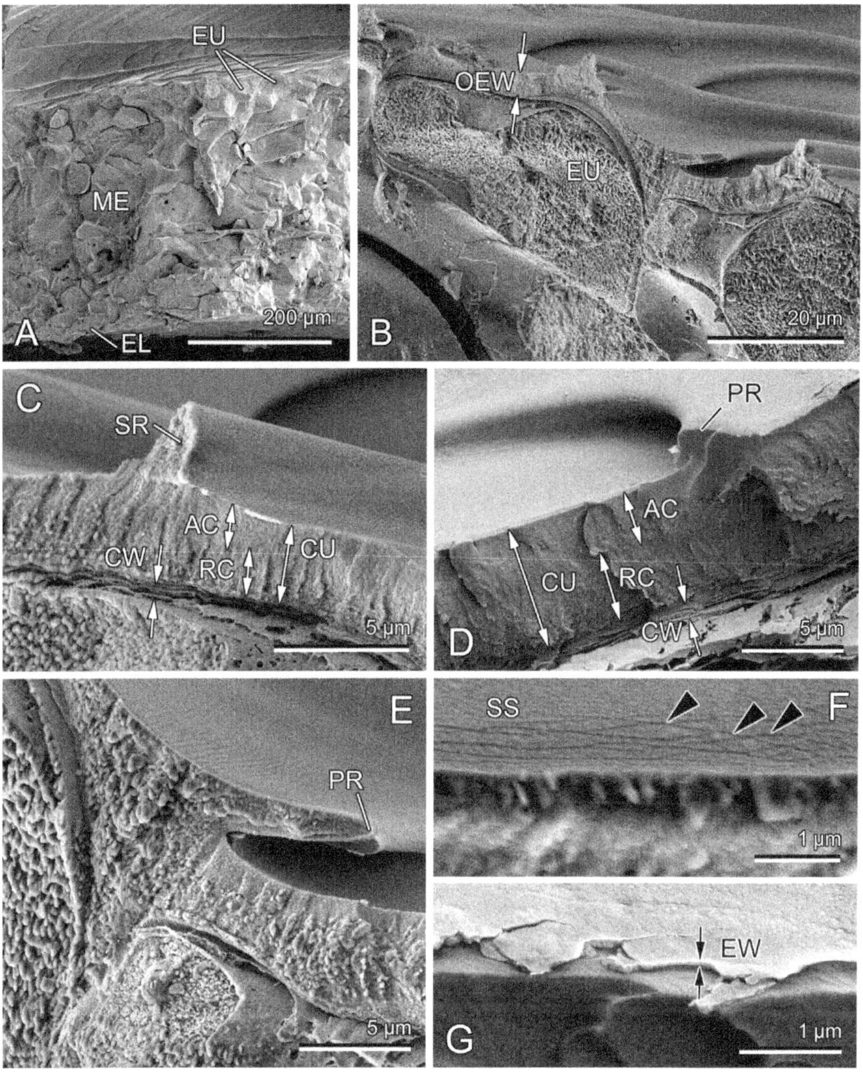

Fig. 8.10 Cryo-SEM micrographs of the inner structure of the peristome, fractured samples. **A**. Fracture of the peristome. **B–G**. Outer epidermal wall structure of the upper peristome surface. *AC*, amorphous cuticle; *CU*, cuticle; *CW*, external cellulose wall; *EL*, lower epidermal cell layer; *EU*, upper epidermal cell layer; *EW*, epicuticular wax layer; *ME*, mesophyll; *OEW*, outer epidermal wall; *PR*, parabolic ledge; *RC*, reticulate cuticle; *SR*, second-order ridge; *SS*, surface sculpturing. *Arrowheads* in **F** show small cracks of the epicuticular wax layer

6.55 µm, n=29), situated in superficial depressions, are similar to those described above in the leaf and other pitcher surfaces. Dendroid trichomes are much shorter (length: 87.84±14.94 µm, n=15) and more abundant than attenuate ones (length: 221.95± 69.99 µm, n=5). Dendroid trichomes show the same shape as those on the upper face of the lid, however, have fewer branches, and often with one branch

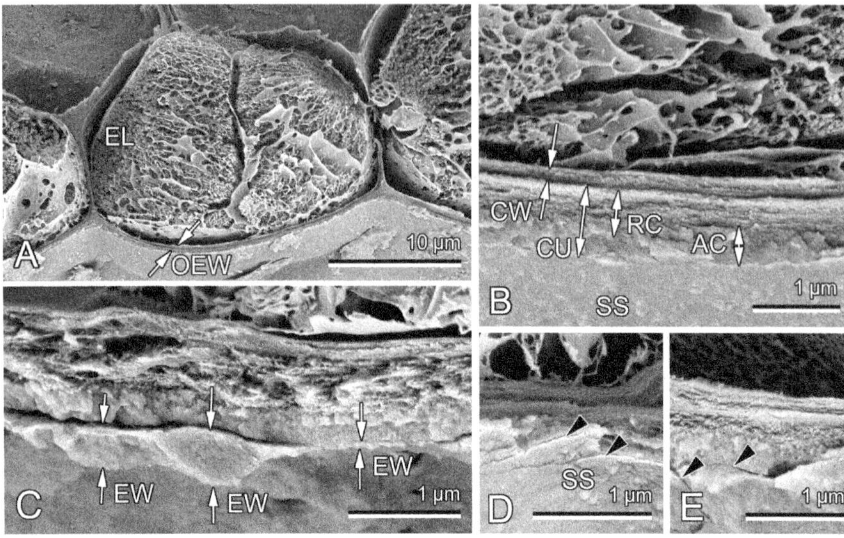

Fig. 8.11 Cryo-SEM micrographs of the inner structure of the lower side of the peristome, fractured samples. *AC*, amorphous cuticle; *CU*, cuticle; *CW*, external cellulose wall; *EL*, lower epidermal cell layer; *EW*, epicuticular wax layer; *OEW*, outer epidermal wall; *RC*, reticulate cuticle; *SS*, surface sculpturing. *Arrowheads* in **D** and **E** show cracks of the epicuticular wax layer

longer than others. Attenuate trichomes possess long gradual tapers, which are round in cross-section. Both dendroid and attenuate trichomes are crowded tightly to the surface and distally (upward) pointed.

The inner (waxy) surface of the slippery zone is extremely rough because of numerous crescent, down-pointed lunate cells, which have overhanging lower margins, creating anisotropic relief of the surface (Fig. 8.12C). The density of these cells becomes reduced and they finally disappear in the direction towards the pitcher bottom. The surface has a matt grayish appearance, since it is additionally densely covered with microscopic epicuticular wax crystals (Fig. 8.12D). The wax coverage emerges first as small separate crystals of irregular shape close to the lower side, in the inner arm of the peristome in the vicinity of the teeth (Fig. 8.12E). Furthermore, it turns into fields of densely placed crystals of the lower wax layer (Fig. 8.12F). This layer resembles a foam (Fig. 8.12G), composed of interconnected membraneous platelets (length: 0.82±0.11 μm, width: 0.53±0.11 μm, thickness: 36.54±7.87 nm, n=20). Cavities between these platelets, forming the cells of the foam, are usually irregular in shape and vary greatly in size (length: 0.65±0.17 μm, width: 0.42±0.11 μm, n=20; measured in the plane, parallel to the pitcher wall surface).

Moving towards the pitcher bottom, platelets of the upper wax layer appear and cover the lower wax layer completely (Fig. 8.12H). They are mostly irregular in shape with pointed extensions, although almost oval or rectangular ones also occur. These separate, clearly discernible irregular platelets are placed extremely densely and mostly contact adjacent ones so that gaps are hardly distinguishable. Platelets'

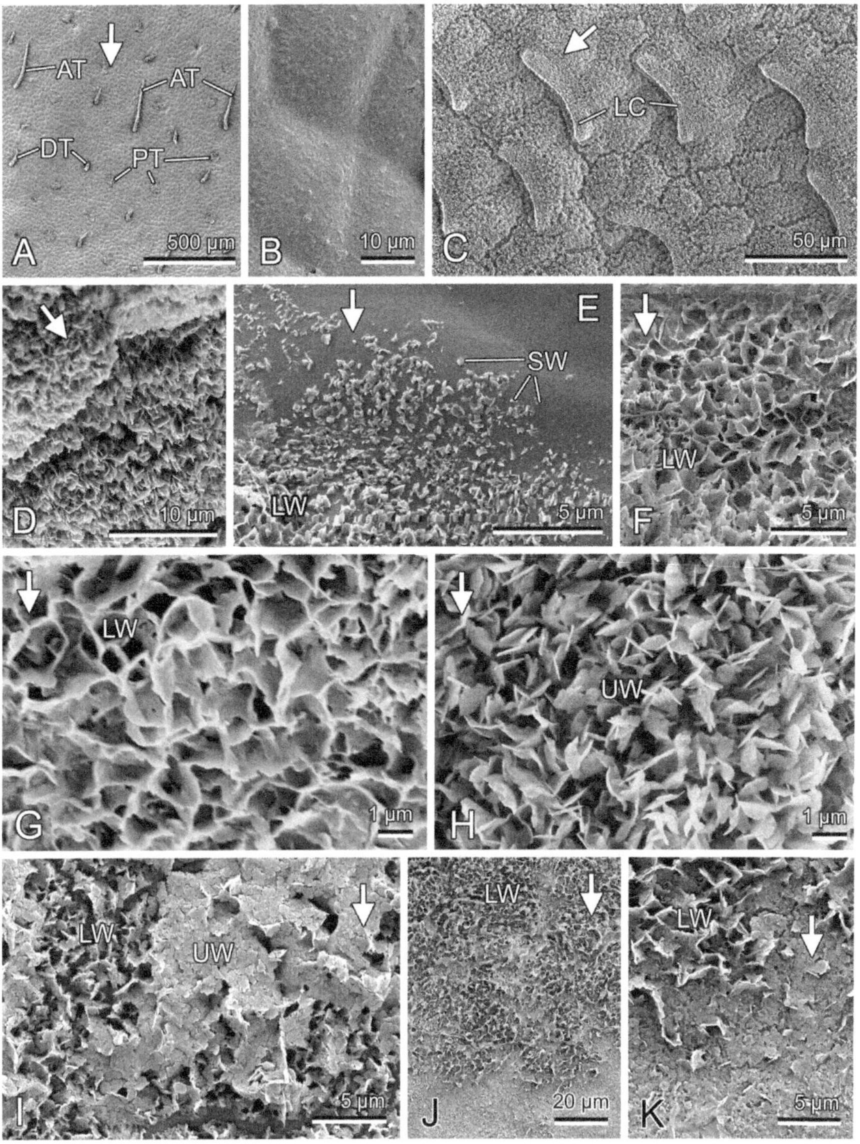

Fig. 8.12 Cryo-SEM micrographs of the external and internal (waxy) surface of the slippery zone. **A**, **B**. External surface. **C–K**. Waxy surface. *AT*, attenuate trichomes; *DT*, dendroid trichomes; *LC*, lunate cells; *LW*, lower wax layer; *PT*, peltate trichomes; *SW*, separate wax crystals; *UW*, upper wax layer. *Arrows* in **A** and **C–K** indicate the direction towards the bottom of the pitcher

planes are orientated more or less perpendicular to the surface of the pitcher wall. Orientation of planes along the axis perpendicular to the surface is rather random; they create no clear pattern on the surface. Upper platelets are larger than those of the lower layer (largest dimension: 1.01±0.19 µm, thickness: 34.95±8.70 nm, n=20) and vary greatly in size and shape from almost round to irregular-shaped with indistinct margins, e.g. oval, leaf-like, triangular, rhombic etc. Close to the boundary between the waxy surface and the transitional zone, upper wax platelets no longer form complete coverage, crowd to the surface and then disappear, exposing the lower wax layer (Fig. 8.12I). The latter becomes sparser and less uniform, and finally vanishes (Fig. 8.12J, K).

The experiment with spraying of water over the waxy surface shows its strong hydrophobicity. Water is present in the form of separate droplets, having a very high contact angle with the surface (Fig. 8.8C, D).

The pitcher wall at the level of the slippery zone is as thick as the peristome (approx. 350 μm) and demonstrates the same inner structure (Fig. 8.13A) as in both

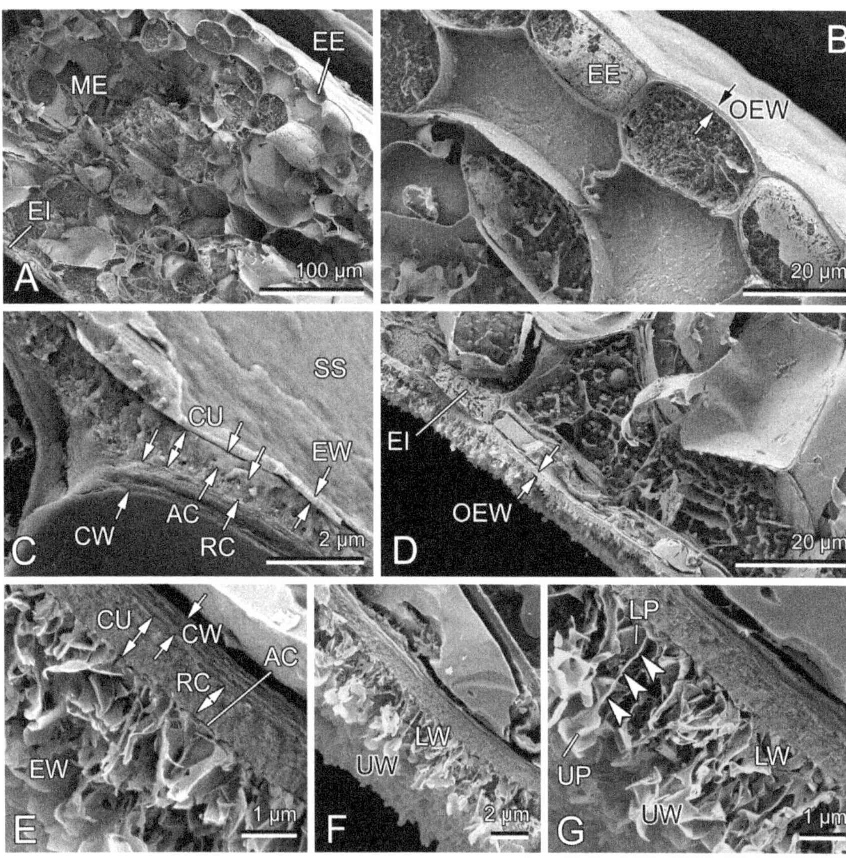

Fig. 8.13 Cryo-SEM micrographs of the inner structure of the slippery zone, fractured samples. **A**. Fracture of the pitcher wall at the level of the slippery zone. **B**, **C**. External surface. **D**–**G**. Waxy surface. *AC*, amorphous cuticle; *CU*, cuticle; *CW*, external cellulose wall; *EE*, external epidermal cell layer; *EI*, inner epidermal cell layer; *EW*, epicuticular wax layer; *LW*, lower wax layer; *ME*, mesophyll; *OEW*, outer epidermal wall; *RC*, reticulate cuticle; *SS*, surface sculpturing; *UW*, upper wax layer. *Arrowheads* show stalks connecting upper (*UP*) and lower platelets (*LP*)

the lid and peristome. On the external side, the outer epidermal wall is relatively thin compared to the lid and especially the peristome (1.23±0.21 μm, n=14) with a rather thick (0.35±0.19 μm, n=14) external cellulose wall (Fig. 8.13B, C). In the cuticle (thickness: 0.85±0.17 μm), the deeply located, coarse reticulate region and upper, very homogeneous, amorphous part of nearly the same thickness are clearly visible (Fig. 8.13C). A smooth, thin epicuticular wax layer completely covers the cuticle and develops irregular indistinct surface sculpturing (Fig. 8.13C). Being significantly variable in thickness (0.18±0.13 μm, n=14), it is rather heavy at some places and clearly shows amorphous texture.

The outer epidermal wall of the internal side, in the slippery zone, is of nearly the same overall thickness (1.27±0.20 μm, n=12) and has almost the same thickness ratio between outer cellulose wall (0.40±0.08 μm, n=12) and cuticle (0.83± 0.12 μm, n=12) (Fig. 8.13D, E) as on the external side. In the cuticle, the reticulate fraction with clearly distinguishable coarse fiber texture increases up to the entire cuticle thickness, whereas the amorphous region is present as a thin, hardly visible stripe in the upper sites (Fig. 8.13E). The specific epicuticular wax coverage on top of the cuticle reaches extremely great thickness (3.59±0.56 μm, n=12) and is comprised of two superimposed layers: the lower (thickness: 1.58±0.30 μm, n=12) and upper (thickness: 1.75±0.40 μm, n=12) (Fig. 8.13F). Lower platelets emerge from the cuticle and remain attached to it through the whole platelet margin. Upper platelets derive from lower ones, namely from their pointed extensions, and bear relatively long (0.52±0.17 μm, n=31) and thin stalks (Fig. 8.13G). Stalks connect lower platelets with upper ones; the latter serve to make up the upper wax layer just above the lower one.

8.3.6 External and Glandular Surfaces of the Digestive Zone

The digestive zone is separated from the slippery one by the transitional zone (Fig. 8.14A). The latter represents a narrow (approx. 1.5–2.0 mm) cross band on the internal side of the pitcher wall and is characterized by extremely smooth surface at both macro- and microscales (Fig. 8.14B).

The external side of the pitcher at the level of the digestive zone bears the same surface structures (stomata, peltate, dendroid, and attenuate trichomes) (Fig. 8.14C) and shows similar surface relief (Fig. 8.14D) as in the slippery zone. The presence of numerous digestive glands, distributed densely and regularly (Fig. 8.14E, F), is the characteristic feature of the internal (glandular) surface of this functional zone. Density of the glands is higher in the upper part of the zone than at the bottom of the pitcher. Glands are large (length: 185.13±23.05 μm, n=25; width: 137.31±19.38 μm, n=24), round-shaped or slightly elongated either in longitudinal (in upper part of the zone) or in transverse directions (in the middle and lower parts).

Each gland is situated in a shallow depression with a slope, slanting towards the bottom of the pitcher, and is partly covered by a small hood, overhanging the upper margin of the depression (Fig. 8.14G). Downward-pointing hoods and slanting slopes of depressions create the anisotropy of the glandular surface profile. The

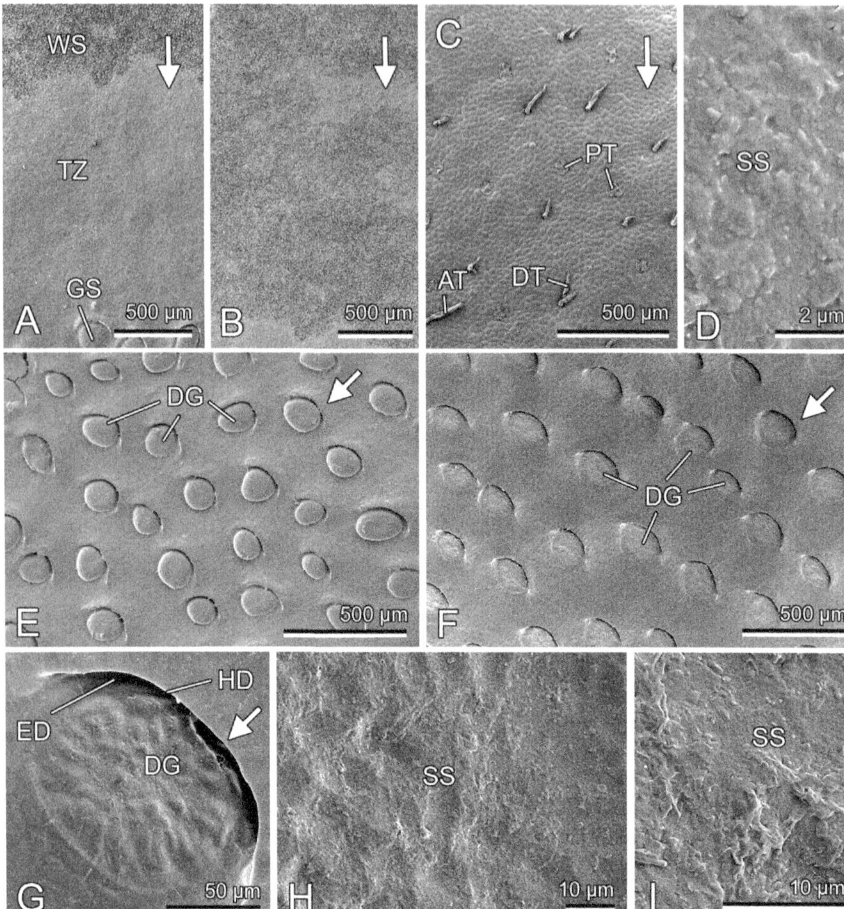

Fig. 8.14 Cryo-SEM micrographs of the transitional and digestive zones. **A**, **B**. Internal surface of the transitional zone. **C**, **D**. External surface of the digestive zone. **E–I**. Internal (glandular) surface of the digestive zone. *AT*, attenuate trichomes; *DG*, digestive glands; *DT*, dendroid trichomes; *ED*, epidermal depressions; *GS*, glandular surface; *HD*, hoods; *PT*, peltate trichomes; *SS*, surface sculpturing; *TZ*, transitional zone; *WS*, waxy surface. *Arrows* in **A–C** and **E–G** indicate the direction towards the bottom of the pitcher

depth of depressions as well as dimensions and prominence of hoods decreases from the top (Fig. 8.14E) to the bottom of the zone (Fig. 8.14F). The surface between glands has a very smooth appearance and demonstrates almost no asperities at high SEM magnifications. On glands, the surface is evidently uneven and possesses certain pitted sculpturing (Fig. 8.14H) together with some irregular microtopography (Fig. 8.14I).

Although the pitcher wall in the lower half of the pitcher, at the level of the digestive zone, is as thick (350–400 µm) as in the leaf lamina and thicker than in

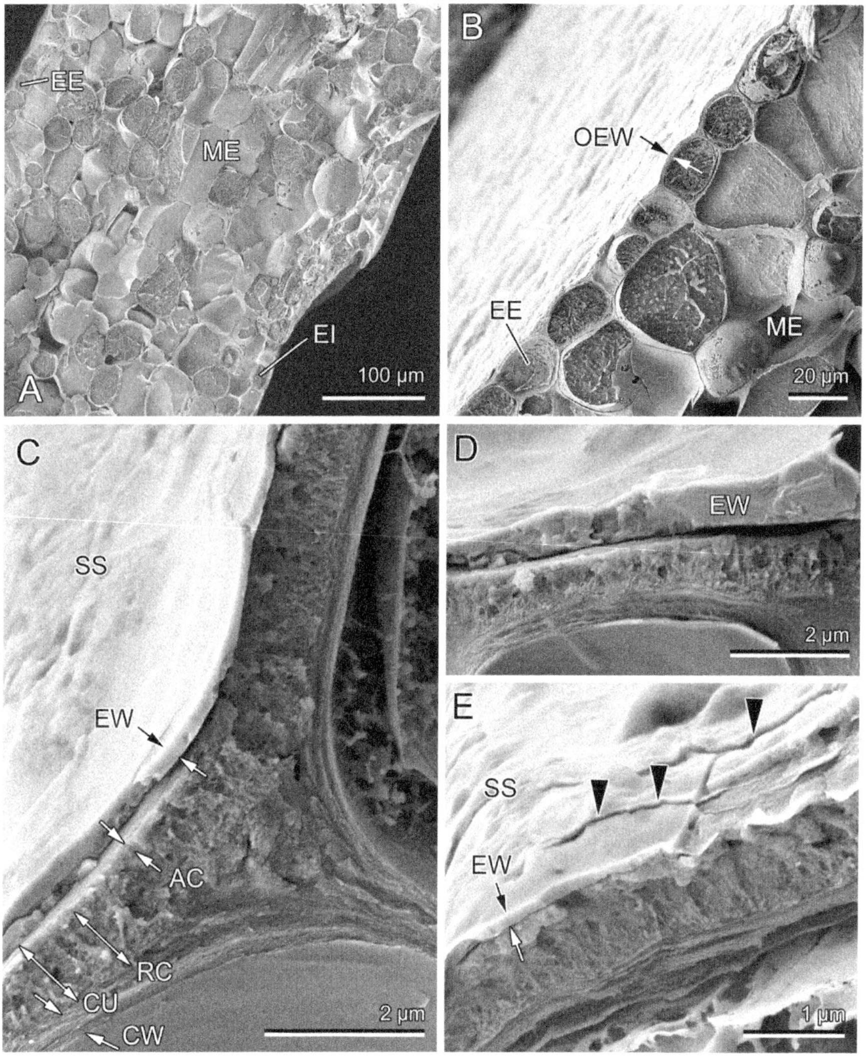

Fig. 8.15 Cryo-SEM micrographs of the inner structure of the external surface of the digestive zone, fractured samples. **A**. Fracture of the pitcher wall at the level of the digestive zone. **B–E**. Outer epidermal wall structure of the external surface. *AC*, amorphous cuticle; *CU*, cuticle; *CW*, external cellulose wall; *EE*, external epidermal cell layer; *EI*, inner epidermal cell layer; *EW*, epicuticular wax layer; *ME*, mesophyll; *OEW*, outer epidermal wall; *RC*, reticulate cuticle; *SS*, surface sculpturing. *Arrowheads* in **E** indicate cracks of the epicuticular wax layer

other pitcher zones, it has typical inner structure (Fig. 8.15A), observed in all other pitcher parts. The outer epidermal wall of the external pitcher side (Fig. 8.15B) is thinner ($1.66 \pm 0.26\,\mu m$, n=11) than in the leaf lamina, but thicker than in other pitcher zones, with the only exception being the peristome. The thickness of the cuticle (1.03±0.16 μm, n=11) exceeds that measured in the external cellulose wall (0.41±0.08 μm, n=11) by about 2.5 times (Fig. 8.15C). The cuticle shows a lower reticulate region and an upper homogeneous amorphous part with the thickness ratio, varying from 1:1 to 4:1. Relatively heavy epicuticular wax coverage of variable thickness (0.39±0.30 μm, n=12), which surpasses 1 μm in a few places, may be identified either as a smooth layer or crust (Fig. 8.15C, D). It has homogeneous amorphous texture, some prominent sculpturing, and cracks rarely in the vicinity of the fracture plane (Fig. 8.15E).

The inner side of the digestive zone demonstrates different morphologies of the coverage in the superficial cell layer, depending on the site, i.e. on glands, hoods or in areas between glands (Fig. 8.16A, B). Between glands and in hoods (Fig. 8.16C, D), the epidermal cell layer is present. The outer epidermal wall is

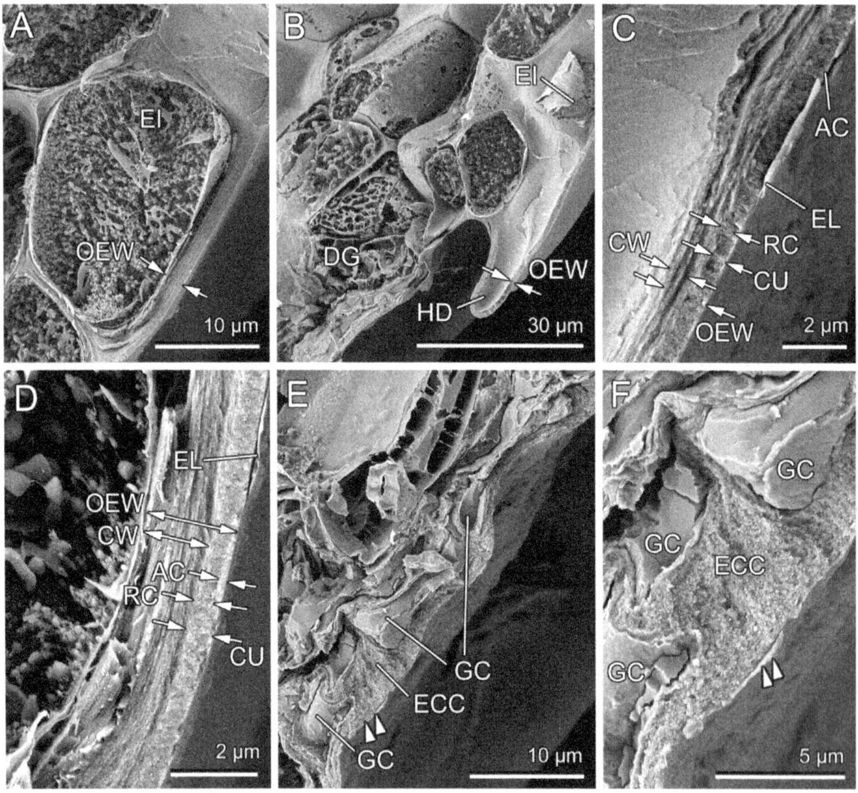

Fig. 8.16 Cryo-SEM micrographs of the inner structure of the internal (glandular) surface of the digestive zone, fractured samples. **A**, **D**. Epidermis between glands. **B**. Gland and hood. **C**. Hood. **D**. Region between glands. **E**, **F**. Gland. *AC*, amorphous cuticle; *CU*, cuticle; *CW*, external cellulose wall; *DG*, digestive gland; *ECC*, extracellular coverage of the gland; *EI*, inner epidermal cell layer; *EL*, epicuticular layer; *GC*, upper gland cell; *HD*, hood; *OEW*, outer epidermal wall; *RC*, reticulate cuticle. *Arrowsheads* in **E** and **F** show the outermost film on top of glands

rather thick (between glands: 1.80±0.31 μm, n=8; hood: 1.15±0.18 μm, n=5) and shows typical layered structure at these sites (Fig. 8.16A–D). The cuticle has the same thickness (hood: 0.61±0.18 μm, n=5) or is half as thick (between glands: 0.59±0.07 μm, n=8) as the external cellulose wall (hood: 0.56±0.04 μm, n=5; between glands: 1.20±0.08 μm, n=8) and consists of the heavy reticulate region and a very thin (hardly visible in some places) amorphous layer in the upper part (Fig. 8.16C, D). The epicuticular coverage, detected in some places, appears either as a thin (0.05±0.02 μm, n=7) film or somewhat thicker (0.11±0.08 μm, n=4) layer. It probably represents the glandular secretion, which causes uniform smooth appearance of the surface. Digestive glands lack epidermis, and the upper cell layer bears extracellular coverage (Fig. 8.16E), differing from that found in all other surfaces studied. This coverage has very variable thickness (1.71±0.86 μm, n=14) that probably results in an uneven surface profile, shows no evident layers, has coarse reticulate structure and consists of extremely inhomogeneous material (Fig. 8.16E, F). The outermost film (thickness: 0.07±0.03 μm, n=4), detected at some places on top of glands, presumably represents the glandular secretion and may be responsible for prominent surface sculpturing (Fig. 8.16F).

8.4 Discussion

Our comparative study shows the difference in inner structure between the leaf lamina and the pitcher in *N. alata*. The lamina tissue contains two different mesophyll layers (palisade and spongy mesophyll), whereas the pitcher mesophyll is composed of undifferentiated parenchyma cells. This structural difference obviously reflects functional specialisation of leaves (photosynthesis, water and gas exchange) and pitchers (predominantly trapping function). The two-layered mesophyll has also been reported in the leaf lamina of *N. khasiana, N. ampularia* and *N. rafflesiana*, while in leaves of *N. gracilis*, the undifferentiated mesophyll has been previously described (Pant and Bhatnagar, 1977). In our previous study on the glandular surface of the digestive zone in *N. ventrata* pitchers, we also discovered that the pitcher wall mesophyll consists of undifferentiated cells (Gorb et al., 2004).

Regarding the surface morphology, we found, here, differences between the leaf lamina and the pitcher as well as between different pitcher parts (Fig. 8.17). External pitcher surfaces (the upper face of the lid, the lower surface of the outer peristome arm, and the external surface of the pitcher tube) differ essentially from inner pitcher surfaces (the inner lid face, upper surface of the peristome, waxy and glandular surfaces). While external surfaces do not show considerable differences, there are substantial distinctions in both the surface topography and coverage between inner surfaces of different pitcher zones, highly specialised to serve different functions. Below, we discuss inner surfaces in separate subsections.

As for external pitcher surfaces, they are covered with a thin, smooth, epicuticular wax layer, similar to that found in both leaf sides. This layer, varying in thickness among surfaces, represents an amorphous wax and creates no distinct

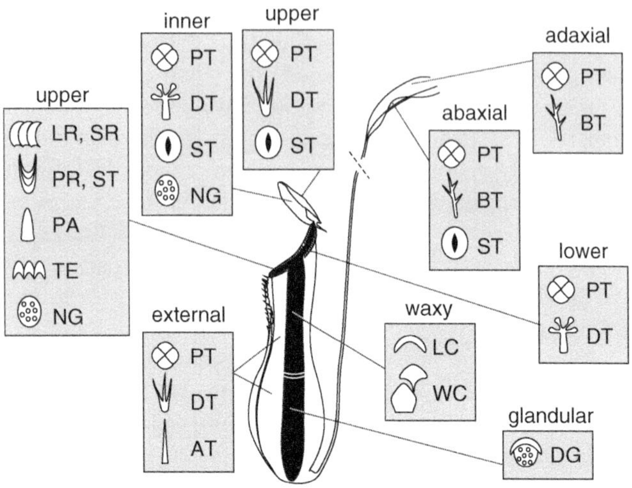

Fig. 8.17 Diagram of distribution of various surface structures on the leaf and different surfaces in *N. alata* pitchers. *AT*, attenuate trichomes; *BT*, bootjack trichomes; *DG*, digestive glands with hoods; *DT*, dendroid trichomes; *LC*, lunate cells; *LR*, large ridges of the first order; *NG*, nectar glands; *PA*, papillae; *PR*, parabolic ledges; *PT*, peltate trichomes; *SR*, small ridges of the second order; *ST*, stomata with guard cells; *TE*, teeth of the inner arm margin; *WC*, wax crystals

surface sculpturing. Stomata are present in all surfaces, being most numerous on the adaxial leaf side, as it has been previously reported for other *Nepenthes* species (Pant and Bhatnagar, 1977; Owen and Lennon, 1999). Contrary to smooth leaf surfaces, bearing regularly placed peltate trichomes and having long branched hairs only at lamina edges, external pitcher surfaces are hirsute. Trichomes of different types (peltate, stellate and dendoid, all branched) have been described on the external pitcher surfaces by previous authors (Pant and Bhatnagar, 1977; Owen and Lennon, 1999; Clarke, 2001). We also found non-branched, long trichomes on the external surface of the pitcher tube. Apart from erect dendriod trichomes on the lid and lower surface in the outer peristome arm, prominent trichomes (bootjack on the lid, dendroid and antennuate on the external surface of the pitcher tube) are crowded to the surface and pointed towards the direction of the pitcher mouth. Although there are some differences in trichome morphology and their distribution among external pitcher surfaces, these surfaces do not seem to serve specialized functions (except erect trichomes; see section "the lid").

We found no information on functional specialization of external pitcher surfaces in literature. We suppose that these surfaces are not involved in animal trapping and do not effect insect attachment. Since the pubescence is rather sparse, it should not preclude insect attachment and locomotion. Macroscopic surface structures, such as stomata and especially trichomes, presumably provide anchorage sites for claws of crawling insects. It has been previously reported that insects can attach well to macroscopically rough substrate by interlocking their claws with surface irregularities (Nachtigall, 1974; Cartmill, 1985; Betz, 2002; Dai et al., 2002). Such an effect of trichomes on insect attachment to hairy plant surfaces has been found in several insect-plant systems (Roberts, 1930; Schwoerbel, 1956; Roberts and Foster, 1983; Kennedy, 1986; Hare and Elle, 2002; Kessler and Baldwin, 2004; Sugiura and Yamazaki, 2006). Besides, crowded trichomes may serve insects as route guides, pointing out the preferable direction towards the pitcher mouth.

In external surfaces, parts between trichomes and stomata are rather smooth and lack distinct surface sculpturing. Successful attachment to various smooth substrates has been previously observed in many insects (Way and Murdie, 1965; Stork, 1980; Edwards, 1982; Eigenbrode, 1996; Eigenbrode et al., 1999; Federle et al., 2000; Gorb and Gorb, 2002). Attachment to smooth surfaces is accomplished by adhesive pads, having evolved independently in many insect lineages (Gorb, 2001; Beutel and Gorb, 2001; Gorb et al., 2002). We suppose that areas between stomata and trichomes on the external pitcher surfaces represent, probably, appropriate substrate for pad adhesion. We have found in our previous comparative study on the physicochemical properties of functional pitcher zones in *N. alata* that external pitcher surfaces are wettable with water and have relatively high surface energy (Gorb and Gorb, 2006). This can, additionally, ensure a high adhesion force, based on the capillary interaction between the plant substrate and the pad surface.

8.4.1 Attractive Zone: The Lid

The lid is closed in very young pitchers and open in mature ones. During pitcher development, the lid hermetically seals the pitcher until it reaches the operative state (Macfarlane, 1908; Stehli, 1934; Lloyd, 1942; Juniper and Burras, 1962). The sealing is ensured by (1) tightly applied surfaces of the lead and peristome due to interlocking of epidermal cells and (2) branched hairs, densely situated at the lid edge and on the lower surface of the outer arm of the peristome (Macfarlane, 1908; Owen and Lennon, 1999). The hairs of different pitcher parts interweave and close a crack when it appears. For mature pitchers it has been hypothesised that, in many species having a large overhanging lid; it prevents rain water from entering the pitcher, which otherwise would cause the pitcher to flood and dilute the digestive fluid (Stehli, 1934; Juniper and Burras, 1962; Slack, 1979; Juniper et al., 1989; Clarke, 1997, 2001). Because of nectar glands, which cover the inner lid face, an attracting function of the lid has been proposed and then fully accepted by later authors (Hooker, 1859; Fenner, 1904; Stehli, 1934; Lloyd, 1942; Juniper and Burras, 1962; Adams and Smith, 1977; Slack, 1979; Juniper et al., 1989; Clarke, 1997, 2001; Barthlott et al., 2004). However, no nectar glands were found on the lid surface in *N. ampularia* and *N. inermis* (Lloyd, 1942).

There are only a few studies dealing with the role of the lid in animal trapping. It has been reported that insects (ants) run, with no risk of capture, on the lid (Lloyd, 1942; Slack, 1979). Our previous data on the wettability and free surface energy of the lid faces in *N. alata* support the assumption that these surfaces are appropriate substrates for insect pad adhesion.

Lid shapes, the structure of nectar glands and variability of their distribution, as well as trichome types and their pattern on the lid surface, have been previously described (Fenner, 1904; Lloyd, 1942; Adams and Smith, 1977; Parkes, 1980; Clarke, 2001). We demonstrate here, for the first time, the presence of the superficial layer on the surface in the inner lid face. We believe that this layer represents the secretion produced by nectar glands. Contrary to previous authors (Fenner, 1904), we think that the nectar flows out of the epidermal depressions, where the glands are located, since the superficial layer was detected not only on gland surfaces, but also in sites between the glands.

8.4.2 Attractive Zone: The Peristome

The peristome, as well as the lid, is a part of the attractive (alluring) zone, since its nectar glands, interposed between teeth at the edge of the inner arm, lure potential prey to enter the pitcher (Hooker, 1859; Lloyd, 1942; Stehli, 1934; Slack, 1979; Juniper and Burras, 1962; Juniper et al., 1989; Clarke, 1997, 2001). It is important to note that the peristome has the greatest density of nectar glands compared to other pitcher parts (Owen and Lennon, 1999; Barthlott et al., 2004). Although there is a lack of detailed information on the chemical composition of this nectar, it has been previously reported that the glands produce a sugar-rich fluid, favored by foraging insects (Martin and Juniper, 1970; Adams and Smith, 1977; Owen and Lennon, 1999).

It has been previously believed that the trapping role of the peristome is fulfilled mostly due to the marginal position of the glands. Attempting to reach the nectar, insects move towards the edge of the inner arm of the rim, where they stay on sites (surfaces) in a precarious position, not being able to maintain their foothold (Stehli, 1934; Adams and Smith, 1977; Juniper and Burras, 1962; Ratsirarson and Silander, 1996; Clarke, 1997, 2001). Such poor attachment at these sites has been explained by effects of surface roughness, anisotropic relief and almost vertical slopes of the peristome, and by the proximity of the waxy pitcher surface. Other authors have reported that the rim is not slippery or unclimable and allows free walking for some insects, such as ants (Lloyd, 1942; Moran et al., 1999; Juniper et al., 1989). Recent study on *N. bicalcarata* and *N. alata* showed that pitchers may capture insects (experiments with ants *Oecophyla smaragdina*) with the aid of slippery peristome if it is wet (Bohn and Federle, 2004). The trapping success was found to be based on the presence of lubricating water (from rain, condensation) or nectar films on well wettable, microstructured peristome surface. It was demonstrated that water lubrication and anisotropic surface topography are the two factors preventing insect attachment to the peristome: the first one disables pad adhesion, and the second one precludes claw interlocking. The authors assumed that the corrugations of the peristome surface are also essential for its complete wettability. Our previous comparative study on the physicochemical properties of the functional pitcher surfaces in *N. alata* also demonstrated that the peristome was best wettable by water surface with the highest surface energy (Gorb and Gorb, 2006).

Different aspects of the peristome morphology in representatives of the genus *Nepenthes*, such as variability in shape, surface relief, arrangement of nectar glands and tissue structure have been extensively studied by previous authors (Lloyd, 1942; Juniper and Burras, 1962; Pant and Bhatnagar, 1977; Adams and Smith, 1977; Juniper et al., 1989; Owen and Lennon, 1999; Bohn and Federle, 2004). Based of our present data, we consider that a very thick cuticle in the upper surface of the peristome contribute not only to mechanical stiffness and robustness of the whole rim (Lloyd, 1942; Juniper et al., 1989), but is also involved in the formation of such surface relief features as second-order ridges and parabolic ledges of steps. In ridges, the cuticle protrudes to a considerable height over the surface, whereas parabolic edges represent cuticle thickening or folds. We also found that these ridges and steps originate from papillae at the edge of the outer peristome arm. The use of the cryo-SEM method made it possible, for the first time, to visualize droplet-like structures on the surface of the inner arm. We interpret these structures as droplets of nectar, produced by marginal nectar glands. Since the cuticle is covered with a very thin film of the epicuticular wax, the water-containing nectar did not completely wet the surface, but forms small droplets. The presence of nectar droplets may lead to an increase of surface hydrophilicity, previously reported (Bohn and Federle, 2004; Gorb and Gorb, 2006). Our experiments with sprayed water show that it covers the surface as a layer, which is held on the peristome by its corrugated relief (between large ridges of the first order). This supports the previous assumption about the role of peristome topography in its complete wettability by water (Bohn and Federle, 2004) and suggests that the particular height of ridges and distance between them may be an adaptation for control of the water layer thickness.

8.4.3 Slippery Zone: The Waxy Surface

The slippery zone (previously called "conductive" after Hooker, 1859), the inner surface of which is covered with microscopic epicuticular wax crystals, is a characteristic feature of traps in most *Nepenthes* species, with the few exceptions of *N. ventricosa, N. bicalcarata,* and *N. ampularia* (Lloyd, 1942; Cheek and Jebb, 2001; Bohn and Federle, 2004). No significant interspecific differences have been seen in the morphology of the zone apart from the variation in the zone's band width (Juniper et al., 1989) and the absence of lunate cells in *N. ampularia* (Pant and Bhatnagar, 1977). However, it has also been reported that in some species, the pitchers differ regarding the presence of the waxy layer (Lloyd, 1942; Di Giusto et al., see contribution in this volume).

The key role of the slippery zone's waxy surface in trapping and retention functions of the pitcher has been reported by many previous authors. The first observations and experiments have shown that, because of the wax coverage, this surface is extremely slippery for insects and offers no hold for their feet (Macfarlane, 1893; Knoll, 1914; Lloyd, 1942). Recent experimental studies with *N. alata* and *N. ventrata* and a series of insect species (flies *Drosophila melanogaster* and *Calliphora vomitoria*, ants *Iridomyrmex humilis*) confirmed that the wax coverage hinders insect locomotion (Gaume et al., 2002, 2004). Our tests with beetles *Adalia bipunctata* and two wax layers in *N. alata* pitchers demonstrated that both layers essentially decreased the attachment force (Gorb et al., 2005). Since insects have been observed grooming their feet after having contact with a waxy surface, it has been suggested that wax crystals contaminate insect attachment organs. Feet contamination by crystals, detached from the upper layer of the wax coverage, was first demonstrated in flies (Juniper and Burras, 1962; Juniper et al., 1989). On the basis of this result, it has been assumed that insect pads with attached wax material are not able to adhere to the surface properly, and that contamination is the main mechanism of the trapping function of the slippery zone. Contamination of adhesive pads by the wax of *N. ventrata* was also observed in tests with flies *Calliphora vomitoria* (Gaume et al., 2004). We have found, for *N. alata,* that only wax platelets of the upper layer contaminated attachment organs of beetles *Adalia bipunctata*, whereas crystals of the lower layer did not (Gorb et al., 2005). On the lower wax layer, the reduction of contact area between the insect pad and the plant surface probably causes the decrease in the attachment force. It has also been suggested by previous authors that lunate cells hinder the insects climbing on the pitcher wall; to provide claw clinging, the cells should be turned the other way (Bobisut, 1889; Knoll, 1914; Lloyd, 1942). Experiments with *N. alata* pitchers, where the wax was removed chemically, and ants *Iridomyrmex humilis* showed that insects indeed can climb an inverted waxy surface, but they failed on naturally oriented pitchers (Gaume et al., 2002). We have also found that the waxy surface is extremely unwettable by water and has very low surface energy compared to all other pitcher surfaces (Gorb and Gorb, 2006). This may additionally cause low attracting force between the plant surface and attachment pads of insects.

Morphology, distribution and origin of lunate cells, considered to be deformed stomata, have been widely discussed in the literature (Lloyd, 1942; Adams and Smith, 1977; Pant and Bhatnagar, 1977; Owen and Lennon, 1999). Regarding the waxy coverage, numerous authors have previously shown that it is composed of densely placed scale-like or plate-like crystals of epicuticular waxes (Lloyd, 1942; Adams and Smith, 1977; Owen and Lennon, 1999; Juniper and Burras, 1962; Juniper et al., 1989; Riedel et al., 2003). From the chemical point of view, these waxes represent a mixture of different aliphatic compounds, where aldehydes, alcohols and fatty acids contribute almost equally to the composition (*N. x williamsii*; Juniper et al., 1989) or long-chained aldehydes dominate (*N. alata;* Riedel et al., 2003; Gorb et al., 2005). It has been found that wax crystals are organised into two morphologically different layers: scales in the upper layer and plates in the lower one (Juniper and Burras, 1962; Juniper et al., 1989). Many layers of wax platelets, slightly differing in shape and chemistry, were reported for *N. alata* (Riedel et al., 2003). Our previous study on the waxy surface in *N. alata* pitchers showed two distinct layers in the waxy coverage that differ in their morphology, chemical composition and mechanical properties (Gorb et al., 2005).

Previous studies were conducted with the use of conventional microscopy methods, requiring diverse surface preparation (solvent or mechanical treatment), and this led to the loss of information on the entire structure of the wax coverage. Thus, the distribution of crystals and connections between different wax layers remained unclear. The cryo-SEM method, used here, allows us, for the first time, to visualise the spatial structure of the wax coverage in a native state. We found that wax crystals, mechanically composing two layers, morphologically constitute one unit. This consists of a lower platelet with irregular upper margins and upper platelets with thin stalks, pointed downwards. The platelets of two layers are connected to each other through thin filamentous outgrowths. The wax coverage, however, behaves mechanically as two separate layers. The upper layer can be easily separated from the lower one under normal load or shear. Applied force results in a breakage of stalks, connecting the upper platelets with the lower ones. The experiment with sprayed water, showing behaviour of small separate water droplets on the waxy surface, visualizes its high hydrophobic properties, reported previously (Gorb and Gorb, 2006).

8.4.4 Digestive Zone: The Glandular Surface

As the digestive zone plays the primary role in prey utilisation, the glandular surface has been previously studied mostly for its function in digestion, absorption and transport of the insect-derived compounds (Fenner, 1904; Lloyd, 1942; Luettge, 1964, 1965; Juniper et al., 1989; Schulze et al., 1999; Owen and Lennon, 1999; Owen et al., 1999; An et al., 2001). Digestive glands, covering the surface, were found to secrete diverse digestive enzymes, organic acids, Cl^- and Na^+ ions and water, and to absorb ions and low molecular mass mixtures associated with prey breakdown. However, this zone also influences the trapping efficiency of the pitcher, since animals are usually captured and drowned in the digestive fluid (Juniper et al., 1989; Adams and Smith, 1977; Owen and Lennon, 1999). In earlier studies it has been assumed that the formic acid, found in the secretion, blocks the spiracles of insects leading to their suffocation and drowning (Fenner, 1904; Stehli, 1934). Later, it was hypothesised that the digestive fluid contains a wetting agent, which prevents struggling prey from floating and swimming, and promotes their sinking in spite of their hydrophobic surface properties (Juniper and Burras, 1962; Juniper et al., 1989; Clarke, 1997). However, the role of the glandular surface itself in insect capturing and retention has only been examined in very few studies.

Once insects fall down into the pitcher and are inside the digestive pool, they try to get out of the fluid by climbing up the unsubmersed glandular surface. Previous authors have pointed out that the glands do not provide support points for insect feet, when they attempt to scale the vertical pitcher wall (Haberland, 1889, cited after Fenner, 1904). It was later suggested that placement and orientation of the hoods, covering digestive glands, may prevent insects from using the glands and epidermal depressions, where the glands are found, as foothold sites for their claws to escape from the pitcher (Lloyd, 1942; Juniper et al., 1989). Moreover, it has been described that insects, trying to climb the glandular surface, always come into contact with the hood and gland itself, to which they stick (Fenner, 1904). Experiments with flies *Drosophila melanogaster* and ants *Iridomyrmex humilis* on epidermal surfaces of the *N. alata* pitchers demonstrated that insect locomotion was strongly impeded on the glandular surface, probably due to glandular secretion, presumably acting like glue (Gaume et al., 2002).

Our previous study on the digestive zone in *N. ventrata* showed that, although surface anisotropy of the glandular surface does not facilitate effective claw interlocking, pad-bearing insects (bugs *Pyrrhocoris apterus* and especially flies *Calliphora vicina*) are presumably able to adhere to smooth areas between glands (Gorb et al., 2004). It was assumed that in general, a rather stiff glandular surface should provide the appropriate substrate for feet attachment and is not responsible for insect capture and retention. It was found that influence of the glandular surface on insect attachment differs depending on insect weight and design of their attachment organs. Smooth areas between glands, having rather low tenacity, allow successful attachment for adhesive pads in middle-sized and large insects, whereas spots with strong adhesion, associated probably with glands, may glue small insects and impede their locomotion. Strong hydrophilicity and high surface energy of the glandular surface (measured in both *N. ventrata* and *N. alata*) should provide high attractive forces between the plant surface and insect adhesive organs and lead to successful attachment for insects (Gorb et al., 2004; Gorb and Gorb, 2006). The glandular surface is constantly covered with a film of water-containing fluids (either gland secretion or fluid from the digestive pool), which may cause the surface to be extremely slippery for insects and result in poor attachment. Such a mechanism of insect trapping due to lubricating fluid was also recently described for ants *Oecophylla smaragdina* on the hydrophilic peristome surface of both *N. bicalcrata* and *N. alata* (Bohn and Federle, 2004).

In literature, there is much data on morphology of the glandular surface. General morphology and fine structure of digestive glands as well as their distribution and variability within the zone have been described (Fenner, 1904; Lloyd, 1942; Adams and Smith, 1977; Pant and Bhatnagar, 1977; Juniper et al., 1989; Owen and Lennon, 1999). Here, we show, for the first time, that the surface is indeed covered with a thin film, which we interpret as a gland secretion. We were not able to detect wax coverage either on glands, or at sites between glands. From the functional point of view, absence of a wax layer leads to stronger wettability by water-containing liquids, such as gland secretion or digestive fluid. Extreme reduction of the epicuticular wax coverage has been previously reported for leaves of submerged aquatic plants, the surface of which is adapted to be readily wetted by water to facilitate gas exchange (Sculthorpe, 1967). The extracellular coverage of glands differs essentially in structure from the cuticle in regions between glands and in other pitcher surfaces. We interpret the coarse reticulate texture and the absence of the amorphous layer in the coverage of upper gland cells as an adaptation to increase surface permeability, which is important for stimulus acceptance as well as for transport and absorption functions of the digestive glands (Juniper et al., 1989; Barthlott et al., 2004).

Acknowledgments Plant material for the study was kindly provided by I. Hans (Botanical Garden at the University of Hohenheim, Stuttgart, Germany). Discussions with D. Voigt (Max Planck Institute for Metals Research, Stuttgart, Germany) on the effect of different plant surfaces on insect attachment are acknowledged. We thank V. Kastner for linguistic correction of the manuscript. This work was supported by the Federal Ministry of Education, Science and Technology, Germany (project InspiRat 01RI0633C), Landesstiftung Baden-Württemberg, Germany (project Plants as Source for Biomimetics of Anti-Adhesive Surfaces), and SPP 1420 priority program of the German Science Foundation (DFG) "Biomimetic Materials Research: Functionality by Hierarchical Structuring of Materials" (project GO 995/9-1).

References

Adams, R.M, and Smith, G.W. (1977) An S.E.M. survey of the five carnivorous pitcher plant genera. *Am. J. Bot.* 64: 265–272.

An, C.-I., Fukusaki, E.-I., and Kobayashi, A. (2001) Plasma-membrane H^+-ATPases are expressed in pitchers of the carnivorous plant *Nepenthes alata* Blanco. *Planta* 212: 547–555.

Barthlott, W., Neinhuis, C., Cutler, D., Ditsch, F., Meusel, I., Theisen, I., and Wilhelmi, H. (1998) Classification and terminology of plant epicuticular waxes. *Bot. J. Linn. Soc.* 126: 237–260.

Barthlott, W., Porembski, S., Seine, R., and Theisen, I. (2004) *Karnivoren*. Stuttgart: Ulmer.

Betz, O. (2002) Performance and adaptive value of tarsal morphology in rove beetles of the genus *Stenus* (Coleoptera, Staphylinidae). *J. Exp. Biol.* 205: 1097–1113.

Beutel, R., and Gorb, S.N. (2001) Ultrastructure of attachment specialization of hexapods (Arthropoda): Evolutionary patterns inferred from a revised ordinal phylogeny. *J. Zool. Syst. Evol. Res.* 39: 177–207.

Bobisut, O. (1889) Ueber den Functionswechsel der Spaltoeffnungen in der Gleitzone der Nepenthes-Kannen (cited after Lloyd, 1942).

Bohn, H.F, and Federle, W. (2004) Insect aquaplaning: *Nepenthes* pitcher plants capture prey with the peristome, a fully wettable water-lubricated anisotropic surface. *PNAS* 101: 14138–14143.

Cartmill, M. (1985) Climbing. In: *Functional vertebrate morphology*, ed. Hildebrand, M., Bramble, D.N, Liem, K.F., and Wake, D.B. Cambridge: The Belknap Press, pp. 73–88.

Cheek, M., and Jebb, M. (2001) Nepenthaceae. In: *Flora Malesiana*, ed. by Nooteboom, H.P. (cited after Di Guisto et al., this book).

Clarke, C. (1997) Nepenthes*of Boneo*. Kota Kinabalu, Sabah, Malaysia: Natural History Publications (Borneo) & Science and Technology Unit.

Clarke, C. (2001) Nepenthes*of Sumatra and peninsular Malaysia*. Kota Kinabalu, Sabah, Malaysia: Natural History Publications (Borneo).

Dai, Z., Gorb, S., and Schwarz, U. (2002) Roughness-dependent friction force of the tarsal claw system in the beetle *Pachnoda marginata* (Coleoptera, Scarabaeidae). *J. Exp. Biol.* 205: 2479–2488.

Danser, B.H. (1928) The Nepenthaceae of the Netherlands Indies. *Bull: du Jardin Botanique* 9: 249–438.

Edwards, P.B. (1982) Do waxes of juvenile *Eucalyptus* leaves provide protection from grazing insects? *Austral. J. Ecol.* 7: 347–352.

Eigenbrode, S.D. (1996) Plant surface waxes and insect behaviour. In: *Plant cuticles – an integral functional approach*, ed. Kerstiens, G. Oxford: BIOS, pp. 201–222.

Eigenbrode, S.D., Kabalo, N.N., and Stoner, K.A. (1999) Predation, behavior, and attachment by *Chrysoperla plorabunda* larvae on *Brassica oleracea* with different surface waxblooms. *Entom. Exp. Applicata* 90: 225–235.

Ellison, A.M., and Gotelli, N.J. (2001) Evolutionary ecology of carnivorous plants. *TREE* 16: 623–629.

Federle, W., Rohrseitz, K., and Hölldobler, B. (2000) Attachment forces of ants measured with a centrifuge: better "wax-runners" have a poorer attachment to a smooth surface. *J. Exp. Biol.* 203: 505–512.

Fenner, C.A. (1904) Beitraege zur Kenntnis der Anatomie, Entwicklungsgeschichte und Biologie der Laubblaetter und Druesen einiger Insektivoren. *Flora* 93: 335–333.

Gaume, L., Gorb, S., and Rowe, N. (2002) Function of epidermal surfaces in the trapping efficiency of *Nepenthes alata* pitchers. *New Phytol.* 156: 476–489.

Gaume, L., Perret, P., Gorb, E., Gorb, S., Labat, J.-J., and Rowe, N. (2004) How do plant waxes cause flies to slide? Experimental tests of wax-based trapping mechanisms in the three pitfall carnivorous plants. *Arth. Struct. Dev.* 33: 103–111.

Goebel, K. (1923) *Organographie* (cites after Pant and Bhatnagar, 1977).

Gorb, E., Haas, K., Henrich, A., Enders, S., Barbakadze, N., and Gorb, S. (2005) Composite structure of the crystalline epicuticular wax layer of the slippery zone in the pitchers of the carnivorous plant *Nepenthes alata* and its effect on insect attachment. *J. Exp. Biol.* 208: 4651–4662.

Gorb, E., Kastner, V., Peressadko, A., Arzt, E., Gaume, L., Rowe, N., and Gorb, S. (2004) Structure and properties of the glandular surface in the digestive zone of the pitcher in the carnivorous plant *Nepenthes ventrata* and its role in insect trapping and retention. *J. Exp. Biol.* 207: 2947–2963.

Gorb, E.V., and Gorb, S.N. (2002) Attachment ability of the beetle *Chrysolina fastuosa* on various plant surfaces. *Entom. Exp. Applicata* 105: 13–28.

Gorb, E.V., and Gorb, S.N. (2006) Physicochemical properties of functional surfaces in pitchers of the carnivorous plant *Nepenthes alata* Blanco (Nepenthaceae). *Plant Biol.* 8: 841–847.

Gorb, S.N. (2001) *Attachment devices of insect cuticle*. Dordrecht, Boston, London: Kluwer Academic Publishers.

Gorb, S.N., Beutel, R.G., Gorb, E.V., Jiao, J., Kastner, V., Niederegger, S., Popov, V., Scherge, M., Schwarz, U., and Vötsch, W. (2002) Structural design and biomechanics of friction-based releasable attachment devices in insects. *Integr. Compar. Biol.* 42: 1127–1139.

Haberlandt, G. (1894) Ueber Bau and Funktion der Hydathoden. *Ber. Dt. Bot. Ges.* 12: 367–378.

Hare, J.D., and Elle, E. (2002) Variable impact of diverse insect herbivores on dimorphic *Datura wrightii*. *Ecology* 83: 2711–2720.

Holloway, P.J. (1982) Structure and histochemistry of plant cuticular membranes: an overview. In: *The plant cuticle*, ed. by Cutler, D.F., Alvin, K.L., and Price, C.E. London: Academic Press, pp. 1–32.

Hooker, J.D. (1859) On the origin and development of the pitcher of *Nepenthes*, with an account of some new Bornean plants of the genus. *Trans. Linn. Soc.* 22:415–424.

Jebb, M., and Cheek, M. (1997) A skeletal revision of *Nepenthes* (Nepenthaceae). *Blumea* 42: 1–106.

Jeffree, C.E. (1986) The cuticle, epicuticular waxes and trichomes of plants, with reference to their structure, functions and evolution. In: *Insects and the plant surface*, ed. by Juniper, B., and Southwood, R. London: Edward Arnold, pp. 21–63.

Jeffree, C.E. (1996) Structure and ontogeny of plant cuticles. In: *Plant cuticles: An integrated functional approach*, ed. by Kerstiens, G. Oxford: BIOS Scientific Publishers, pp. 33–82.

Juniper, B.E. (1986) The path to plant carnivory. In: *Insects and the plant surface*, ed. by Juniper, B., and Southwood, R. London: Edward Arnold, pp. 195–318.

Juniper, B.E., and Burras, J.K. (1962) How pitcher plants trap insects. *New Sci.* 269: 75–77.

Juniper, B.E., Robins, R.J., and Joel, D.M. (1989) *The carnivorous plants*. London: Academic Press.

Kennedy, C.E.J. (1986) Attachment may be a basis for specialization in oak aphids. *Ecol. Entomol.* 11: 291–300.

Kessler, A., and Baldwin, I.T. (2004) Herbivore-induced plant vaccination. Part I. The orchestration of plant defences in nature and their fitness consequences in the wild tobacco *Nicotiana attenuata*. *Plant J.* 38: 639–649.

Knoll, F. (1914) Ueber die Ursache des Ausgleitens der Insektenbeine an wachsbedeckten Pflanzenteilen. *Jarb. Wiss. Bot.* 54: 448–497.

Lloyd, F.E. (1942) *The carnivorous plants*. New York: The Ronald Press Company.

Luettge, U. (1964) Untersuchungen zur Physiologie der Carnivoren-Druesen. I. Die an den Verdauungsvorgangen beteiligten Enzyme. *Planta* 63: 103–117.

Luettge, U. (1965) Untersuchungen zur Physiologie der Carnivoren-Druesen. II. Ueber die Resorption verschiedener Substanzen. *Planta* 66: 331–334.

Lyshede, O.B (1982) Structure of the outer epidermal wall in xerophytes. In: *The plant cuticle*, ed. by Cutler, D.F., Alvin, K.L., and Price, C.E. London: Academic Press, pp. 87–98.

Macfarlane, J.M. (1893) Observations on the pitchered insectivorous plants, I (cited after Lloyd, 1942).

Macfarlane, J.M. (1908) Nepenthaceae (cited after Lloyd, 1942).

Martin, J.T, and Juniper, B.E. (1970) *The cuticle of plants*. London: Edward Arnold.

Moran, J.A, Booth, W.E., and Charles, J.K. (1999) Aspects of pitcher morphology and spectral characteristics of six Bornean *Nepenthes* pitcher plant species: Implications for prey capture. *Ann. Bot.* 83: 621–528.

Nachtigall, W. (1974) *Biological mechanisms of attachment*. Berlin, Heidelberg, New York: Springer.

Owen, T.P., and Lennon, K.A. (1999) Structure and development of the pitchers from the carnivorous plant *Nepenthes alata* (Nepenthaceae). *Am. J. Bot.* 86: 1382–1390.

Owen, T.P., Lennon, K.A., Santo, M.J., and Anderson, A.Y. (1999) Pathways for nutrient transport in the pitchers of the carnivorous plant *Nepenthes alata*. *Ann. Bot.* 84: 459–466.

Pant, D.D., and Bhatnagar, S. (1977) Morphological studies in *Nepenthes* (Nepenthaceae). *Phytomorphology* 27: 13–34.

Parkes, D.M. (1980) Adaptive mechanisms and glands in some carnivorous plants (cited after Juniper et al., 1989).

Payne, W.W. (1978) A glossary of plant hair terminology. *Brittonia* 30: 239–255.

Ratsirarson, J., and Silander, J.A. (1996) Structure and dynamics in *Nepenthes madagascariensis* pitcher plant micro-communities. *Biotropica* 28: 218–227.

Riedel, M., Eichner, A., and Jetter, R. (2003) Slippery surfaces of carnivorous plants: composition of epicuticular wax crystals in *Nepenthes alata* Blanco pitchers. *Planta* 218: 87–97.

Roberts, J. I. (1930) The tobacco capsid (*Engytatus volucer*, Kirk.) in Rhodesia. *Bull. Ent. Res.* 21: 169–183.

Roberts, J.J.; and Foster, J E. (1983) Effect of leaf pubescence in wheat on the bird cherry oat aphid (Homoptera: Aphidae). *J. Econ. Entomol.* 76: 1320–1322.

Schulze, W., and Schulze, E.D. (1990) Insect capture and growth of the insectivorous *Drosera rotundifolia*. *Oecologia* 82: 427–429.

Schulze, W., Frommer, W.B., and Ward, J.M. (1999) Transporters for ammonium, amino acids and peptides are expressed in pitchers of the carnivorous plant *Nepenthes*. *Plant J.* 17: 637–646.

Schulze, W., Schulze, E.D., Pate, J.S., and Gillison, A.N. (1997) The nitrogen supply from soils and insects during growth of the pitcher plants *Nepenthes mirabilis, Cephalotus follicularis* and *Darlingtonia californica*. *Oecologia* 112: 464–471.

Schwoerbel, W. (1956) Beobachtungen und Untersuchungen der Biologie einiger einheimischer Wanzen (Heteroptera: *Pyrrhocoris* Fall., *Coptosoma* Lap., *Corizus* Fall., *Gampsocoris* Fuss, *Rhinocoris* Hhn.). *Zool. Jb. Syst.* 84: 329–354.

Sculthorpe, C.D. (1967) *Aquatic vascular plants*. New York: St. Martin's Press.

Slack, A. (1979) *Carnivorous plants*. Cambridge, Massachusetts: The MIT Press.

Stehli, G. (1934) *Pflanzen auf Insektenfang – Schilderungen aus dem Leben von fleischfressenden und insektenfangenden Pflanzen*. Stuttgart: Kosmos.

Stern, K. (1916) Beitraege zur Kenntnis der Nepenthaceen (cited after Lloyd, 1942).

Stork, N. E. (1980) Experimental analysis of adhesion of *Chrysolina polita* (Chrysomelidae, Coleoptera) on a variety of surfaces. *J. Exp. Biol.* 88, 91–107.

Sugiura, S., and Yamazaki, K. (2006) Consequences of scavenging behaviour in a plant bug associated with a glandular plant. *Biol. J. Linn. Soc.* 88: 593–602.

Thum, M. (1988) The significance of carnivory for the fitness of *Drosera* in its natural habitat. 1. The reactions of *Drosera intermedia* and *Drosera rotundifolia* to supplementary feeding. *Oecologia* 75: 427–480.

Troll, W. (1932) Morphologie der schildfoermigen Blaetter. *Planta* 17: 153–314.

Way, M.J., and Murdie, G. (1965) An example of varietal resistance of Brussel sprouts. *Ann. Appl. Biol.* 56: 326–328.

Color Plates

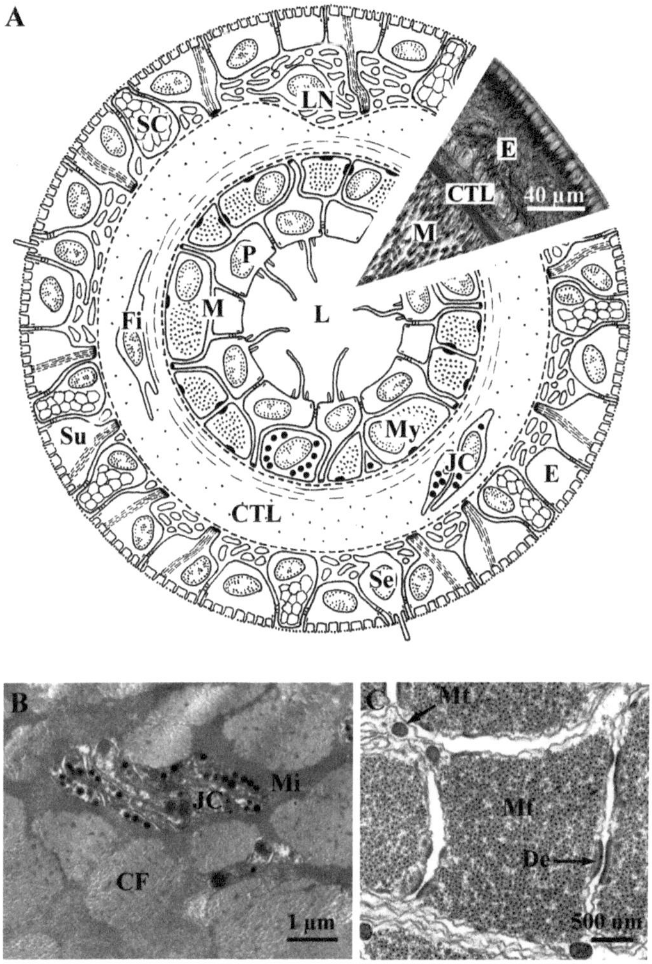

Plate 1 Morphology and ultrastructure of the echinoderm tube foot stem. (**A**) Reconstruction of a transverse section through a typical stem of asteroid or echinoid disc-ending tube foot (modified from Flammang 1996). The inset is a light micrograph of a histological section through the stem wall in the sea star *Asterias rubens* (original). (**B**) Transverse section through the outer sheath of the connective tissue layer (original TEM picture from the asteroid *Marthasterias glacialis*). (**C**) Transverse section through a myocyte of the mesothelium (original TEM picture from the holothuroid *Holothuria forskali*). Abbreviations: CF, collagen fiber; CTL, connective tissue layer; De, desmosome; E, epidermis; Fi, fibrocyte; JC, juxtaligamental cell; L, ambulacral lumen; LN, longitudinal nerve; M, mesothelium; Mf, myofibril; Mi, microfibrillar network; Mt, mitochondria; My, myocyte; P, peritoneocyte; SC, secretory cell; Se, sensory cell; Su, support cell

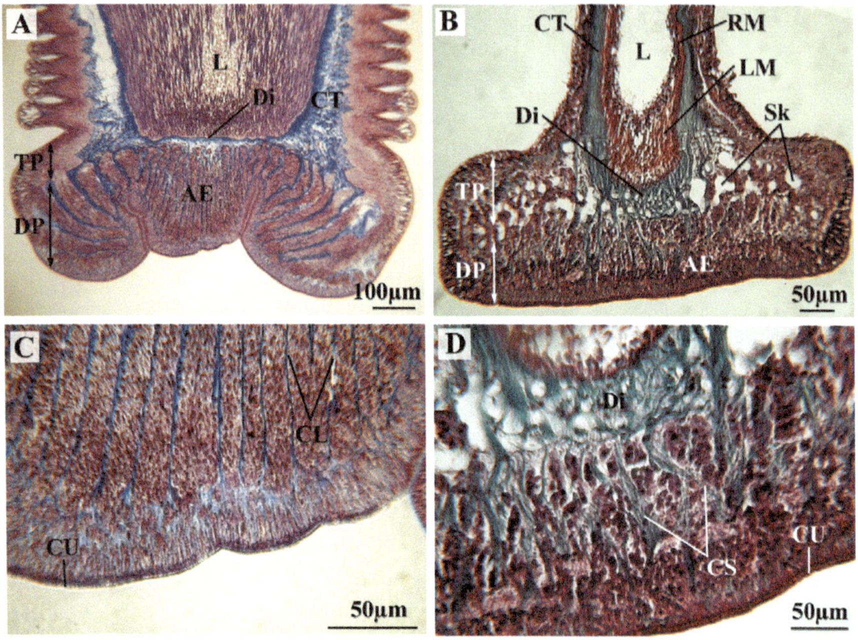

Plate 2 Longitudinal histological sections through the discs of the tube feet of the asteroid *Asterias rubens* (**A**,**C**; the section goes through the margin of the disc to show the connective tissue radial lamellae) and the echinoid *Paracentrotus lividus* (**B**,**D**) (from Santos et al. 2005a). Abbreviations: AE, adhesive epidermis; CL, connective tissue radial lamellae; CS, connective tissue septa; CT, connective tissue; CU, cuticle Di, diaphragm; DP, distal pad; L, lumen; LM, levator muscle; RM, retractor muscle; Sk, skeleton; TP, terminal plate

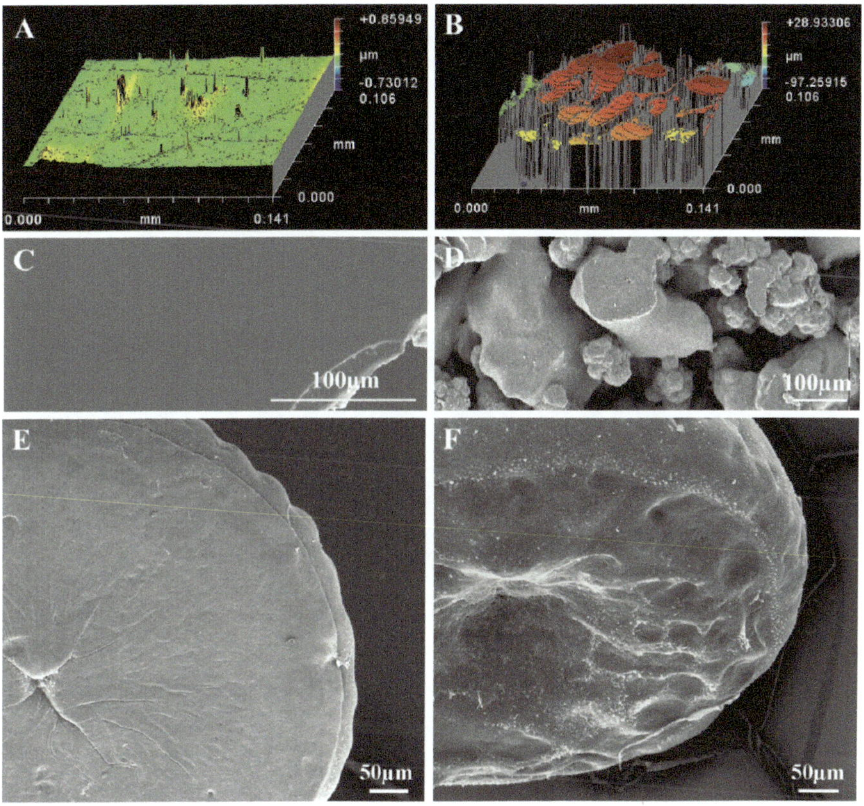

Plate 3 3-D (scanning white light interferometer) and SEM images of the surface of smooth (**A**,**C**) and rough (**B**,**D**) polypropylene samples, and of the distal surfaces of tube foot discs from the echinoid *Paracentrotus lividus* attached to each of these substrata (**E**,**F**) (from Santos et al. 2005a)

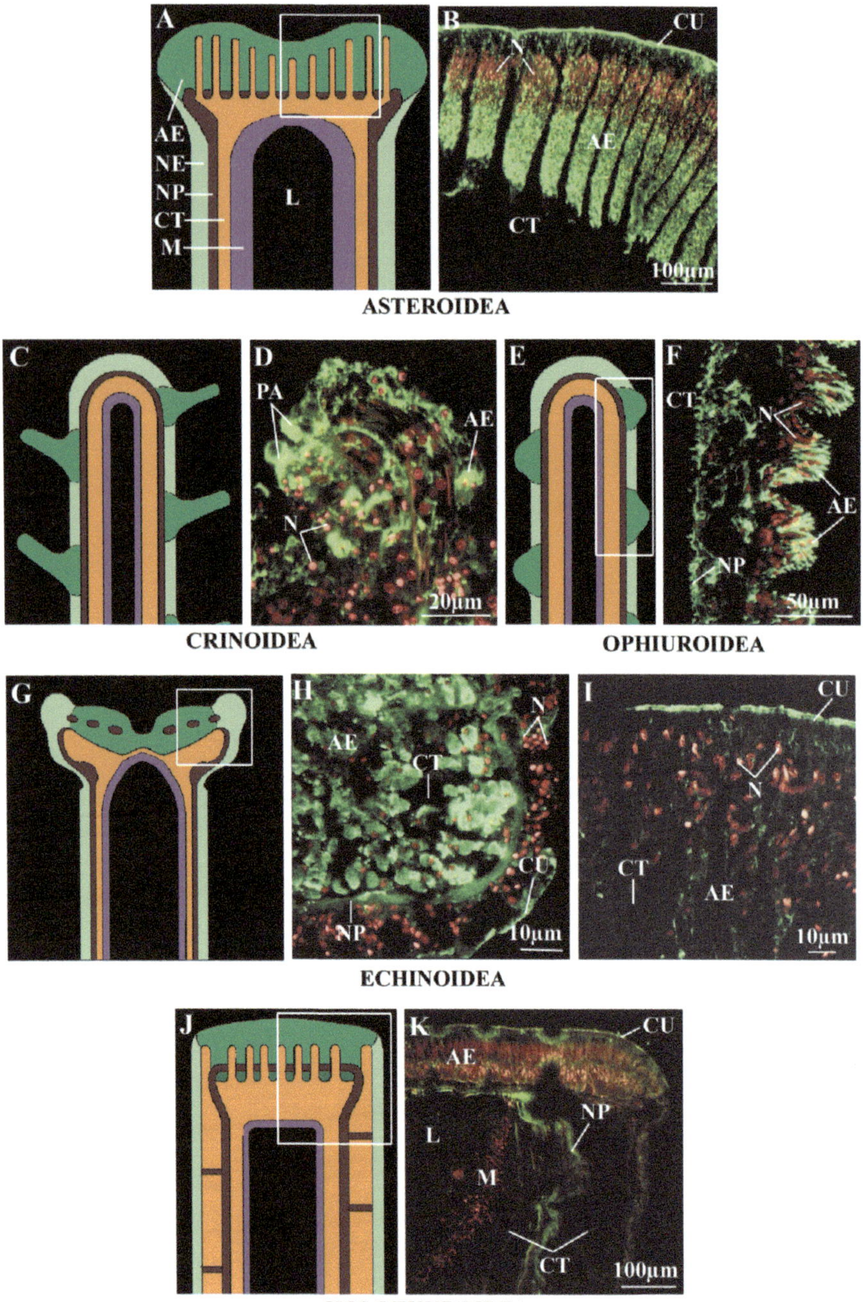

Plate 4 Schematic representations of longitudinal sections through the tube feet of the different echinoderm classes (**A**, **C**, **E**, **G**, **J**) and immunofluorescent labelling of corresponding tube foot sections with antibodies raised against the adhesive material of the asteroid *Asterias rubens* (immunoreactive structures are labelled in green while nuclei appear in red) (**B**, **D**, **F**, **H**, **I**, **K**) (originals; see text for details). Abbreviations: AE, adhesive epidermis; CU, cuticle; CT, connective tissue; L, lumen; N, nuclei; NE, non-adhesive epidermis; NP, nerve plexus; M, mesothelium

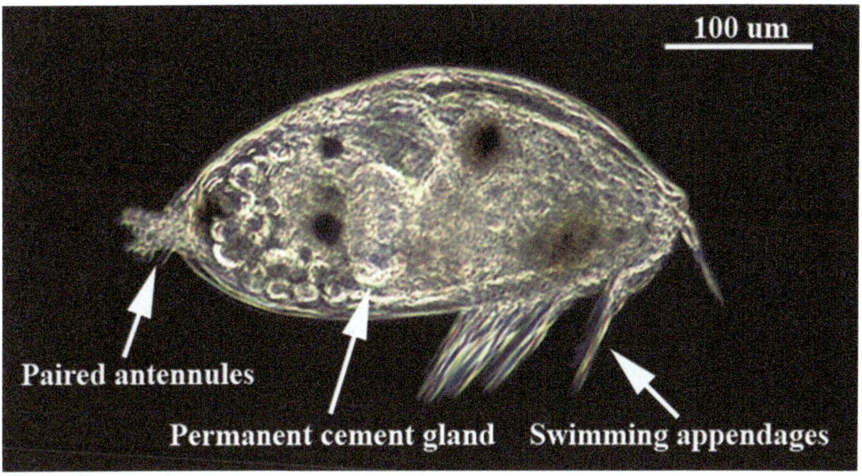

Plate 5 A light micrograph of a *Balanus amphitrite* cypris larva, imaged using dark field optics

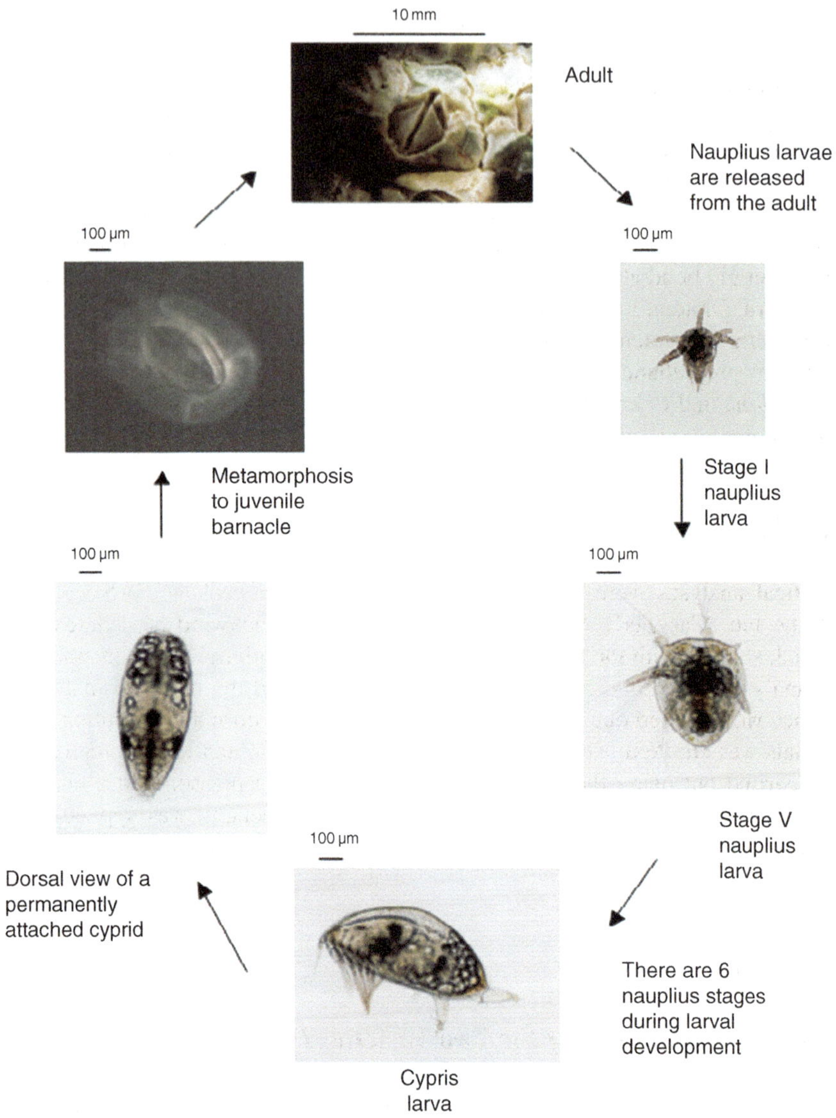

Plate 6 A typical balanid barnacle life history (credit, Dr. M. Kirby, Newcastle University)

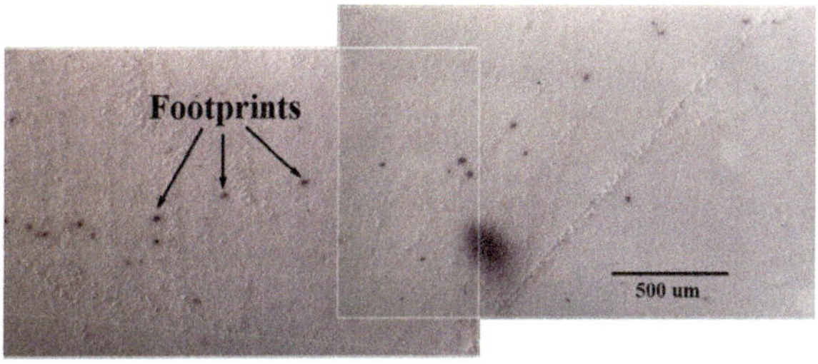

Plate 7 *B. amphitrite* footprints, stained on nitrocellulose membrane using an antibody specific to a 76 kDa subunit of the SIPC

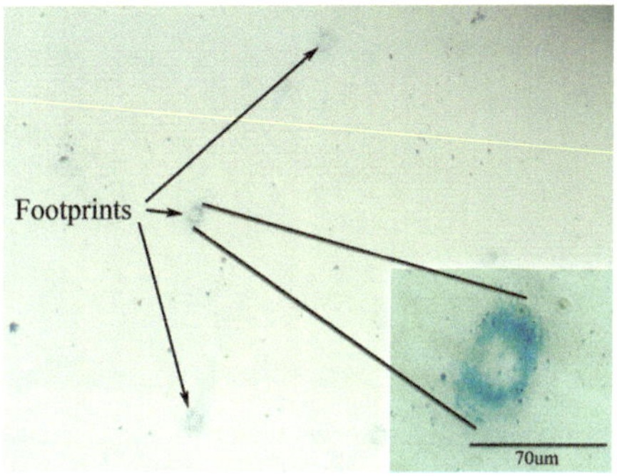

Plate 8 *S. balanoides* footprints on glass could be used for AFM studies. Staining using Coomassie Brilliant Blue protein dye reagent allows visualisation

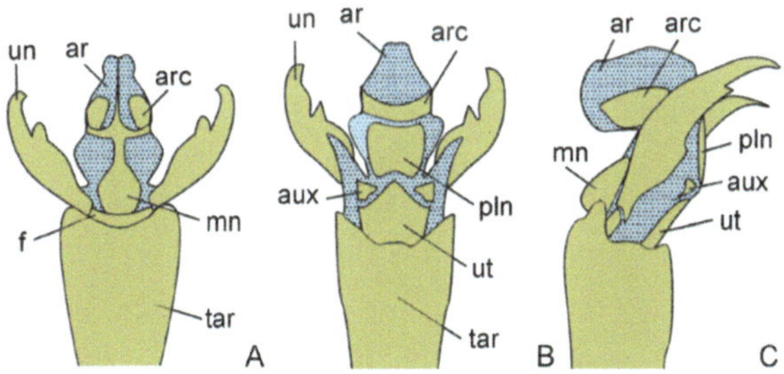

Plate 9 Terminology of arolium structures (*Apis mellifera*). **A**. Dorsal aspect. **B**. Ventral aspect. **C**. Lateral aspect. ar, arolium; arc, arcus; aux, auxiliae; f, marginal flange of the terminal tarsomere; mn, manubrium; pln, planta; tar, tarsal segments; un, claw; ut, unguitractor plate

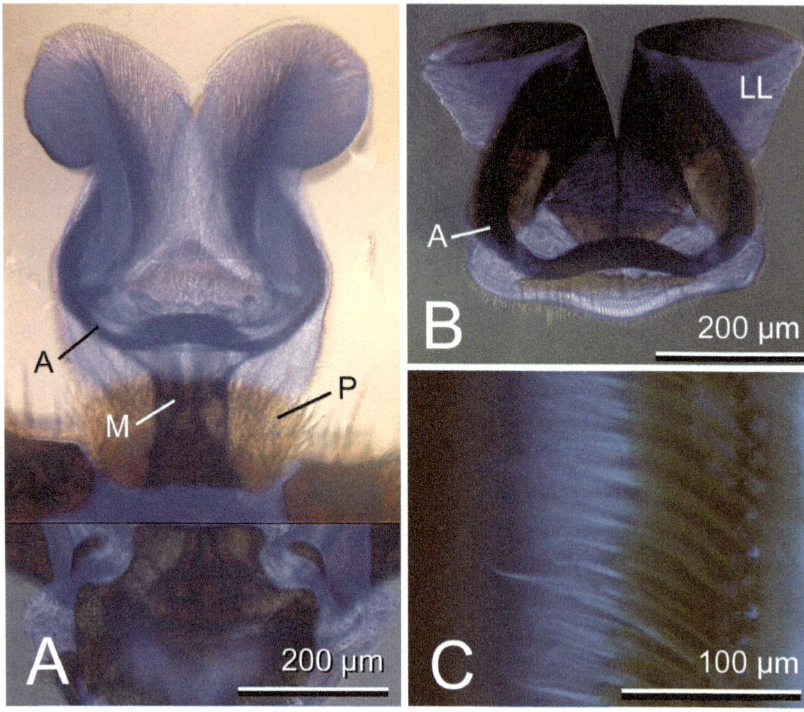

Plate 10 Resilin-bearing structures in the hornet pretarsus. **A**. Arolium, ventral aspect. The distal part of the pretarsus with arolium and its proximal part with unguitractor are shown in the *upper and lower panels*, respectively, divided by the *dark line*. **B**. Arolium, dorsal aspect. **C**. Setae located on the ventral side of tarsomeres, lateral aspect. A, arcus; LL, lateral lobe; M, manubrium; P, planta

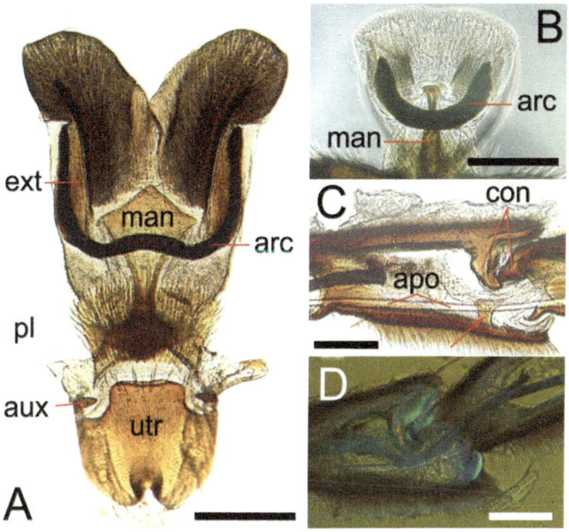

Plate 11 Pretarsus and tarsomeres of hymenopteran legs. **A**. Arolium in *Vespa crabro* (Vespidae). **B**. Arolium in *Apis mellifera* (Apidae). **C**. Intertarsomere articulation in *V. crabro*. **D**. Same, fluorescence of resilin-bearing tendons. apo, apodeme; arc, arcus; aux, auxilia; con, condyli of neighbouring tarsomeres; ext, external area; fix, fixator of the apodeme, man, manubrium; pl, planta; tr, trachea; utr, unguitractor. Scale bars – 200 μm

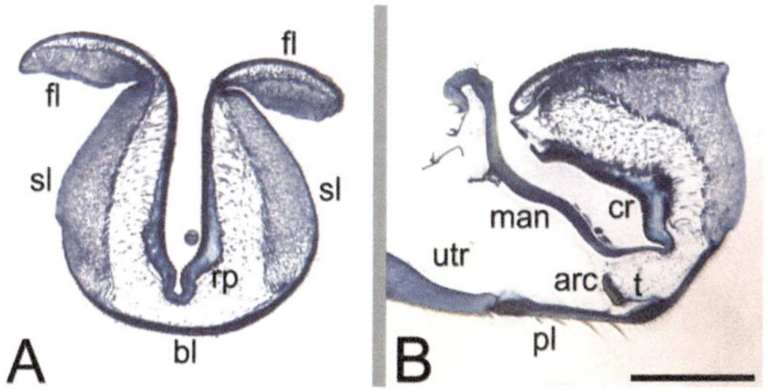

Plate 12 Semi-thin sections of the folded arolium in a hornet, *Vespa crabro*. A. Frontal plane. B. Sagittal plane. Scale bar 200 μm. arc, arcus; bl, bottom lobe; cr, system of ridges; fl, flap lobe; man, manubrium; pl, planta; rp, resilin pillow; sl, side lobe; t, trace of invagination; utr, unguitractor

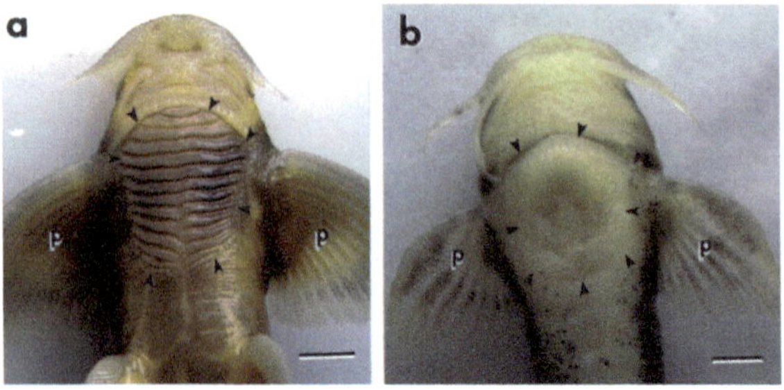

Plate 13 AO (*arrowheads*) of the mountain-stream catfishes, *Pseudocheneis sulcatus* (**a**, standard length: 13.7 cm) and *Glyptothorax pectinopterus* (**b**, standard length: 14 cm) located between the base of the pectoral fins (p). Bars, 5 mm. [(Figure **a** is from Das and Nag (2005) and shown with permission of the Blackwell Science Publishers, Oxford WHICH?)]

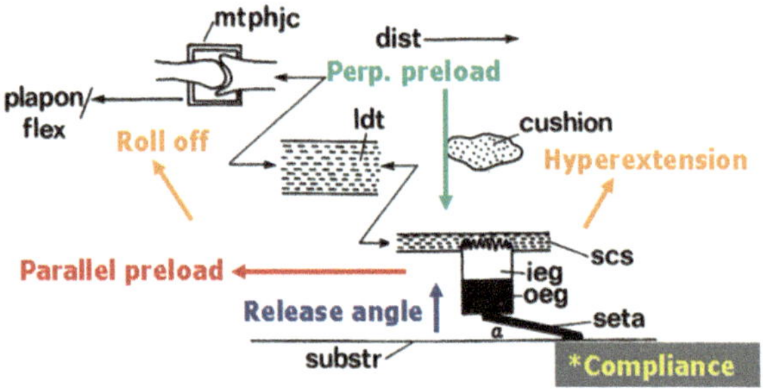

Plate 14 Diagrammatic representation of the relationships between the seta/substrate contact pattern, the mechanical linkages from seta to skeleton and the application of forces bringing about and releasing the adhesive grip. The setal stalk changes its angle with respect to the substratum and ventral surface of the foot as substrate contact is made and released. At contact, the stalk angle (α) is depressed below that critical for maximal contact to be made. Perpendicular preloading (Perp. Preload) is applied by unfurling of the digit from its hyperextended position distally, or by rolling onto the more proximal scansors proximally, and is assisted by a distal cushion (such as a vascular sinus network). The parallel preload is applied and maintained by a combination of active loading through the tensile skeleton of the digit, mediated by flexor muscles acting through the plantar aponeurosis, and by gravitational loading (on vertical surfaces). Detachment is facilitated by hyperextension distally and roll off proximally, both of which raise the setal shaft above the critical angle threshold. The parallel and perpendicular preloads are diminished as a result of pressure changes and unloading of the tensile skeleton. Abbreviations: α, angle of setal stalk with substratum; dist., distal; ieg, inner epidermal generation; ldt, lateral digital tendon; mtphjc, metatarsophalangeal joint capsule; oeg, outer epidermal generation; plapon/flex, connection to the plantar aponeurosis and crural flexor muscles; scs, stratum compactum of the dermis of the scansor; substr, substratum (modified from Russell, 2002)

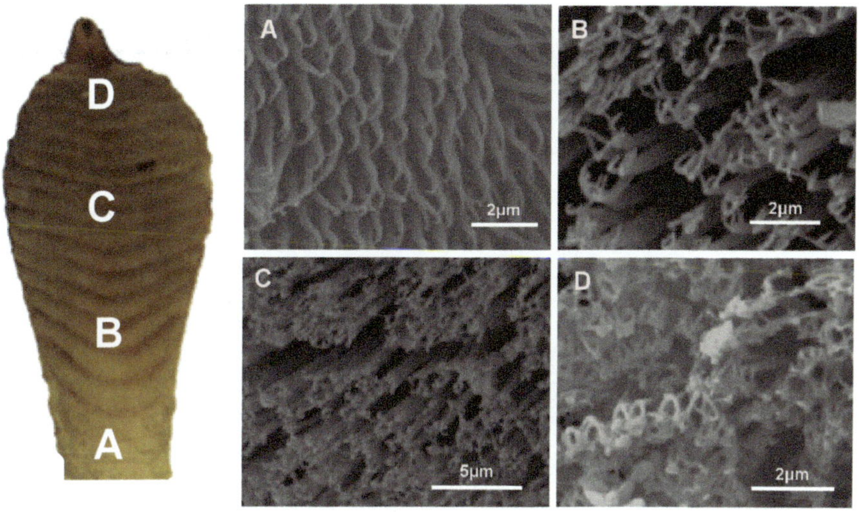

Plate 15 Variation in setal size and complexity from the proximal (A) to the distal end of the digit (D) of the Tokay gecko (*Gekko gecko*). **A** curved spines carried on the basal scales of the digit **B** branched prongs carried on proximal lamellae **C** setae carried on proximally located scansors **D** setae carried on the distalmost scansors

Plate 16 The site of collection and habitat of *Rhoptropus* cf *biporosus*. **A** The location of the site of Gai'As where specimens were collected. **B** A specimen resting on a vertical sandstone rock face (photo courtesy of Tony Gamble) **C** The terrain at Gai'As, which consists of isolated piles of sandstone boulders and outcrops

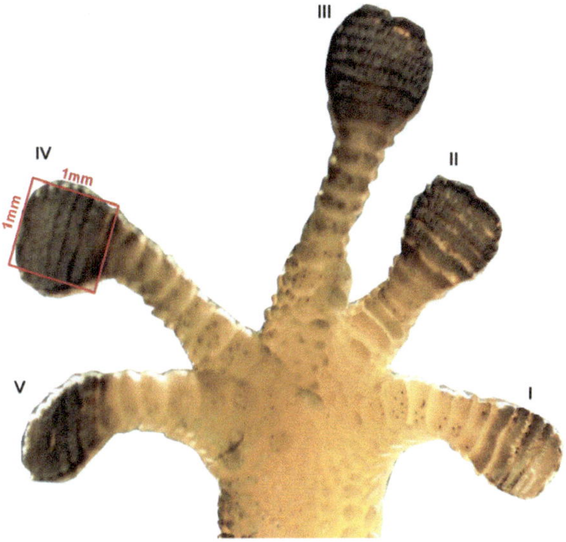

Plate 17 The right manus of *Rhoptropus* cf *biporosus* showing the subdigital pad of digit IV and its association with the 1 mm^2 area that was examined on each of the sample surfaces. Digits are numbered I–V

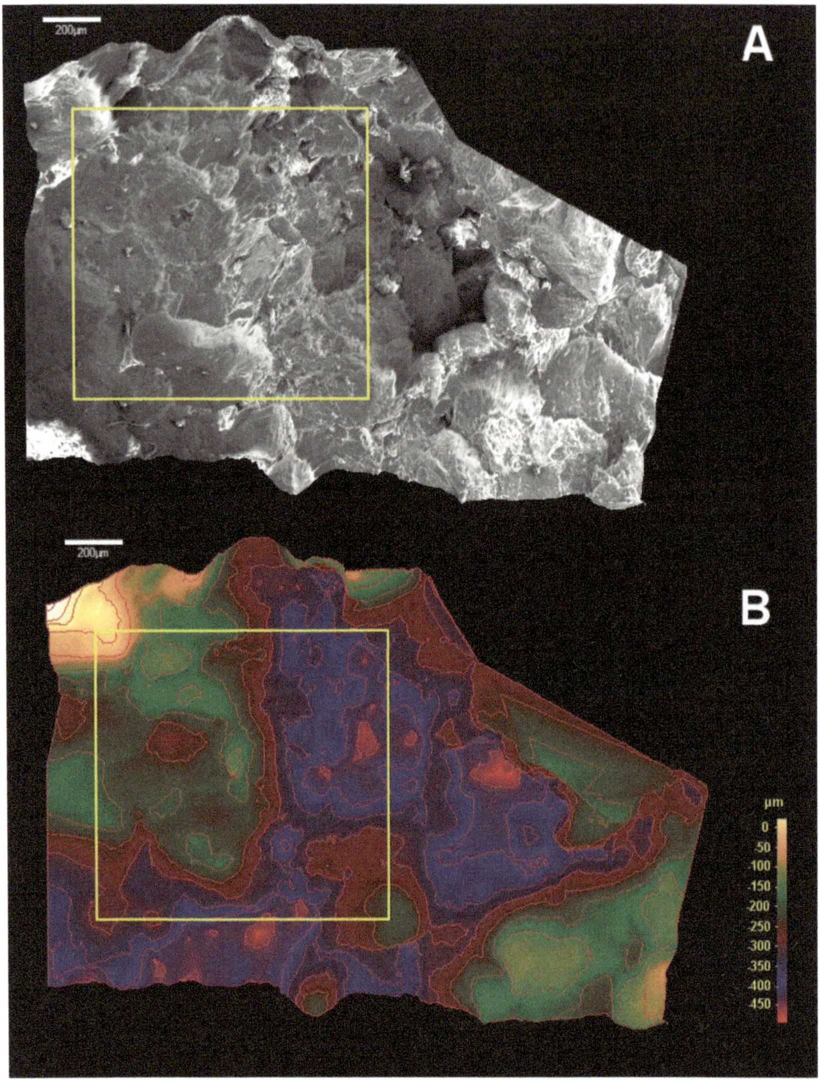

Plate 18 The surface topology of one of the sandstone surfaces (sandstone 5, Fig. 5.5). **A** Digital Elevation Model of the surface showing its general topographical features. **B** Depth-coloured image of the surface showing the surface area available for contact in 50 μm increments from the highest to the lowest point of the sandstone. Both A and B show a 1 mm^2area of the surface analyzed as representing the approximate area of a subdigital pad of *Rhoptropus* cf *biporosus*

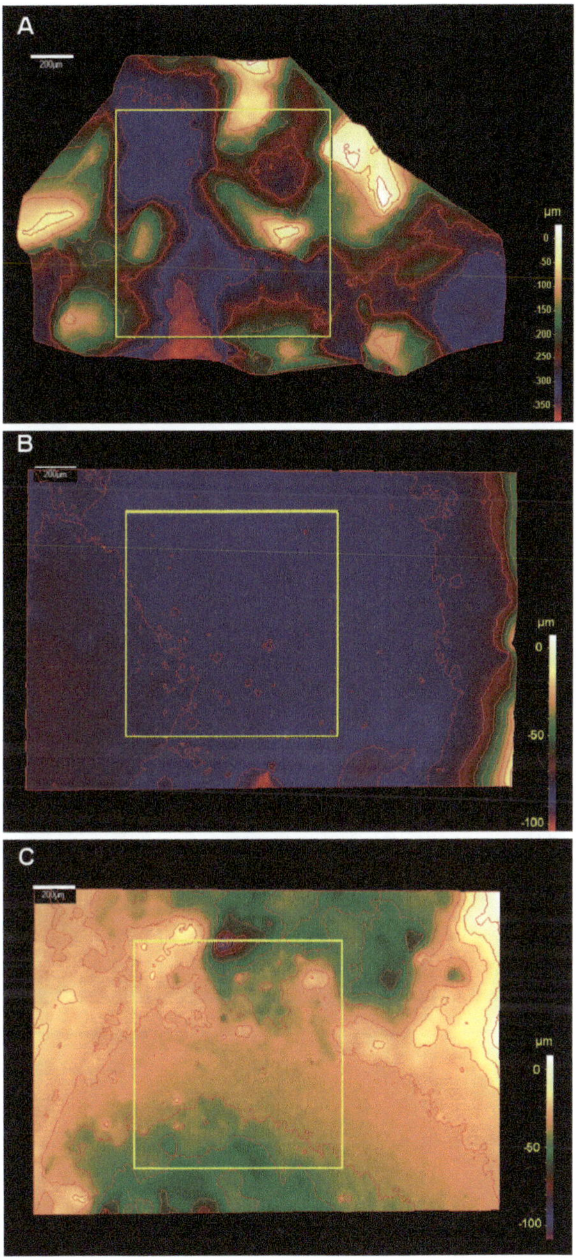

Plate 19 Depth coloured images showing the topology and the area available for contact in 50 µm increments from the highest to the lowest point for three example surfaces **A** 60 Grit Sandpaper **B** Acetate Overhead Transparency **C** 0.5 µm Polishing Film. A, B, and C are all overlain with a 1 mm^2 area representing the approximate area of a subdigital pad of *Rhoptropus* cf*biporosus*

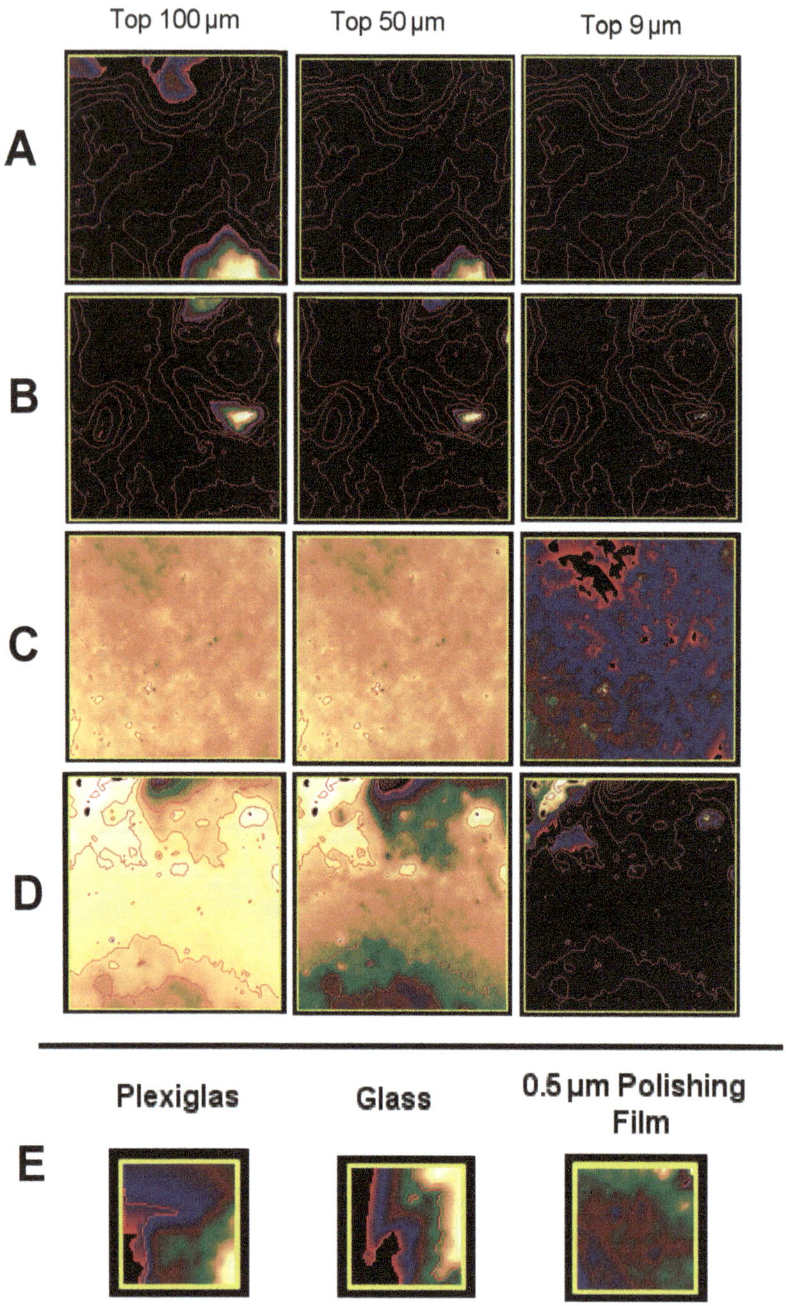

Plate 20 Depth coloured images for a 1 mm^2 area of four example surfaces **A** Sandstone 6 (Fig. 5.5) **B** 60 grit sandpaper **C** Acetate and **D** 0.5 µm polishing film showing the area available for contact within the uppermost 100 µm, 50 µm and 9 µm of each surface. **E** Depth coloured images for a 0.04 mm^2 area of three smooth surfaces: Plexiglas, glass and 0.5 µm polishing film, showing the area available for contact within the uppermost 9 µm of each surface. These images show that the evaluation of 1 mm^2 areas for these surfaces may have resulted in an underestimation of the surface area available for contact at this level. Areas that are available for contact within each range are coloured and those that are not available are *black*

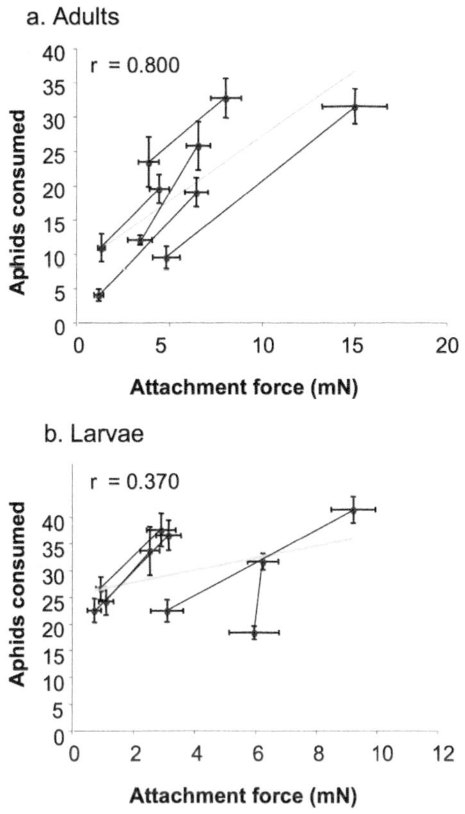

Plate 21 Plot of mean attachment force as given in Fig. 6.1 and mean aphid consumption rate as given in Fig. 6.2 for the coccinellid species tested. *Open circles* are data for adult females and *closed circles* are data for fourth-instar larvae. Correlation coefficients were calculated for the data within each developmental stage. Points connected by lines are for the same species tested on wild-type CEW or reduced CEW pea plants. The point with the higher consumption rate in each connected pair of points is always for the reduced EW plants

Plate 22 Some of the lowland species of *Nepenthes* distributed in Brunei Darussalam, showing the diversity of pitcher types within the genus. The pitchers framed in blue possess a waxy zone (*indicated by the arrow*) while the pitchers framed in red lack such a zone. The knife (9 cm long) gives the scale. The pitcher of *N. gracilis* has the same length as the knife

Plate 23 The three varieties of *Nepenthes rafflesiana* found in Brunei with upper and lower pitchers shown in the upper and lower parts of the figure, respectively. Note the presence of a waxy layer (pale area) in the *variety elongata* (*long arrow*), whose basal limit can be assessed by the presence of a hip (*short arrow*) on the outer part of the pitcher

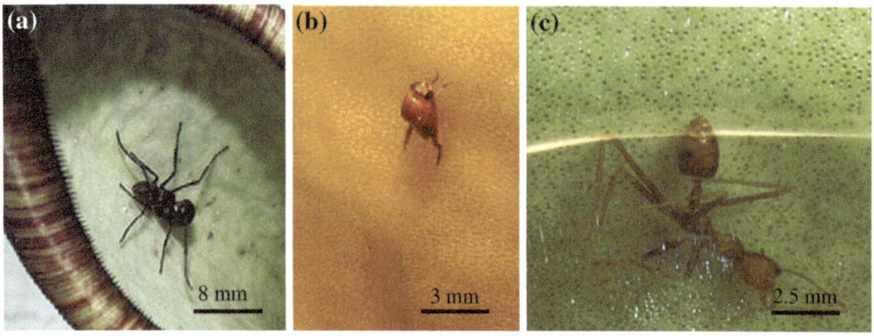

Plate 24 Insect behaviour within the pitchers of *N. rafflesiana*. (**a**) A worker of *Polyrhachis* sp. trying with great difficulty to climb up the waxy surface of a lower pitcher of the variety *elongata*. Note that it applies entire tarsa against the waxy wall to increase the contact area (**b**) A *Drosophila melanogaster* fly that had succeeded in moving its body from the digestive liquid of the variety *typica* and that is slowly climbing up the glandular surface by walking. Its wings have been wetted by the digestive liquid and are adhering to its body. (**c**) An *Oecophylla smaragdina* ant, which has begun to sink inside a pitcher of the variety *typica* after several unsuccessful attempts to emerge from the liquid

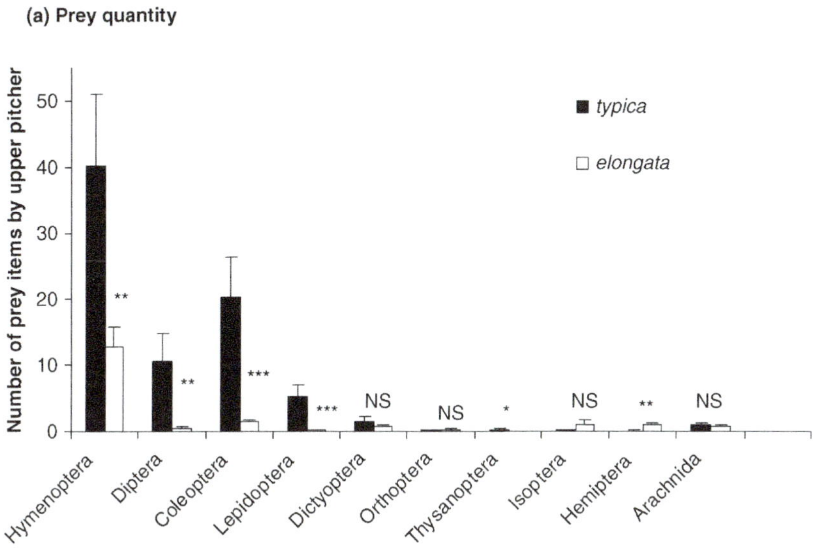

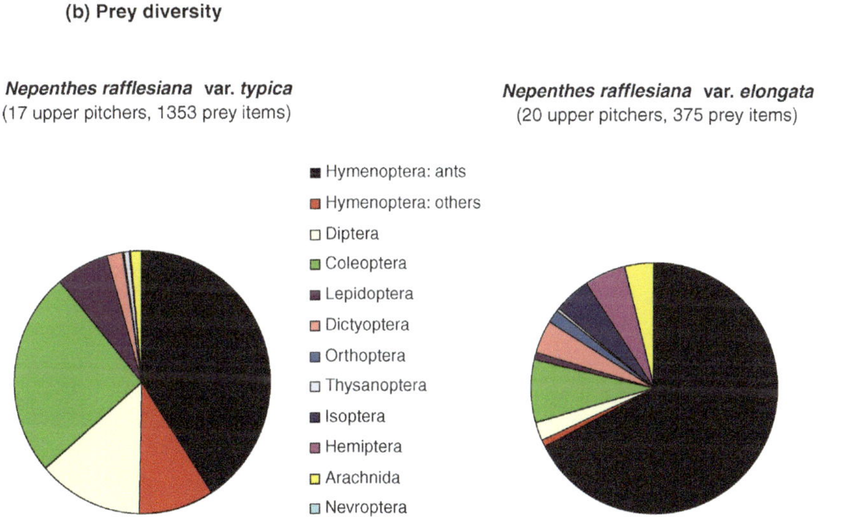

Plate 25 Prey spectra of the two studied *Nepenthes* varieties for their (**a**) quantity and (**b**) diversity. Results of the one-way ANOVAs performed during the MANOVA procedure are given in Fig. 7.8a: NS, means not significant from each others; *, $p < 0.1$; **, $p < 0.001$; ***, $p < 0.0001$. The errors bars refer to standard deviations

Index

R

S

T

GPSR Compliance

The European Union's (EU) General Product Safety Regulation (GPSR) is a set of rules that requires consumer products to be safe and our obligations to ensure this.

If you have any concerns about our products, you can contact us on ProductSafety@springernature.com

In case Publisher is established outside the EU, the EU authorized representative is:

Springer Nature Customer Service Center GmbH
Europaplatz 3
69115 Heidelberg, Germany

Batch number: 10816149

Printed by Printforce, the Netherlands